Dictionary of Remote Sensing and Geoinformatics

Dictionary of Remote Sensing and Geoinformatics

Anand Malik

Rawat Publications

Jaipur • New Delhi • Bangalore • Guwahati • Kolkata

ISBN 978-81-316-0773-2

Disclaimer: While the author of this book have made every effort to avoid any mistake or omission and have used his skill, expertise and knowledge to the best of his capacity to provide accurate and updated information. The author and Rawat Publication does not give any representation or warranty with respect to the accuracy or completeness of the contents of this publication and are selling this publication on the condition and understanding that they shall not be made liable in any manner whatsoever. Rawat Publication and the author expressly disclaim all and any liability/responsibility to any person, whether a purchaser or reader of this publication or not, in respect of anything and everything forming part of the contents of this publication. Rawat Publication shall not be responsible for any errors, omissions or damages arising out the use of the information contained in this publication.

Further, the appearance of the personal name, location, place and incidence, if any; in illustrations used herein is purely coincidental and work of imagination. Thus the same should in no manner be termed as defamatory to any individual.

Published by

Prem Rawat for **Rawat Publications**
Satyam Apts, Sector 3, Jawahar Nagar, Jaipur 302 004 (India)
Phone: 0141 265 1748 / 7006 Fax: 0141 265 1748
E-mail: info@rawatbooks.com
Website: www.rawatbooks.com

New Delhi Office
4858/24, Ansari Road, Daryaganj, New Delhi 110 002
Phone: 011 2326 3290

Also at Bangalore, Guwahati and *Kolkata*

Typeset by Rawat Computers, Jaipur
Printed at Chaman Enterprises, New Delhi

Acknowledgements

A dictionary is a very important tool for anyone who is learning a new language or a subject. The present work is also meant for anyone who has interest in this subject. I started collecting new words during the initial education in Geoinformatics from Indian Institute of Remote Sensing (IIRS), Dehradun, India and International Institute for Geo-information Science and Earth Observation (ITC), The Netherlands. From there the idea of spreading this knowledge through book was conceived. I will always be grateful to the faculties of both institutes and their rich libraries.

In the production of this book, I am indebted to a large number of people. I express my gratitude to the teachers, researchers and organizations who have contributed to the quality and information content of this book. I am grateful to all the authors of the numerous books, research publications and web links for their classic and innovative contribution to the field of remote sensing and GIS. I have tried to consult the literature and put them in organized manner so that user gets benefited.

I would like to thank my wife Renuka for her encouragement and constant support, son Akhil and daughter Aishwarya for their love and inspiration during the writing of this book.

Rawat Publications, for their understanding in conceptualizing the idea of this book and finally publishing it.

Anand Malik

A

AARS Asian Association for Remote Sensing.

Abscissa In a rectangular coordinate system, the distance of the x-coordinate along a horizontal axis from the vertical or y-axis. For example, a point with the coordinates (4, 3) has an abscissa of 4.

Abscissa

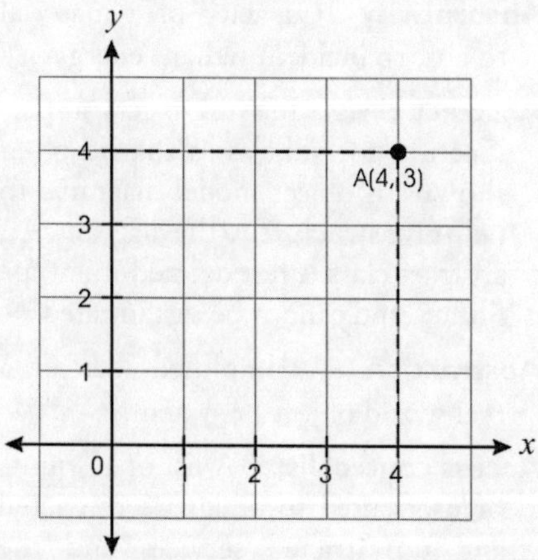

Absolute accuracy The degree to which the position of an object on a map conforms to its correct location on the earth according to an accepted coordinate system.

Absolute coordinates Coordinates that are referenced to the origin of a given coordinate system.

Absolute mode A way of using a digitizing tablet in which locations on the tablet are mapped to specific locations on the screen. Moving the digitizer puck on the tablet surface causes the screen pointer to move to precisely the same position on the screen.

Absolute Coordinates

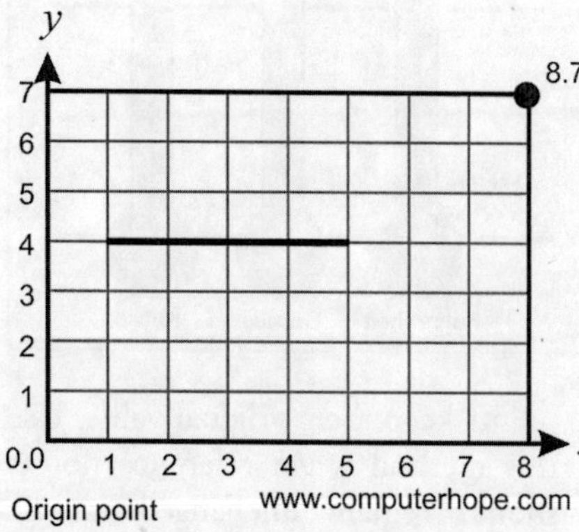

Absolute positioning mode in which a position is identified with respect to a well-defined coordinate system, commonly a geocentric system (i.e. a system whose point of origin coincides with the centre of mass of the earth).

Absolute temperature Temperature of the ability of a material to absorb electromagnetic energy at a specific wavelength. Ratio of the absorbed to incident electromagnetic radiation on surface. Also called absorptivity.

Absolutely constrained adjustment In Survey Analyst for field measurements, one of two possibilities for performing a constrained adjustment. In the absolutely constrained adjustment, the coordinates of the reference

Absolute Temperature

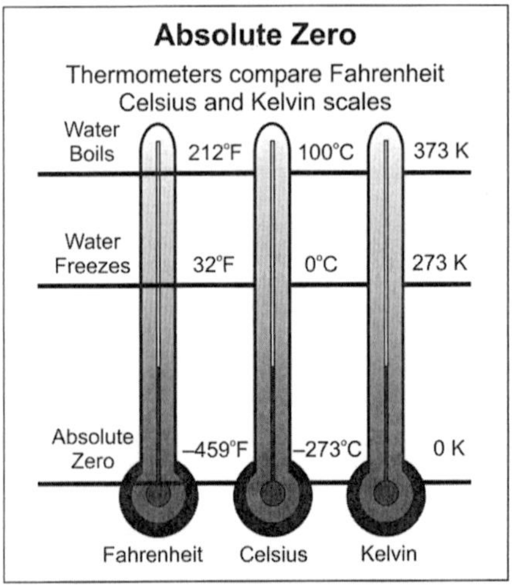

points keep their original value. Use this method when reference points should remain unchanged in the survey dataset.

Absorbed energy the incident radiant energy absorbed during absorption.

Absorptance A measure of the ability of a material to absorb EM energy at a specific wavelength.

Absorption The process by which radiant energy is absorbed and converted into other forms of energy.

Absorption band Wavelength interval within which electromagnetic radiation is absorbed by the atmosphere or by other substances.

Absorption feature A dip in a *spectrum* caused by the absorption of *photons* by molecules interacting with a beam of *light*. The feature can be as simple as a gaussian, or can be much more complicated. It may be due to a reflecting opaque object, or by transmission of the light beam through a partly transparent atmosphere.

Absorption spectrum A continuum *spectrum* that displays *absorption features.*

Absorptivity Capacity of a material to absorb incident radiant energy.

Abstract class In ArcObjects, a specification for subclasses that is often shown on object model diagrams to help give structure to the diagram. An abstract class is not defined in a type library and cannot be instantiated.

Abstraction A simplified idea of a real-world object or system.

Access control list A list of accounts or users used to designate restricted and unrestricted services and the

Absorption

Over view of Electromagnetic Radiation Absorption (using visible light).

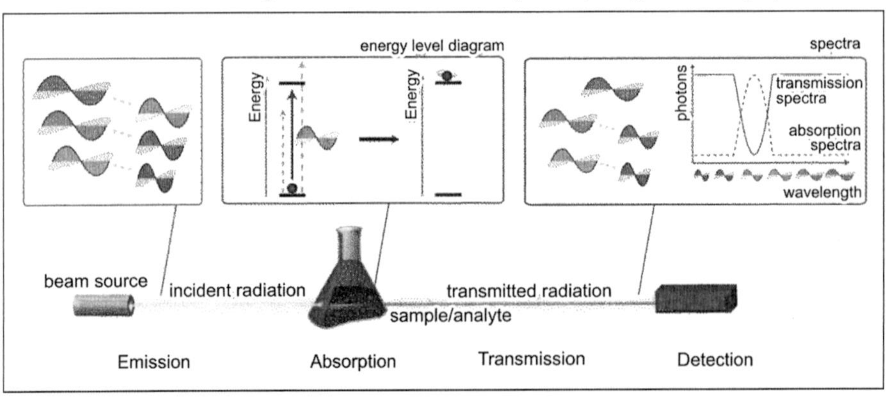

authentication criteria required to access an object.

Access key A keyboard shortcut that allows a user to access the contents of the Main menu by holding down the Alt key and pressing the underlined letter on the menu or menu command item. An access key is created by placing an ampersand (&) in front of the appropriate letter in the command's caption.

Accommodation the change of focus of the eye for distance. The tension of the muscle that changes the focal length of the lens of eye. Thus it brings into focus objects at different distances.

Accuracy Degree of conformity with a standard. Accuracy relates to the quality of a result and is distinguished from precision which relates to the quality of the operation by which the result is obtained.

Acetate In ArcWeb Services, circles, lines, polygons, points, or markers that become transparent when not active. Acetate features are overlaid on other map layers and can be independently annotated.

Achromatic vision The perception by the human eye of changes in brightness, often used to describe the perception of monochrome or black and white scenes.

Acknowledgment file In ArcGIS, an XML file that contains an acknowledgement message.

Acknowledgment message In Arc GIS, a message from a replica to its relative acknowledging the data changes received by the sending replica.

ACL Acronym for *access control list*. A list of accounts or users used to designate restricted and unrestricted services and the authentication criteria required to access an object.

Acoustic device Acoustics is the interdisciplinary science that deals with the study of all mechanical waves in gases, liquids, and solids including vibration, sound, ultrasound and infrasound. The word 'Acoustic' is derived from the Greek work '*akoustikos*' meaning 'of or for hearing, ready to hear'. Acoustic device is a device for amplifying or transmitting sound for a particular purpose.

Acquisition (1) An image captured by a satellite sensor. (2) The process of searching for and locking on to a received signal.

Acquisition date The date the image was acquired. Format: yyyy/mm/dd

Acquisition of signal The time a site receives a signal from a spacecraft.

Across-track An across-track sensor is one that uses a mirror system that moves from side to side in the range to obtain remote sensing data.

Across – Track Scanner

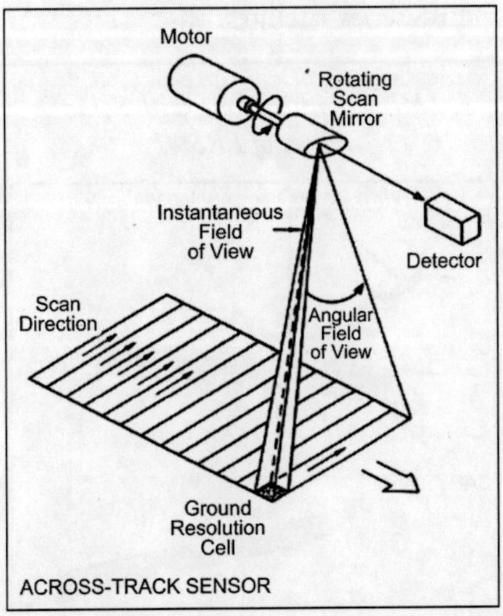

Active Remote Sensing

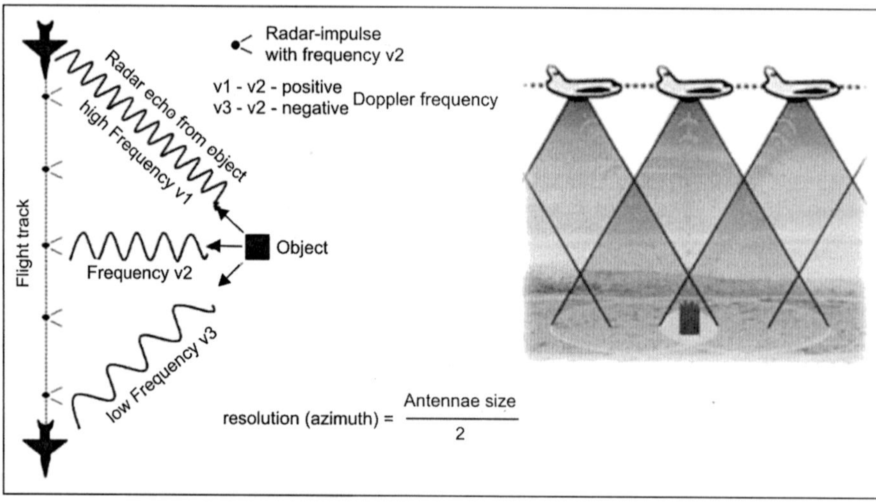

Active data frame The data frame currently being worked on – for example, the data frame to which layers are being added. The active data frame is highlighted on the map, and its name is shown in bold text in the table of contents.

Active microwave instrument (AMI) ordinarily referred to as radar, this is the term commonly used to describe the radar instrument onboard the ERS series satellites.

AMI Antennae Long with Solar Array of ERS - 1

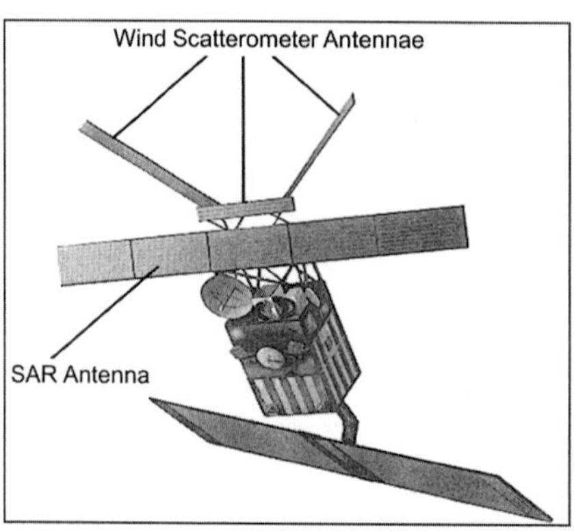

Active network In Survey Analyst for field measurements, the focus of the analyses applied. In an active network, users may list a network's datum points in the Survey Explorer, detect and solve breaks in the network's computation sequence, and find and repair cycles in the network.

Active remote sensing Remote sensing methods that provide their own source of electromagnetic radiation to illuminate the terrain. Radar is one example.

Active sensors A remote sensing device that operates by emitting a beam of energy, and then analysing the beam after it strikes the study object and returns.

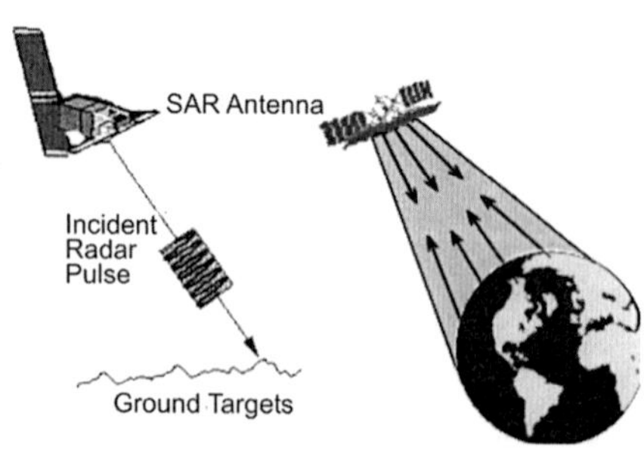

Active System

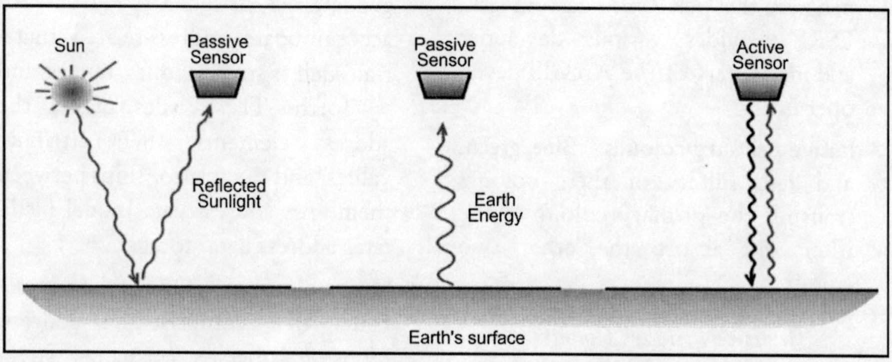

Active server pages A Microsoft server-side scripting technology that can be used to create and run dynamic, interactive Web applications, which are typically coded in JScript, JavaScript or VBScript. An ASP file contains not only the text and HTML tags that standard Web documents contain, but also commands written in a scripting language, which can be carried out on the server or the client.

Active system (sensor or Remote sensing) a remote sensing system that provides its own source of energy and records the energy reflected or refracted back to the sensor.

Active template library A set of C++ template classes, developed by Microsoft for use in building Windows COM objects.

Active theme In ArcView 3.x, the theme in a view document to which button and tool actions or corresponding menu commands are applied. An active theme appears raised in a view's table of contents.

ActiveX connector A type of ArcIMS Application Server Connector that is a Component Object Model (COM) dynamic link library (DLL) which can be used in a COM application such as Microsoft Active Server Pages (ASP).

Acuity A measure of human ability to perceive spatial variations in a scene. It varies with the spatial frequency, shape, and contrast of the variations, and depends on whether the scene is coloured or monochrome.

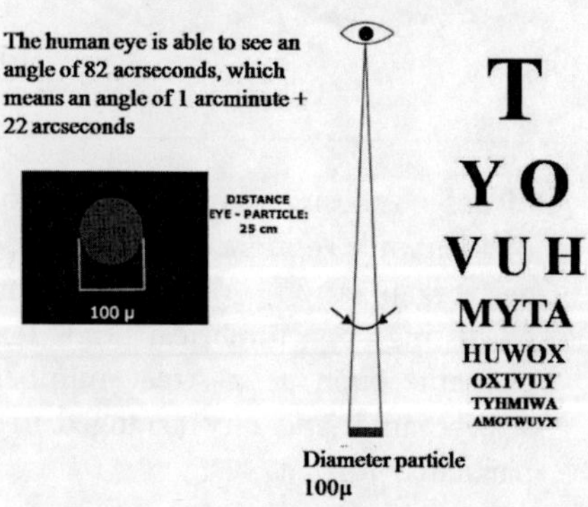

Acutance A measure, using a microdensitometer or other instrument, of how well a photographic system shows sharp edges between contiguous bright and dark areas.

Add ArcWeb data An ArcWeb Services tool that allows ArcWeb data to be added to ArcGIS. Add ArcWeb Data is available as part of ArcWeb Toolbar for ArcGIS.

Add-in An extension to a software program that performs a custom task. ESRI provides various developer add-ins as part of the ArcGIS developer kit.

Additive primary colours Blue, green, and red. Filters of these colours transmit the primary colour of the filter and absorb the other two colours.

Additive Primary Colours

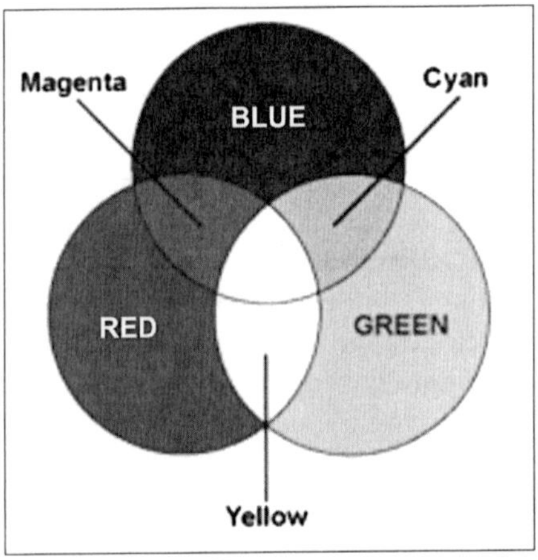

Address A designation of the location of a person's residence or workplace, an organization, or a building, consisting of numerical and text elements such as a street number, street name, and city arranged in a particular format.

Address data Data that contains address information used for geocoding. Address data may consist of one individual address or a table containing many addresses.

Address data format The arrangement of address information in a database, most often consisting of such address elements as house number, street direction, street name, street type, city, and postal code.

Address data model The rules of a geodatabase designed specifically to accommodate address-related material, such as streets, zones, ranges, and so forth. These rules define the address elements, their attribute values, and the relationships between them. An address data model facilitates address data storage.

Address element One of the components that comprise an address. House numbers, street names, street types, and street directions are examples of address elements.

Address event In ArcGIS, features that can be located based on address matching with a street network or other address identifier, such as ZIP Codes or lot numbers.

Address event table In ArcGIS, a table containing addresses but no spatial reference information. Using GIS software, address event tables can be geocoded to create a spatial data layer.

Address field A column in a table that stores one or some address elements. An address field can be present in reference data, address data, or both.

Address finder web service A SOAP ArcWeb service used to find geographical coordinates of an address and to find an address based on geographical coordinates.

Address format The particular structure and arrangement of address elements and a corresponding method of matching that can be used for a specific application. The address format may vary based on locale or country.

Address Locator

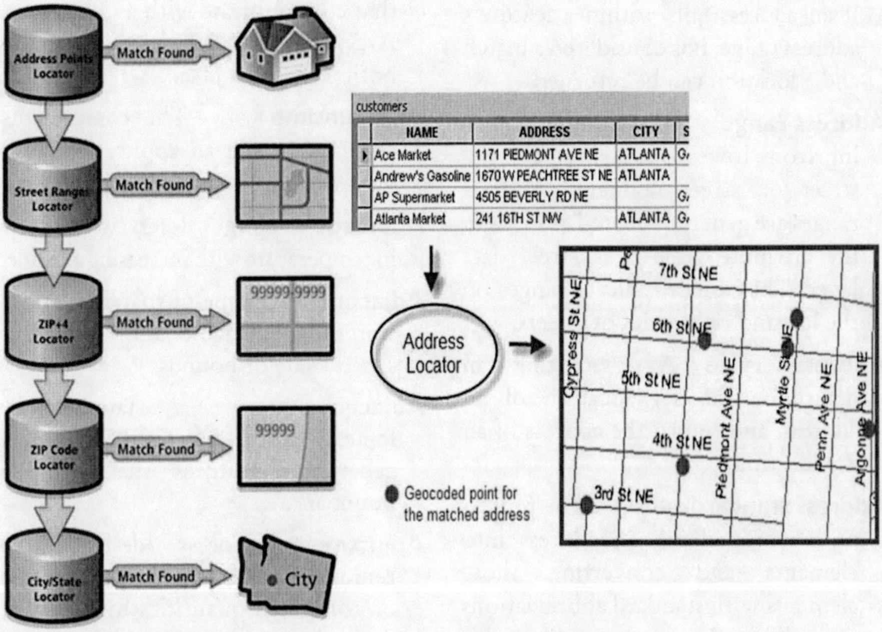

Source: http://www.esri.com

Address geocoding A GIS operation for converting street addresses into spatial data that can be displayed as features on a map, usually by referencing address information from a street segment data layer.

Address locator A dataset in ArcGIS that stores the address attributes, associated indexes, and rules that define the process for translating non spatial descriptions of places, such as street addresses, into spatial data that can be displayed as features on a map. An address locator contains a snapshot of the reference data used for geocoding, and parameters for standardizing addresses, searching for match locations, and creating output. Address locator files have a .loc file extension. In ArcGIS 8.3 and previous versions, an address locator was called a geocoding service.

Address locator property A parameter in an address locator that defines the process of geocoding.

Address locator style A template on which an address locator is built. Each template is designed to accommodate a specific format of address and reference data, and geocoding parameters. The address locator style template file has a .lot file extension.

Address manager web service A SOAP ArcWeb service used to batch geocode a custom set of point addresses. Address Manager Web Service is available as a SOAP interface and a Web interface.

Address matching A process that compares an address or a table of addresses to the address attributes of a reference data set to determine whether a particular address falls within an address range associated with a feature in the reference data set. If an address falls within a feature's address range, it is considered a match and a location can be returned.

Address range Street numbers running from lowest to highest along a street or street segment. Address ranges are generally stored as fields in the attribute table of a street data layer. They often indicate ranges on the left and right sides of streets.

Address service A service that can determine the x,y location of an address, and return the address of an x,y location.

Address standardization The process of breaking down an address into elements and converting those elements with standard abbreviations or spellings. For best practices, this process applies to preparing the reference data and address data for matching.

Address standardizer A tool that prepares and breaks down an address into elements that can be used for geocoding. The process may translate some elements into standard keywords or abbreviations; for example, 'Avenue' may be translated to 'Ave'.

Address style A set of guidelines that define the format of addresses and methods of matching that can be used for a specific application.

Adds table The geodatabase system table, created when a feature class or table is registered as versioned, that maintains information on all rows that have been inserted or updated.

ADF Acronym for *Application Developer Framework*. The set of custom Web controls and templates that can be used to build Web applications that communicate with a GIS server. ArcGIS Server includes an ADF for both .NET and Java.

ADF runtime The components required to run an application built with the ADF.

Adiabatic cooling Refers to decrease in temperature with increasing altitude.

Adjacency A type of spatial relationship in which two or more polygons share a side or boundary.

Adjacency query A statement or logical expression used to select geographic features that share a boundary.

Adjustment Process designed to remove inconsistencies in measured or computed quantities by applying derived corrections to compensate for random or accidental errors.

Adjustment level In Survey Analyst for cadastres, a number that corresponds to a specific set of displacement vectors in a series of adjustments to the cadastral fabric. The adjustment level is generally identified by the date and time of the adjustment.

Adjustment, land-line Positioning land lines on a map to indicate their true, theoretical, or approximate location relative to the adjacent terrain and culture, by reconciling the information shown on Bureau of Land Management plats and field records with the ground evidence of the location of the lines.

Adjustment, standard accuracy Adjustment of a survey resulting in values for positions and (or) elevations that comply with the National Map Accuracy Standards.

Adiabatic Change

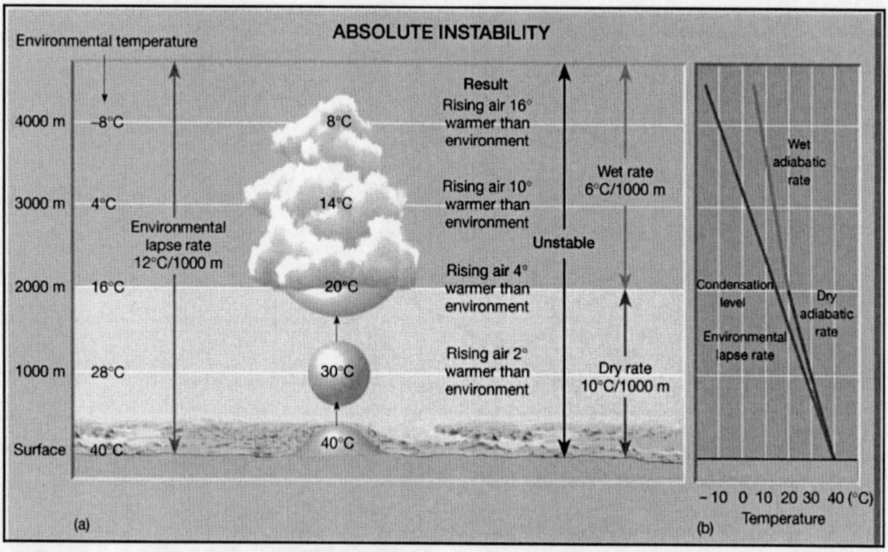

Adoption The process of appropriating a technology and putting it into use for one's own purposes; the act by an individual, organization, or community of choosing a technology and putting it into effect.

ADT Acronym for *abstract data type*. For Open LS, a data type and structure for location information that is shared by two or more services. ADTs are application schemas that are encoded in XML for location services (XLS).

Advanced synthetic aperture radar (ASAR) This the newest SAR instrument from *ESA*. The Advanced Synthetic Aperture Radar (ASAR) instrument onboard the *ENVISAT* satellite extends the mission of the *Active Microwave Instrument* (AMI) *Synthetic Aperture Radar* (SAR) instruments flown on the *European Remote Sensing* (ERS) Satellites ERS-1 and ERS-2. ASAR uses an *active* phased-*array antenna*, with *incidence angles* between 15 and 45 degrees. Applications for this sensor will include the study of ocean waves, sea ice extent and motion, and land surface studies such as deforestation and desertification, to name a few.

The basic idea of Advanced Synthetic Aperture Radar (ASAR); a side-looking case, two example scatterers, A and B, are shown in the ground scene. L_e = maximum flight path length for effective SAR processing.

Advanced very high resolution radiometer (AVHRR) Cross track multispectral scanner on a NOAA polar-orbiting satellite that acquires five spectral bands of data (0.55 to 12.50 Fm) with a ground resolution cell of 1.1 by 1.1 km.

Advanced Synthetic Aperture Radar (ASAR)

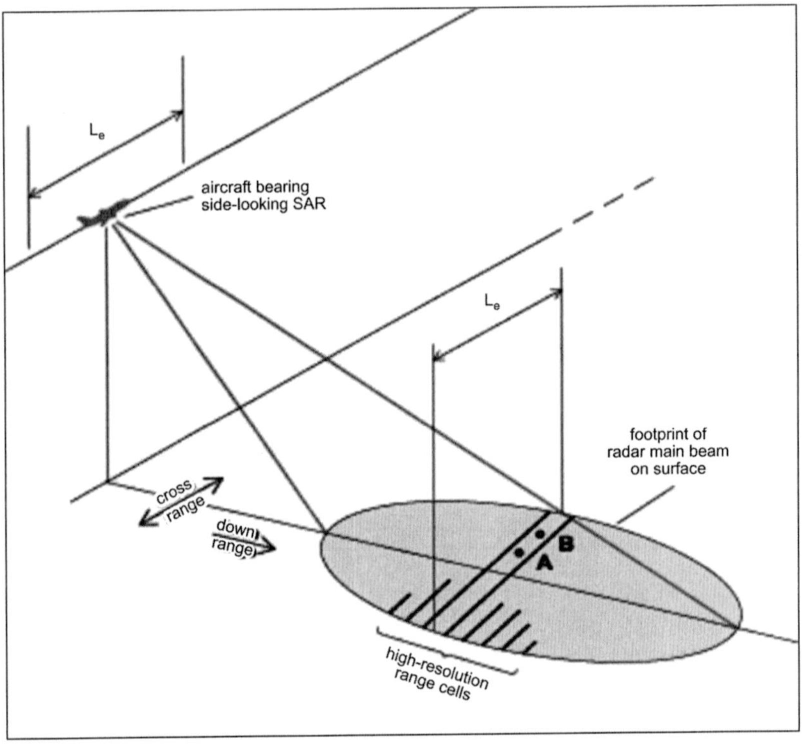

The NOAA satellite with AVHRR on board.

Aerial magnetic survey Survey that records variations in the earth's magnetic field.

Aerial photograph A photograph of the earth's surface taken from a platform flying above the surface but not in orbit, usually an aircraft. Aerial photography is often used as a cartographic data source for base mapping, locating geographic features and interpreting environmental conditions.

Aero triangulation The process of developing a network of horizontal and or vertical positions from a group of known positions using direct or indirect measurements from aerial photographs and mathematical computations. It is the process of densifying and extending ground

Aerial Photograph

Aero Triangulation

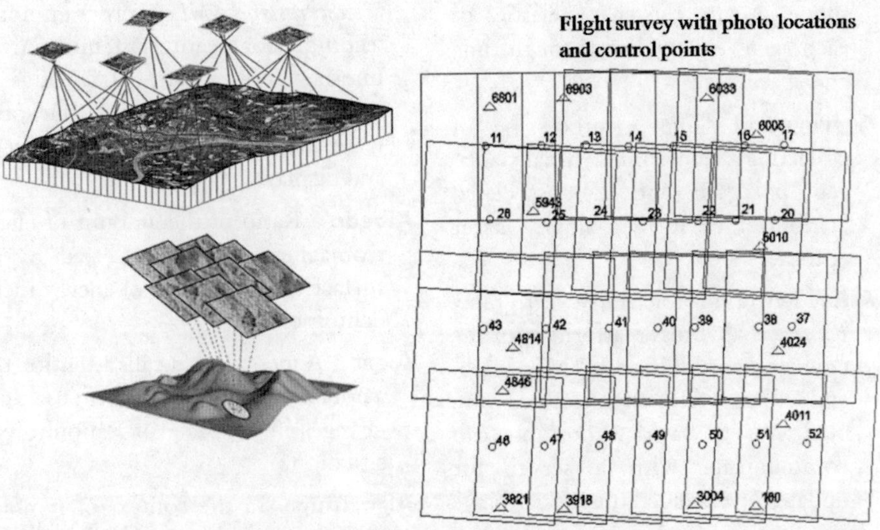

Sources: 1. http://iphonedroneimagery.com/tag/aertriangulation/
2. http://photo.geo.tuwien.ac.at
3. http://geogenesis.net

control through computational means. AT extends horizontal and vertical control from relatively few ground survey control points are called tied points, and they are used to control subsequent photogrammetric mapping. Each stereo model is scaled and levelled using the adjusted coordinate values of the pass points located in the stereo model.

Aero triangulation (flight survey with photo locations and control points.

Affine transformation a geometric transformation method commonly used in GIS, which allows rotation, translation, skew and differential scaling on the rectangular object, while preserving line parallelism.

A minimum of three ticks is required to calculate an affine transformation.

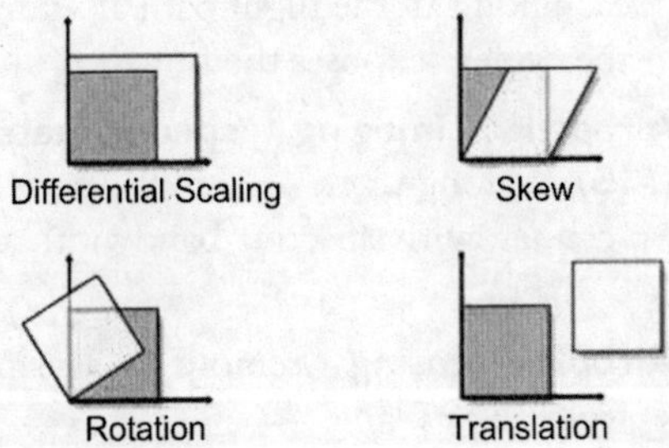

Agent In modelling, an entity within a model that conducts transactions to simulate the actions of a human, group of humans, animal, or other actor.

Agent-based model A simulation of the large-scale consequences of the decisions and interactions of individual members of a population. An agent-based model consists of an environment or framework that defines the scope and rules of actions, along with a number of agents representing one or more actors whose parameters and behaviours are defined. When the model is run, the characteristics of each agent are tracked through time and space.

Aggregation The process of collecting a set of similar, usually adjacent, polygons (with their associated attributes) to form a single, larger entity.

AIR Acronym for *Adobe Integrated Runtime*. A cross-platform runtime environment built on HTML, Flash, and Flex. AIR applications are deployed on a desktop and can communicate with a server for updates or extended functionality.

Air base Ground distance between optical centres of successive overlapping aerial photographs.

Air station In aerial photography, each point in the flight path at which the camera exposes the film.

Airborne imaging spectrometer (AIS) Along-track multispectral scanner with spectral bandwidth of 0.01 Fm.

Airborne sensing remote sensing from an airplane.

Airborne visible and infrared imaging spectrometer (AVIRIS) Experimental airborne along-track multispectral scanner under development at JPL to acquire 224 images in the spectral region from 0.4 to 2.4 Fm.

AIXM Acronym for *Aeronautical Information Exchange Format*. An XML format used to describe aeronautical data transactions created and maintained by EUROCONTROL (European Organisation for the Safety of Air Navigation), in the process of being adopted worldwide.

AJAX Acronym for *Asynchronous JavaScript and XML*. A programming technique for creating fast, interactive Internet applications. AJAX adds a small application to part of the software user's browser for fast loading and display.

Albedo Ratio of the amount of electromagnetic energy reflected by a surface to the amount of energy incident upon it.

Alert A message that calls attention to a notable situation or informs users of changes in the state of a monitored situation.

Algorithm In the context of remote sensing, algorithms generally specify how to determine higher-level data products from lower-level source data. For example, algorithms prescribe how atmospheric temperature and moisture profiles are determined from a set of radiation observations originally sensed by satellite sounding instruments.

Alias An alternative name specified for fields, tables, files, or datasets that is more descriptive and user-friendly than the actual name. On computer networks, a single e-mail alias may refer to a group of e-mail addresses.

Aliasing The jagged appearance of curves and diagonal lines in a raster image. Aliasing becomes more apparent as the size of the raster pixels is increased or the resolution of the image is decreased.

Alidade Instrument or part of an instrument, for determining direction, either horizontal or vertical. In its simplest form, a peep sight or telescope mounted on a straightedge and used for plotting directions graphically. In such instruments as transits and theodolites, the alidade is the part containing the telescope and its attachments.

Align fields A task that identifies the fields required for geocoding, such as address and city, when uploading data.

Aligned dimension A drafting symbol that runs parallel to the baseline and indicates the true distance between beginning and ending dimension points.

Allocation In network analysis, the process of assigning entities or edges and junctions to features until the feature's capacity or limit of impedance is reached. For example, streets may be assigned to the most accessible fire station within a six-minute radius, or students may be assigned to the nearest school until it is full.

Almanac A data file that contains orbit information on all satellites, clock corrections, and atmospheric delay parameters. It is transmitted by a GPS satellite to a GPS receiver, where it facilitates rapid satellite vehicle acquisition within GPS receivers. Almanac data must be acquired before GPS navigation can begin.

Along-track (scanner) An along-track sensor is made up of a linear detector array of CCDs (Charge Coupled Device) that obtains data in the platform's direction of motion (azimuth or along-track dimension). The sensor's instantaneous field of view extends the length of the swath width. The dimension parallel to the path of the platform carrying the sensor. The along-track dimension is

Almanac

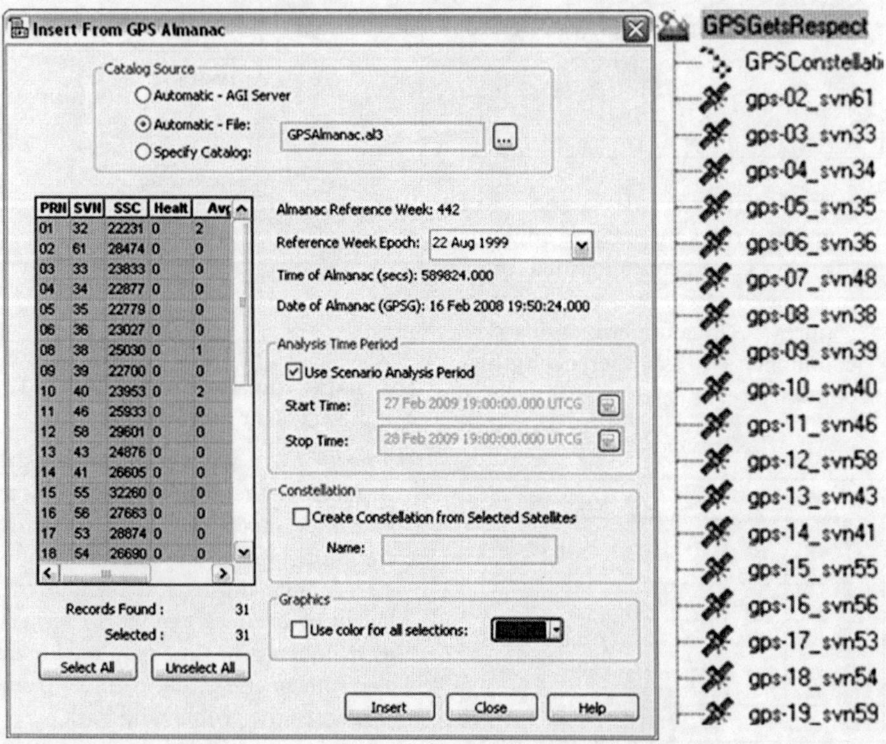

the imaging direction of the sensor that is parallel to the direction in which the satellite or aircraft is moving. For side-looking radars (SLARs), this dimension is sometimes called the cross range. The typical two-dimensional remotely sensed image is created by the movement of the platform in the along-track direction, while the sensor scans or aims at the orthogonal direction.

Along - Track Scanner

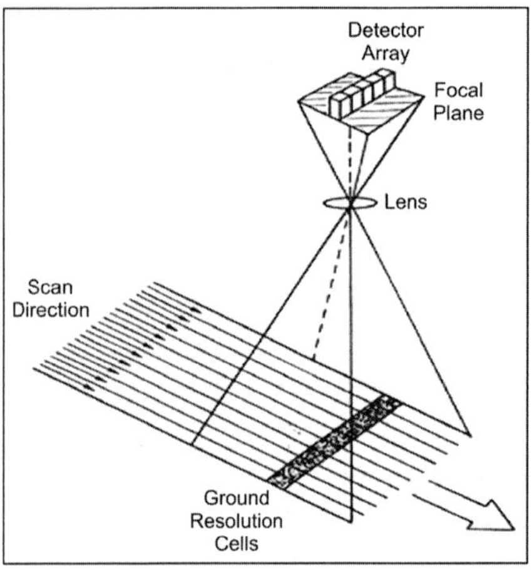

Alphanumeric grid A grid of numbered rows and lettered columns (or vice versa) superimposed on a map, used to find and identify features. Alphanumeric grids are commonly used as a reference system on local street maps.

Alphanumeric Grid

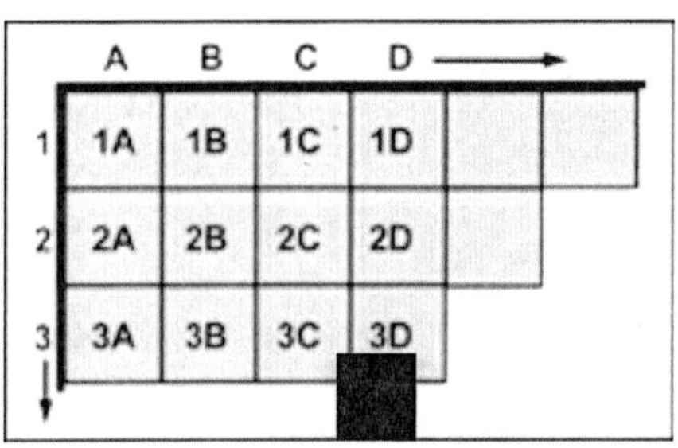

Alternate key An attribute or set of attributes in a relational database that provides a unique identifier for each record and could be used as an alternative to the primary key.

Alternate name A name for an address element, usually a street name, that is different from the official or most common name. For example, a highway number might be an alternate name for a street name.

Alternating polarization (AP) These products are similar to Image Mode products but include a second image acquired using a second polarisation combination. The raw data is acquired in bursts of alternating polarization. See Products glossary AP_MODE.

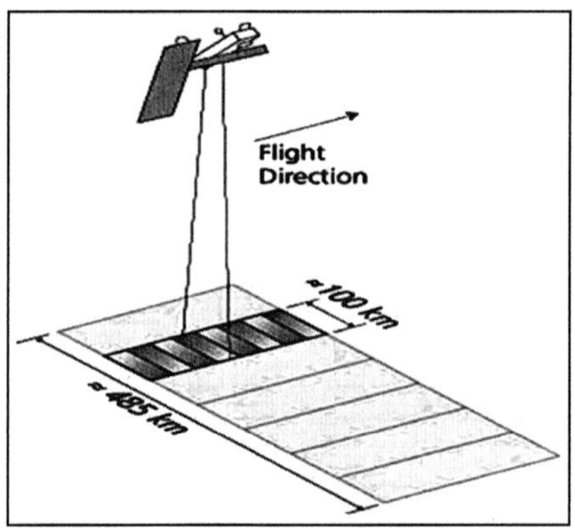

Alternating polarization in ENVIST ASAR slant range image.

Altimeter An active instrument used to measure the altitude of an object above a fixed level. The most common type is an aneroid barometer. A radar altimeter determines the height of an aircraft above the terrain by measuring the time required for an electromagnetic pulse to travel from aircraft to the ground and back.

Altimeter

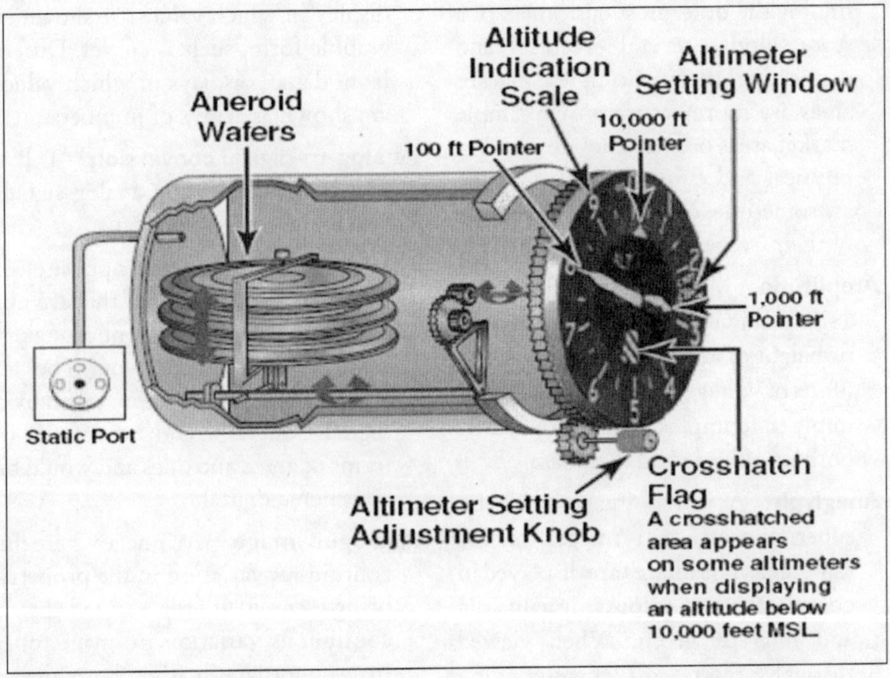

Altitude The angle, in degrees, above the level horizon where an object in the sky appears (the height in space).

AM/FM Acronym for *automated mapping/facilities management*. GIS or CAD-based systems used by utilities and public works organizations for storing, manipulating, and mapping facility information such as the location of geographically dispersed assets.

Ambiguity the certainty in the measurement of the number of wave cycles of the GPS carrier between the satellite and the receiver. For example, coastal areas experiencing tidal fluctuations may be dry land at some times and under water at other times. Ambiguity may be caused by changeable conditions in reality, by incomplete or conflicting definitions of attributes, or by subjective differences in the evaluation of data. It may also be caused by disputes, as when two parties claim ownership of the same tract of land.

American national standards institute The private, nonprofit organization that develops US industry standards through consensus and public review.

American standard code for information interchange The de facto standard for the format of text files in computers and on the Internet that assigns a 7-bit binary number to each alphanumeric or special character. ASCII defines 128 possible characters.

AML Acronym for *ARC Macro Language*. A proprietary, high-level programming language created by ESRI for generating end-user applications in ArcInfo Workstation.

Amoeba An area calculated by finding the outermost customers of a store along several vectors and connecting them. Complex market areas are more accurate than simple market areas because they respond to physical and cultural barriers. They are sometimes called amoebas because of their irregular shapes.

Amplitude Measure of the strength of a signal, and in particular the strength or height of an electromagnetic wave (units of voltage). The amplitude may imply a complex signal, including both magnitude and the phase.

Anaglyph A stereo image made by superimposing two images of the same area. The images are displayed in complementary colours, usually red and blue or green. When viewed through filters of corresponding colours, the images appear as one three-dimensional image.

Anaglyph 3-D Glasses

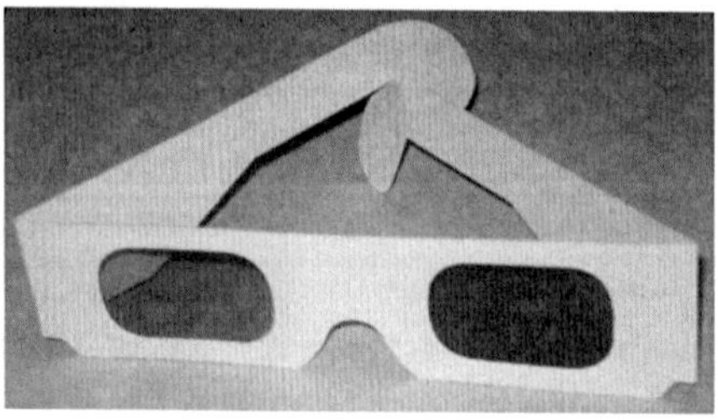

Analog A form of transmitting information characterized by contin- uously variable quantities, as opposed to digital transmission, which is characterized by discrete bits of information in numerical steps. An analog signal is used to transmit audio (such as voice, radio, stereo, and control tones) and is responsive to changes in light, sound, heat and pressure.

Analog display A form of data display in which values are shown in graphic form, such as curves. Differs from digital displays in which values are shown as arrays of numbers.

Analog-to-digital conversion The process of converting analog signals to a digital representation.

Analogue This is the opposite of digital. It refers to things that are not made up of numbers. A photo taken with a film camera would be an analogue picture. A photo taken by a digital camera would be defined in terms of zeros and ones and would be considered digital.

Analogue image An image where the continuous variation in the property being sensed is represented by a continuous variation in image tone. In a photograph this is achieved directly by the grains of photosensitive chemicals in the film; in an electronic scanner, the response in, say, millivolts is transformed to a display on a cathode-ray tube where it may be photographed.

Analysis A systematic examination of a problem or complex entity in order to provide new information from what is already known.

Analysis extent The geographic bounding area within which spatial analysis will occur. The bounding area is set by defining the x,y coordinates of opposite corners, usually the bottom-left and top-right corners of results.

Analysis mask In digital cartography, a means of covering or hiding features on a map to enhance cartographic representation. For example, masking is often used to cover features behind text to make the text more readable.

Analysis of variance A statistical procedure used to evaluate the variance of the mean values for two or more data sets in order to assess the probability that the data comes from the same sample or statistical population.

Ancillary data In digital image processing, data from sources other than remote sensing, used to assist in analysis and classification or to populate metadata.

Ancillary source A supplementary source of information.

Angular beam width In radar, the angle subtended in the horizontal plane by the radar beam.

Angular field of view Angle subtended by lines from a remote sensing system to the outer margins of the strip of terrain that is viewed by the system.

Angular resolving power Minimum separation between two resolvable targets, expressed as angular separation.

Angular unit The unit of measurement on a sphere or a spheroid, usually degrees. Some map projection parameters, such as the central meridian and standard parallel, are defined in angular units.

Animation In ArcMap, ArcScene, and ArcGlobe, a collection of animation tracks that define the dynamic property changes to associated objects. An animation allows for navigation through the display, visualization of temporal changes, or alteration of layer and scene properties, such as layer transparency or the scene background.

Animation manager In ArcMap, ArcScene, and ArcGlobe, the interface in which the keyframes, tracks and time-scale properties of an animation can be edited and an animation can be timed and previewed.

Anisotropic Having a nonuniform spatial distribution of movement or properties, usually across a surface.

Anisotropy A property of a spatial process or data in which spatial dependence (autocorrelation) changes with both the distance and the direction between two locations.

Annotation In cartography, text or graphics on a map that provide information for the map reader. Annotation may identify or describe a specific map entity, provide general information about an area on the map, or supply information about the map itself.

Annotation class A subset of annotation in a standard or feature-linked geodatabase annotation feature class that contains properties that determine how the subset of annotation will display. A standard or feature-linked geodatabase annotation feature class may contain one or more annotation classes.

Annotation construction method One of a number of procedures that dictate what type of annotation feature is created and the number of points required to create new annotation features. Construction methods include horizontal, straight, curved, leader line, and follow feature.

Annotation feature class A geodatabase feature class that stores text or graphics that provide information about features or general areas of a map (annotation). An annotation feature class may be linked to another feature class, so that edits to the features are reflected in the corresponding annotation (feature-linked annotation). Annotation in a geodatabase is edited during an edit session, using the tools on the Annotation toolbar.

Annotation group A container within a map document for organizing and managing text or graphics that provide additional information about features or general areas of a map. Annotation groups allow control of the display of different sets of annotation. Annotation stored in a map document is edited with the tools on the Drawing toolbar.

Annotation layer A layer that references annotation. Information stored for annotation includes a text string, a position at which it can be displayed, and display characteristics.

Annotation target In ArcMap, the annotation group or feature class in a map document where new annotation will be stored when created when using the New Text tools on the Draw toolbar or when copying and pasting annotation. Annotation created with the Annotation Edit tools is stored in the current Editing target, not in the annotation target.

Anomoly (1) A deviation from the normal position. (2) The angular distance between the position of a planet and its last perihelion (point nearest to the sun), or between that of a satellite and its last perigee (point nearest to the centre of the earth). In the image, an area which differs from the surrounding, normal area. For example, a concentration of vegetation within a desert scene constitutes a anomaly.

ANOVA A statistical procedure used to evaluate the variance of the mean values for two or more datasets in order to assess the probability that the data comes from the same sample or statistical population.

ANSI Acronym for *American National Standards Institute*. The private, nonprofit organization that develops US industry standards through consensus and public review.

Ant An open-source, Java-based tool from the Apache Software Foundation that is used to manage the build procedure of applications.

Antenna (Array Antenna) A device for transmitting and receiving radio waves (mainly used in Radar System).

Antenna array An arrangement of several individual antennas so spaced and phased that their individual contributions add in the preferred direction and cancel in other directions. SAR systems, employ a short physical antenna, but through modified data recording and processing techniques, they synthesise the effect of a very long antenna. The result of this mode of operation is a very narrow effective antenna beam width, even at far ranges, without requiring a physically long antenna or a short operating wavelength. For example, in a SAR system, a 2m antenna can be made effectively 600 m long. In the figure shown, an antenna array with its elements along the z axis such that the distance between each two successive elements is equal to d.

Radar Antenna Array

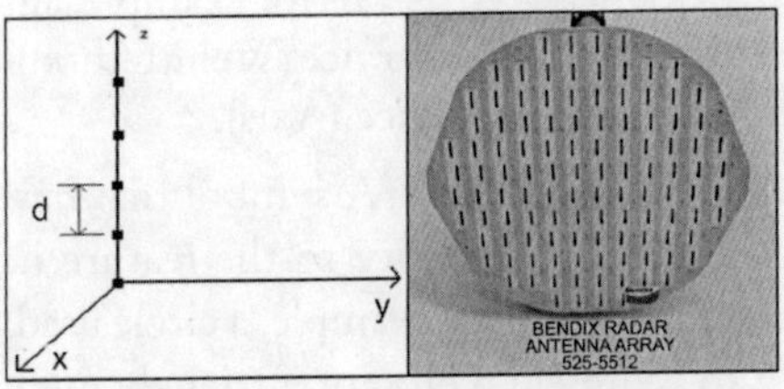

Antipode Any point on the surface of a sphere that lies 180 degrees (opposite) from a given point on the same surface, so that a line drawn between the two points through the centre of the sphere forms a true diameter.

Any-vertex connectivity In network datasets, a type of edge connectivity policy that states that an edge may connect to another edge or junction where they have coincident vertices.

Anywhere fix The ability of a receiver to start position calculations without being given an approximate location and time or a position that a GPS receiver can calculate without knowing its own location or the local time.

AOI The extent used to define a focus area for either a map or database production.

Apartment In Microsoft's COM component programming model, a group of threads, working within a process, that work within the same context.

Aperture The diameter of an opening; the diameter of the primary lens or mirror of a telescope. A cross-sectional area of the antenna that is exposed to the satellite signal.

Aphylactic projection A projection that does not have equal area, conformal, or equidistant characteristics. The compromise projection is an attempt at balance between these characteristics, and is often used in thematic mapping.

API Acronym for *application programming interface*. A set of interfaces, methods, protocols, and tools that application developers use to build or customize a software program. APIs make it easier to develop a program by providing building blocks of prewritten, tested, and documented code that are incorporated into the new program. APIs can be built for any programming language.

API key In ArcWeb Services, an encryption-based authentication method for the ArcWeb Explorer JavaScript API. An API key is valid only for registered URLs.

Apogee In an orbit path, the point at which the object in orbit is farthest from the centre of the body being orbited.

Apollo US lunar exploration programme of satellites with crews of three astronauts.

APP-6A A military symbology specification published by NATO (North Atlantic Treaty Organization). APP-6A is based on MIL-STD-2525A, the predecessor to MIL-STD-2525B.

Apparent thermal inertia (ATI) An approximation of thermal inertia calculated as one minus albedo divided by the difference between daytime and night-time radiant temperatures.

Appending Adding features from multiple data sources of the same data type into an existing data set.

Applet A small program that usually executes from within a Web browser.

Applets are compatible with most platforms, and can also be used within applications or devices that support applets.

Application The end purpose for which remote sensing is used. Most often remote sensing is used to measure, map or monitor features of our environment. In the GIS, to solve problems, automate tasks, or generate information within a specific field of interest. For example, a common agricultural application of GIS is determining fertilization requirements based on field maps of soil chemistry and previous crop yields.

Application developer framework The set of custom Web controls and templates that can be used to build Web applications that communicate with a GIS server. ArcGIS Server includes an ADF for both .NET and Java.

Application programming interface A set of interfaces, methods, protocols, and tools that application developers use to build or customize a software program. APIs make it easier to develop a program by providing building blocks of prewritten, tested, and documented code that are incorporated into the new program. APIs can be built for any programming language.

Application server A computer program that receives user requests through a client application and returns results to the client.

Application Web service A Web service that solves a particular problem; for example, a Web service that finds all of the hospitals within a certain distance of an address. An application Web service can be implemented using the native Web service framework of a Web server; for example, an ASP.NET Web service (WebMethod) or Java Web service (Axis).

Arbitrary symbol A symbol that has no visual similarity to the feature it represents – for example, a circle used to represent a city, or a triangle used to represent a school.

Arc On a map, a shape defined by a connected series of unique x,y coordinate pairs. An arc may be straight or curved.

ARC macro language A proprietary, high-level programming language created by ESRI for generating end-user applications in ArcInfo Workstation.

Arc second An angle equal to one sixtieth of a minute of latitude or longitude.

ArcGIS online A set of web-based base maps, globes and other data and services created by ESRI for use inside ArcGIS products and GIS applications on the Internet.

ArcGIS server web service A Web service processed and executed from within an ArcGIS server. Each Web service has a distinct HTTP location (URL). Web access is enabled by default for all ArcGIS Server services, but can be turned off by an administrator.

ArcGIS spatial analyst An ArcGIS extension that provides spatial modelling and analysis features. It allows the creation, querying, mapping, and analysis of cell-based raster data and integrated vector-raster analysis.

Arcgisant The command, provided with the Java ADF, that starts the

Apache Ant tool that builds and deploys Web applications.

Architecture The internal design of an application or software package; the way software or hardware components are organized into a functioning unit.

Archive A collection of information or data that is stored on a permanent medium such as CDs, discs, or tapes. Information is archived to ensure its security or persistence.

Archiving In ArcGIS and ArcSDE, a procedure that allows a geodatabase to capture and store updates to features and records as the version is posted or edits are saved directly. Archiving builds a lineage of historical information that can be viewed and queried.

ArcIMS ESRI software that allows for centrally hosting and serving GIS maps, data, and applications for use on the Internet. The administrative framework lets users' author configuration files, publish maps, design Web pages, and administer ArcIMS spatial servers. ArcIMS supports Windows, Linux, and UNIX platforms and is customizable on many levels.

ArcIMS administrator The ArcIMS component that allows users to manage ArcIMS services, servers, virtual servers, and folders.

ArcIMS application server The ArcIMS component that handles the distribution of incoming requests. It tracks which services are running on which ArcIMS spatial servers and hands off a request to the appropriate ArcIMS spatial server.

ArcIMS application server connector A component used to connect the Web server to the ArcIMS application server. Types of connectors include ActiveX Connector, ColdFusion Connector, Java Connector, .NET Link, and Servlet Connector. Connectors must be installed on the same computer as the Web server.

ArcIMS architecture A multitier framework that includes ArcIMS components: the ArcIMS Manager/manager applications, application server, application server connectors, monitor, tasker, spatial servers, and viewers. The complete architecture also includes a Windows or UNIX operating system, a Web server, a servlet engine, and client-side Web browsers.

ArcIMS author The ArcIMS component that allows users to organize data into a configuration file that can be used to create a service. A configuration file specifies the map content: which data layers will be displayed and how they will look (colour, symbols, labels, etc.).

ArcIMS designer The ArcIMS component that guides users in designing Web pages based on at least one service and one of the ArcIMS viewers. Users choose from a variety of options including toolbar functions, scale bar properties, and visible layer settings.

ArcIMS manager A suite of Web pages, deprecated in the 9.0 release, that guides users through the process of authoring configuration files, publishing services, designing Web pages, and administering sites. ArcIMS Manager combines the three independent applications (ArcIMS Author, ArcIMS Designer, and ArcIMS Administrator) into one wizard-driven framework. ArcIMS

Manager resides on the Web server computer and can be accessed remotely. It is also referred to as the Web-based Manager and Remote Manager.

ArcIMS manager application Any one of these stand-alone applications: ArcIMS Author, ArcIMS Administrator, or ArcIMS Designer. Each of these applications guides users through a process: authoring configuration files (ArcIMS Author), creating services and administering sites (ArcIMS Administrator), or designing Web pages (ArcIMS Designer).

ArcIMS monitor An ArcIMS component that tracks the state of the ArcIMS spatial server. When a computer system reboots, ArcIMS Monitor restarts services automatically by restarting the site configuration. This configuration is based on the setting saved in a serialized file ending in the file extension .sez.

ArcIMS service A service that allows the content of a configuration file to be published on the Internet. The configuration file provides data layer content and symbology that the service registers to the ArcIMS spatial server and Web server for processing.

ArcIMS service administrator A Web-based administration application that allows users to manage ArcIMS services and ArcSDE services remotely.

ArcIMS tasker An ArcIMS component that removes temporary image files generated by the image and ArcMap image services at a user-defined time interval.

ArcIMS viewer Any one of the three Web site designs that come as standard options in ArcIMS Designer: the HTML, Java Custom, and Java Standard. They provide the functionality and graphic look for Web sites. The Java viewers require a one-time Web download and are only compatible with Web browsers that support Java 2 plug-in functionality.

ArcIMS virtual server A grouping of one or more spatial servers into a single unit for administrative purposes. All of the following are ArcIMS virtual servers: the ArcMap server, extract server, feature server, geocode server, image server, metadata server, query server, and route server. The ArcMap server and route server are optional extensions to ArcIMS.

ArcIMS web site directory The directory that stores the files that make up the ArcIMS Internet GIS application and other files. It is the directory specified as the Working Directory during the installation process. The default location for Windows is C:\ArcIMS. The default location for UNIX is $home.

ArcIMSFolders.sez A serialization file containing ArcIMS folders and any submitted MapNotes and EditNotes. It is stored in the ArcIMS AppServer directory. This file replaces the EsriMapCookies.ser file used in ArcIMS 3.1.

ArcIMSSite.sez A serialization file created in the ArcIMS AppServer directory that saves site parameters, including which services are running and which servers are started. This file replaces the EsriMapCatalog.ser file used in ArcIMS 3.1.

ArcInfo A software package that provides great ability to modify and edit your GIS data.

ArcInfo interchange file A file format, also known as an export file, used to enable a coverage, grid or TIN and an associated INFO table to be transferred between different machines which are not connected by any type of file sharing network. ArcInfo interchange files have a .E00 extension, which increments to .E01, .E02, and so on, if the interchange file is composed of several separate files.

ArcInfo workspace A file-based collection of coverages, grids, TINs, or shapefiles stored as a directory of folders in the file system.

ArcMap server A public ArcIMS virtual server that allows an ArcGIS user to create maps in ArcMap, rather than ArcIMS Author or ArcIMS Manager, and publish them on the Internet. The ArcMap server is an optional extension to ArcIMS.

Arc-node topology The data structure in a coverage used to represent linear features and polygon boundaries and to support analysis functions, such as network tracing. Nodes represent the beginning and ending vertices of each arc. Arcs that share a node are connected, and polygons are defined by a series of connected arcs. An arc that intersects another arc is split into two arcs. Each arc that defines all or part of a polygon boundary, records the number of the polygon to its left and to its right, giving it a direction of travel.

ArcObjects A library of software components that make up the foundation of ArcGIS. ArcGIS Desktop, ArcGIS Engine, and ArcGIS Server are all built using the ArcObjects libraries.

ArcSDE Technology for managing geographic information in a relational database management system (RDBMS). ArcSDE is part of the ArcGIS platform, and is the data server between ArcGIS and relational databases. It is widely used to enable geographic information to be shared by many users across a network and to scale in size from personal, to workgroup, to enterprise use.

ArcSDE administrative user The user who administers ArcSDE geodatabases. The ArcSDE administrative user can be the SDE user, but for DBO-schema ArcSDE geodata bases in SQL Server it is any user whose login is mapped to DBO in the database, and for user-schema geodatabases in Oracle, the ArcSDE administrative user is the user in whose schema the geodatabase is stored.

ArcSDE client application An application or program that communicates with ArcSDE to query, store and manage spatial data. Examples include ArcGIS Desktop, ArcGIS Server, and ArcIMS.

ArcSDE database server In ArcCatalog, an instance of SQL Server Express used to store ArcSDE geodatabases.

ArcSDE geodatabase A geodatabase stored in an RDBMS served to client applications using ArcSDE technology. An ArcSDE geodatabase can support long transactions and versioned workflows, be used as a workspace for geoprocessing tasks, and provide the benefits of a relational database such as security, scalability, backup and recovery, and SQL access.

ArcSDE Personal Edition geodatabase A single-user ArcSDE geodatabase that is created on Microsoft SQL Server Express.

ArcSDE system tables A collection of tables that store metadata about user tables in a geodatabase managed using ArcSDE technology. ArcSDE system tables are owned by an ArcSDE administrative user.

ArcSDE workgroup edition geodatabase A Microsoft SQL Server Express database that uses ArcSDE technology to store, query, and modify spatial data. Workgroup geodatabases accept up to 10 non-Web client connections and unlimited Web client connections to the SQL Server Express instance and are licensed through ArcGIS Server Workgroup edition.

ArcToolbox A user interface in ArcGIS used for accessing, organizing, and managing a collection of geoprocessing tools, models, and scripts.

ArcToolbox window A dockable window used to display, manage, and use the contents of toolboxes in ArcGIS. It provides a shortcut to frequently used tools contained within toolboxes that may be stored in folders or geodatabases on disk.

ArcView A GIS mapping software package suitable for desktop computers.

ArcView project In ArcView 3, a file for creating and storing documents for GIS work. All activity in ArcView 3 takes place within project files, which use five types of documents to organize information: views, tables, charts, layouts, and Avenue scripts. A project file organizes its documents and stores their unique settings in an ASCII format file with the extension .apr.

ArcView project file In ArcView 3, a file for creating and storing documents for GIS work. All activity in ArcView 3 takes place within project files, which use five types of documents to organize information: views, tables, charts, layouts, and Avenue scripts. A project file organizes its documents and stores their unique settings in an ASCII format file with the extension .apr.

ArcWeb explorer The client-side API of ArcWeb Services. It is based on Adobe Flex, a cross-platform development framework for creating rich Internet applications. Developers can use the ArcWeb Explorer Flex API, the JavaScript to Flex Bridge, or the JavaScript API to create their own applications. </SPAN>*ArcWeb Explorer* is sometimes used to refer to the customizable demo application.

ArcWeb mobile toolkit A J2ME-based toolkit for building mobile applications that use ArcWeb Services.

ArcWeb services ESRI-hosted Web services that include map data and on-demand geospatial capabilities needed to add real-time locations, addresses, points of interest, dynamic maps, and routing directions to Web and wireless applications.

ArcWeb services account A Web site for accessing information about an ArcWeb account. This site allows users to activate an ArcWeb Services account, view usage of ArcWeb services, manage POI and address records, and manage groups and alerts.

ArcWeb site An ESRI Web site for building custom services, using applications, and learning about ArcWeb Services.

ArcXML Acronym for *Arc Extensible Markup Language*. A file format that provides a structured method for communication between all ArcIMS components. ArcXML defines content for services and is used for requests and responses between clients, the business logic tier, and servers.

Are A metric areal unit of measure equal to 100 square metres. One are is equal to 1,076.39 square feet, or 0.025 acres.

Area A closed, two-dimensional shape defined by its boundary or by a contiguous set of raster cells.

Area chart A chart that emphasizes the difference between two or more groups of data; for example, the changes in a population from one year to the next. The area of interest is usually shaded in a solid colour.

Area of adjustment In Survey Analyst - Cadastral Editor, a continuous set of parcels that have been selected for adjustment by least-squares.

Area of interest The extent used to define a focus area for either a map or database production.

Areal scale The ratio or relationship between a distance or area on a map and the corresponding distance or area on the ground, commonly expressed as a fraction or ratio. A map scale of 1/1,00,000 or 1:1,00,000 means that one unit of measure on the map equals 1,00,000 of the same unit on the earth.

Argos A French-American system for positioning that uses a small transmitter which sends a signal from the surface to US weather satellites, then back down to ground-based processing centres in France and the US Doppler – shift distortion of the signal is used to determine position which is then transmitted back to the user via telephone, microwave links or other conventional communication channels.

Argument In computing, a value or expression passed to a function, command, or program.

Arithmetic expression A number, variable, function, or combination of these, with operators or parentheses, or both, that can be evaluated to produce a single number.

Arithmetic function A type of mathematical function that performs a calculation on the values of cells in an input raster. There are six arithmetic functions in ArcGIS Spatial Analyst: Abs, Int, Float, Round up (Ceil), Round down (Floor) and Negate.

Arithmetic operator The symbolic representation of a process or operation performed against one or more operands in an expression, such as '+' (plus, or addition) and '>' (greater than). When evaluated, operators return a value as their result. If multiple operators appear in an expression, they are evaluated in order of their operator precedence.

Array In computing, a fundamental data structure consisting of a variable with multiple, sequentially indexed, cells that can each store a value of the same type. Each cell of the array acts as a variable, and the cells are referenced by an index value for each array dimension. One-dimensional arrays, called vectors, and two-dimensional arrays, called matrices, are most common, but arrays may have more dimensions.

Artefact A feature on an image which is produced by the optics of the system or by digital image processing, and sometimes masquerades as a real feature.

Artificial neural network A computer architecture modelled after the human brain and designed to solve problems that human brains solve well, such as recognizing patterns and making predictions from past performance. Neural networks are composed of interconnected computer processors that calculate a number of weighted inputs to generate an output. For example, an output might be the approval or rejection of a credit application. This output would be based on several inputs, including the applicant's income, current debt, and credit history. Some of these inputs would count more than others; cumulatively, they would be compared to a threshold value that separates approvals from rejections. Neural networks 'learn' to generate better outputs by adjusting the weights and thresholds applied to their inputs.

Arundel method a combination of graphical and analytical methods, based on radial triangulation, for point-by-point topographic mapping from aerial photographs.

ASA index Index of the American Standards Association designating film speed, or sensitivity to light. Higher values indicate higher sensitivity. The ASA index has been replaced by the ISO index.

Ascending node (AN) Direction a satellite is travelling relative to the Equator. An ascending node implies a northbound Equatorial crossing.

ASCII Acronym for *American Standard Code for Information Interchange*. The de facto standard for the format of text files in computers and on the Internet that assigns a 7-bit binary number to each alphanumeric or special character. ASCII defines 128 possible characters.

ASP Acronym for *Active Server Pages*. A Microsoft server-side scripting technology that can be used to create and run dynamic, interactive Web applications, which are typically coded in JScript, JavaScript or VBScript. An ASP file contains not only the text and HTML tags that standard Web documents contain, but also commands written in a scripting language, which can be carried out on the server or the client.

ASP.NET A Microsoft-created programming framework built on top of the common language runtime (CLR) that can be used on a Windows server to create Web applications in a variety of programming languages.

Aspatial data Data without inherently spatial qualities, such as attributes.

Aspatial query A request for records of features in a table based on their attribute values.

Aspect The compass direction that a topographic slope faces, usually measured in degrees from north.

Aspect can be generated from continuous elevation surfaces. For example, the aspect recorded for a TIN face is the steepest downslope direction of the face, and the aspect of a cell in a raster is the steepest downslope direction of a plane defined by the cell and its eight surrounding neighbours.

Aspect ratio The ratio of the width of an image to its height. The aspect ratio of a standard computer monitor is 4:3 (rectangular).

Assembly A package of software and its associated resources. For example, an ArcGIS Win32 assembly will typically include executables, DLLs, object libraries, registry files, and help files for a unit of software.

Assignment operator A type of operator that assigns the result of an expression to an output, usually a raster, for storage.

Associated feature class In Survey Analyst - Cadastral Editor, a feature class that uses the cadastral fabric as a basemap, and has been associated with cadastral fabric. Associated feature classes can be corrected to maintain alignment with cadastral fabric parcels after least-squares adjustments of the cadastral fabric.

Association In UML, the relationship between two classes. In an association, instances of the classes in question usually exist together, but can exist on their own.

Assumed bearing A bearing measured from an arbitrarily chosen reference line called an assumed meridian.

ASTER (advanced space borne thermal emission and reflection radiometer) ASTER is one of five remote sensory devices on board the Terra satellite launched into Earth orbit by NASA in 1999. The instrument has been collecting data since February 2000. ASTER provides high-resolution images of the Earth in 14 different wavelengths of the electromagnetic spectrum, ranging from visible to thermal infrared light. The resolution of images ranges between 15 to 90 metres. ASTER data is used to create detailed maps of surface temperature of land, emissivity, reflectance, and elevation.

Astrolabe An instrument that measures the vertical angle between a celestial body and the horizontal plane at an observer's position. The astrolabe was replaced by the sextant in the fifteenth century for navigation, but modern versions are still used to determine local time and latitude.

Asychronous Not synchronized.

Asynchronous Not synchronous; that is, not occurring together or at the same time.

Asynchronous request In programming, a set of actions or events that may occur simultaneously. For example a program that launches another program, then continues execution while the other program is still running is said to be asynchronous.

ATI Apparent thermal inertia.

ATL Acronym for *Active Template Library*. A set of C++ template classes, developed by Microsoft for use in building Windows COM objects.

Atlas A collection of maps usually related to a particular area or theme and presented together. Examples of atlases include world atlases, historical atlases, and biodiversity atlases.

Atlas grid A grid of numbered rows and lettered columns (or vice versa) superimposed on a map, used to find and identify features. Alphanumeric grids are commonly used as a reference system on local street maps.

Atmosphere Layer of gases that surrounds some planets.

Atmospheric absorption The process whereby some or all of the energy of sound waves or electromagnetic waves is transferred to the constituents of the atmosphere.

Atmospheric classification A process whereby data (usually image data) are analysed and classified according to automatic or semi-automatic classification algorithms as opposed to purely manual methods.

Atmospheric correction Image-processing procedure that compensates for effects of selectivity scattered light in multispectral images.

Atmospheric shimmer An effect produced by the movement of masses of air with different refractive indices, which is most easily seen in the twinkling of stars. Shimmer results in blurring on remotely sensed images, and is the ultimate control over the resolution of any system.

Atmospheric transmission A process by which radiation is propagated through a medium – the atmosphere. Measured as transmittance, it is the ratio of the transmitted radiation to the total radiation incident upon the medium.

Atmospheric window Wavelength interval within which the atmosphere readily transmits electromagnetic radiation.

Atomic clock A clock that keeps time by the radiation frequency associated with a particular atomic reaction. Atomic clocks are used in official timekeeping.

Attenuation The dimming and blurring effects in remotely sensed images caused by the absorption and scattering of light or other radiation that passes through the earth's atmosphere. Decrease in the strength of a signal. The decrease in the strength of a signal is usually described by a multiplicative factor in the mathematical description of signal level. A signal is attenuated by application of a gain less than unity. Common causes of attenuation of an electromagnetic wave include losses through absorption and by volume scattering in a medium as a wave passes through.

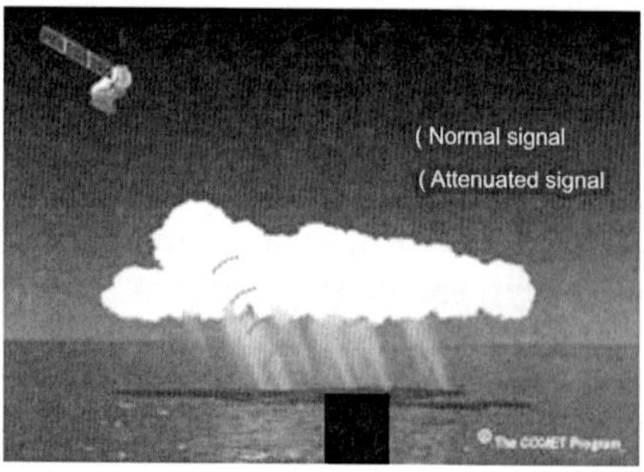

Attitude The angular orientation of a remote sensing system with respect to a geographical reference system. The orientation of the sensor along with information about the accuracy and precision with which this orientation is known. This data is required to perform proper calibration of instrument data. The attitude is usually stored in Euler angle or quaternion form and may be 1) calculated by the on-board computer and telemetered to the ground or 2) calculated by ground processing facilities using a variety of attitude sensor data.

Attractiveness A measure of the combined attributes of a centre or site that are considered positive features or that draw in potential customers or tenants.

Attribute A characteristic which describes a Feature. Attributes can be thought of as questions which are asked about the Feature. Attribute data describe the characteristics of spatial features.

Attribute data Tabular or textual data describing the geographic characteristics of features.

Attribute domain In a geodatabase, a mechanism for enforcing data integrity. Attribute domains define what values are allowed in a field in a feature class or nonspatial attribute table. If the features or nonspatial objects have been grouped into subtypes, different attribute domains can be assigned to each of the subtypes.

Attribute key An attribute or set of attributes in a database that uniquely identifies each record. A primary key allows no duplicate values and cannot be null.

Attribute query A request for records of features in a table based on their attribute values.

Attribute table A database or tabular file containing information about a set of geographic features, usually arranged so that each row represents a feature and each column represents one feature attribute. In raster datasets, each row of an attribute table corresponds to a certain zone of cells having the same value. In a GIS, attribute tables are often joined or related to spatial data layers, and the attribute values they contain can be used to find, query, and symbolize features or raster cells.

Attributes dialog box In ArcMap, a dialog box that displays attributes of selected features for editing.

Attribution The process of assigning attributes to features.

Authalic projection A projection in which the whole of the map as well as each part has the same proportional area as the corresponding part of the earth. An equal-area projection may distort shape, angle, scale, or any combination thereof. No flat map can be both equal-area and conformal.

Authentication The process of validating the identity of a user who logs on to a computer system, network, or Web site.

Authentication web service A SOAP ArcWeb service used to validate access to ArcWeb Services.

Authorization Completion of the software registration process. During authorization, the single-use or server product and/or extensions have been installed and registered, and an authorization file has been generated through the appropriate registration wizard and sent to the software user.

Authorization file A file that contains single use or server product authorization data. Each authorization file contains information regarding the feature name, version number, time-out date, registration number, and authorization code.

Autocorrelation The correlation or similarity of values, generally values that are nearby in a dataset. Temporal data is said to exhibit serial autocorrelation when values measured close together in time are more similar than values measured far apart in time. Spatial data is said to exhibit spatial autocorrelation when values measured nearby in space are more similar than values measured farther away from each other.

Automated cartography The process of making maps using computer systems that carry out many of the tasks associated with map production.

Automated digitizing The creation of vector data from raster data through automated tracing of pixels that are in close proximity and of the same or similar value.

Automated feature extraction The identification of geographic features and their outlines in remote-sensing imagery through post-processing technology that enhances feature definition, often by increasing feature-to-background contrast or using pattern recognition software.

Automated mapping/facilities management GIS or CAD-based systems used by utilities and public works organizations for storing, manipulating, and mapping facility information such as the location of geographically dispersed assets.

Automated text placement An operation in which text is automatically placed on or next to features on a digital map by software application according to rules set by the software user.

Automatic vehicle location (See also IVHS) any system using any sort of technology to locate or track a vehicle.

Automation The automatic functioning of a machine, system, or process, without the need for human interaction.

Automation scale The scale at which non digital data is made digital; for example, a map digitized at a scale of 1:24,000 has an automation scale of

Autovectorization

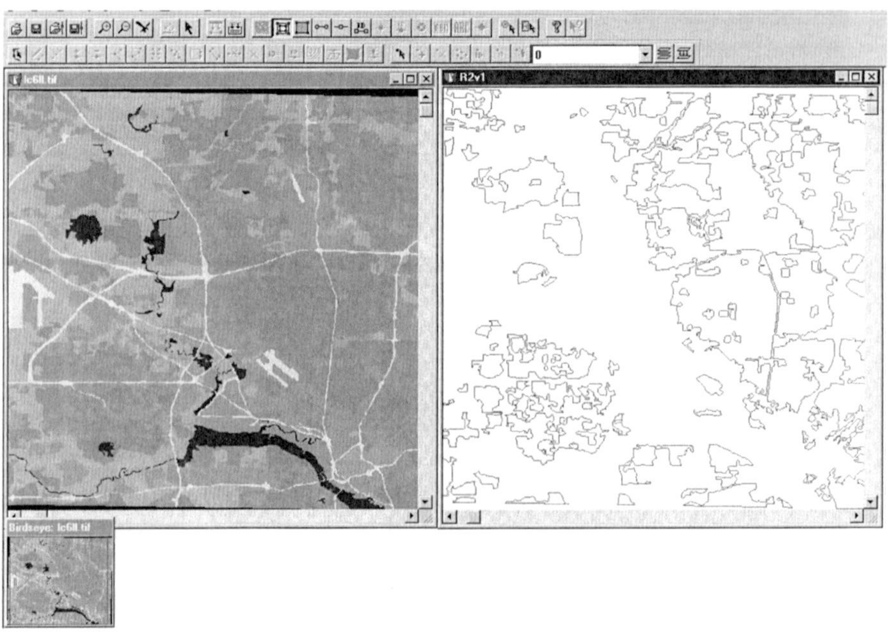

1:24,000. The data can be rendered at different display scales.

Autovectorization The creation of vector data from raster data through automated tracing of pixels that are in close proximity and of the same or similar value.

Availability The number of hours per day that a particular location has sufficient satellites (above the specified elevation angle and less than the specified PDOP value) to make a GPS position fix.

Avenue The object-oriented programming language on which ArcView 3.x is based. Avenue provides tools for customizing ArcView 3.x and developing ArcView 3.x applications.

Average point spacing The average distance separating sample points in a point dataset. A terrain dataset uses the average point spacing of a dataset to define a horizontal tiling system into which to divide input source measurements.

AVHRR Advanced Very High Resolution Radiometer, a multispectral imaging system carried by the TIROS-NOAA series of meteorological satellites. A scanner flown on National Oceanic and Atmospheric Administration (NOAA) polar-orbiting satellites for measuring visible and infrared radiation reflected from vegetation, cloud cover, shorelines, water, snow, and ice. AVHRR data is often used for weather prediction and vegetation mapping.

AVIRIS Airborne visible and infrared imaging spectrometer.

AWS ESRI-hosted Web services that include map data and on-demand geospatial capabilities needed to add real-time locations, addresses, points of interest, dynamic maps, and routing directions to Web and wireless applications.

AWX The client-side API of ArcWeb Services. It is based on Adobe Flex, a cross-platform development framework for creating rich Internet applications. Developers can use the ArcWeb Explorer Flex API, the JavaScript to Flex Bridge, or the JavaScript API to create their own applications. *ArcWeb Explorer* is sometimes used to refer to the customizable demo application.

Axis A line along which measurements are made in order to determine the coordinates of a location. It is also A Java-based toolkit for accessing SOAP Web services. Axis is developed by The Apache Software Foundation.

Azimuth Geographic orientation of line given as an angle measured in degrees clockwise from North.

Azimuth ambiguity A form of ghosting that occurs when the sampling of returned signals is too slow.

Azimuth bandpass filtering Bandpass filtering selects a certain band of frequency components in the signal. Azimuth band-pass filtering refers to filtering in the azimuth direction of the two-dimensional SAR signal. The location of signal energy in the azimuth frequency domain depends on the antenna pointing angle (Doppler centroid 2.6.1.2.2.), so band-pass filtering is necessary to maximize the signal energy in the processed image.

Azimuth beam inversion The azimuth Impulse Response Function (IRF) has amplitude which is modulated by the position of the echo in the azimuth beam pattern as the instrument passes the target. This modulation is removed by modifying the reference function by the inverse of the azimuth beam pattern (expressed as a function of Doppler frequency). This equalizes the echo data across the Doppler bandwidth but exaggerates the noise – thereby degrading the noise figure by the so-called Beamshaping loss.

Azimuth compression In the SAR signal domain, the raw data is spread out in the range and azimuth directions and must be coherently compressed to realize the full-resolution potential of the instrument. Azimuth compression consists of coherently correlating the received signal with the azimuth replica function. (See Azimuth Beam Inversion). The appropriate Hamming weighting is applied also to the reference function. Subsequent correlation has the effect of modulating both signal and noise by similar amounts and hence the signal-to-noise ratio is unchanged by this process

Azimuth descalloping Radiometric correction (descalloping) is performed with a vector multiply in the azimuth direction. See Descalloping in products glossary.

Azimuth direction In radar images, the direction in which the aircraft is heading. Also called flight direction.

Azimuth frequency see Doppler Frequency

Azimuth projection One type of map projection that retains certain accurate directions. Azimuthal projection also refers to one type of map projection that uses a plane as the projection surface.

Azimuth resolution In radar images, the spatial resolution in the azimuth direction.

Azimuth time The time along the flight path.

B

Background Area on an image or the terrain that surrounds an area of interest, or target. In ArcScene or ArcGlobe, the backdrop of the view. The colour of the background can be set to suggest sky, empty space, or any colour that improves visualization.

Background image A satellite image, aerial photograph, or scanned map over which vector data is displayed. Although a background image can be used to align coordinates, it is not linked to attribute information and is not part of the spatial analysis in a GIS.

Backscatter In radar, the portion of the microwave energy scattered by the terrain surface directly back toward the antenna. Or it is the portion of the outgoing radar signal that the target redirects directly back towards the radar antenna. Backscattering is the process by which backscatter is formed. The scattering cross section in the direction toward the radar is called the backscattering cross section; the usual notation is the symbol sigma. It is a measure of the reflective strength of a radar target. The normalized measure of the radar return from a distributed target is called the backscatter coefficient, or sigma nought, and is defined as per unit area on the ground. If the signal formed by backscatter is undesired, it is called clutter. Other portions of the incident radar energy may be reflected and scattered away from the radar or absorbed.

Backscatter coefficient A quantitative measure of the intensity of energy returned to a radar antenna from the terrain.

Backscattering Energy, when hitting a target, can be scattered in many directions. The part of the energy that is scattered back in the exact direction where it came from, is 'backscattered'.

Backshore Part of a beach that is usually dry and is reached only by the highest tides; by extension, a narrow strip of relatively flat coast bordering the sea.

Backup A copy of a file, a set of files, or a disk for safekeeping in case the original is lost or damaged.

BAM Acronym for best available map. The most suitable data source for a map.

Band (channel) A wavelength interval in the electromagnetic spectrum or a slice of wavelengths from the electromagnetic spectrum. Landsat ETM+ has eight bands that collect radiation from different parts of the electromagnetic spectrum. Of the eight bands, three bands are visible light, one band is panchromatic, three bands are infrared, and one band is thermal infrared.

Band gain The band gain condition detected at the start of a WRS scene. H indicates band acquired in high gain mode. L indicates band acquired in low gain mode.

Backshore

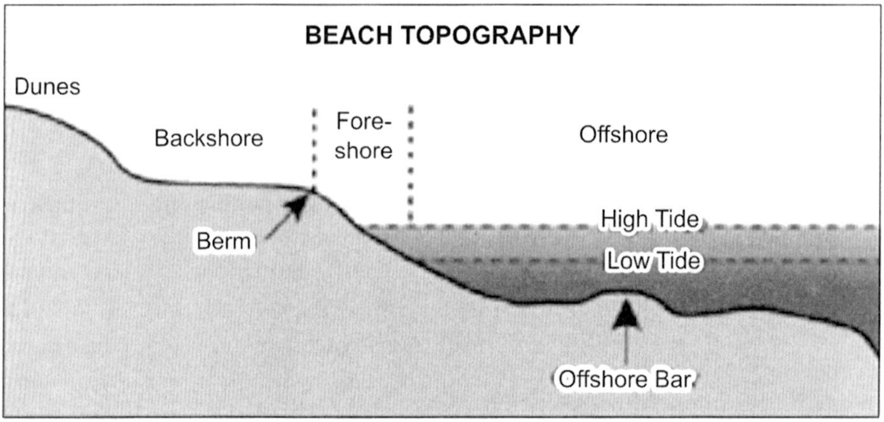

Band ratio A digital image processing technique that enhances contrast between features by dividing a measure of reflectance for the pixels in one image band by the measure of reflectance for the pixels in the other image band.

Band separate An image format that stores each band of data in a separate file.

Band-pass filter A wave filter that allows signals in a certain frequency to pass through, while blocking or attenuating signals at other frequencies.

Bandwidth A measure of the span of frequencies available in the signal or passed by the band limiting stages of the system. Bandwidth is a fundamental parameter of any imaging system and determines the ultimate resolution available, usually measured in bits per second (bps).

Bar scale A map element used to graphically represent the scale of a map. A scale bar is typically a line marked like a ruler in units proportional to the map's scale.

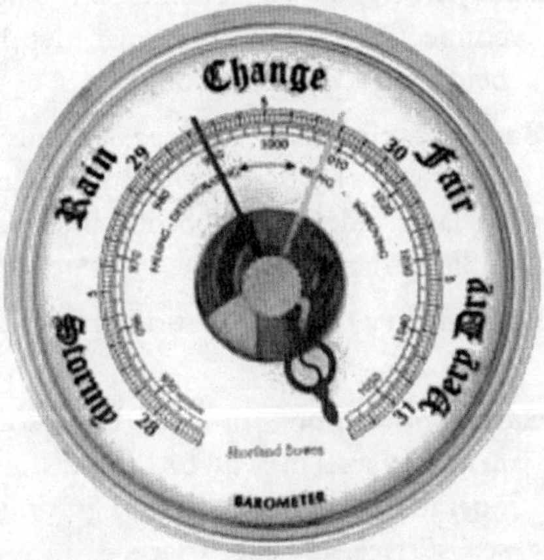

Barometer an instrument used to measure atmospheric pressure.

Barrier In network analysis, an entity that prevents flow from traversing a network edge or junction.

Base data Map data over which other thematic information is placed.

Base height In aerial photography, the height or altitude from which a photograph is taken.

Base layer A data layer in a GIS to which all other layers are geometrically referenced.

Base map A map depicting background reference information such as landforms, roads, landmarks, and political boundaries, onto which other thematic information is placed. A base map is used for locational reference and often includes a geodetic control network as part of its structure.

Base Station Also called a reference station. A receiver that is set up on a known location specifically to collect data for differentially correcting rover files. The base station calculates the error for each satellite and, through differential correction, improves the accuracy of GPS positions collected at unknown locations by a roving GPS receiver.

Base symbol In ArcGIS Tracking Analyst, the default symbol used to represent an event or a feature on a map.

Base table In a geodatabase, the spatially-enabled DBMS table that holds the main attribute values of a dataset. A business table with a spatial column is a feature class, and a business table with a raster column is a raster dataset or a raster catalogue. In the database, the business table name is the dataset name.

Base tag A text formatting tag that allows control of how the ESRI Maplex Labeling Engine places labels based on multiple fields relative to a feature. The field identified with the base tag is placed nearest the feature, and other fields are placed relative to the position of the base field.

Base-height ratio Air base divided by aircraft height. This ratio determines vertical exaggeration on stereo models.

Baseline A baseline consists of a pair of stations for which simultaneous GPS data has been collected.

Batch file A text file containing commands that is sent to the CPU to be executed automatically. A batch file allows the central processing unit (CPU) to process the commands at off-peak times or at a regularly scheduled time, rather than on demand from the user.

Batch geocoding The process of geocoding many address records at the same time.

Batch mode operation A procedure which uses a given ArcToolbox tool to process a set of information, or batch, rather than applying the tool to one piece at a time. Batch mode operation is available through tools in ArcGIS 8.3 and previous versions.

Batch processing Method of data processing in which data and programs are entered into a computer that carries out the entire processing operation with no further instructions.

Batch table In ArcToolbox, a table which displays the input name, user-selected parameters, and output name, where applicable, for all entries pertaining to a group, or batch, of jobs. Batch tables are available through tools in ArcGIS 8.3 and previous versions.

Batch vectorization An automated process that converts raster data into vector features for an entire raster or a portion of it based on user-defined settings.

Bathymetric curve A line on a map connecting points of equal depth below a hydrographic datum.

Bathymetric map A map representing the topography of a sea floor or lake bed, using contour lines to indicate depth.

Bathymetry Configuration of the sea floor.

Battle dimension In MOLE, the primary area in which a force unit operates, such as air, space, ground, sea, surface, and subsurface.

Battleships grid A grid of numbered rows and lettered columns (or vice

Bathymetry

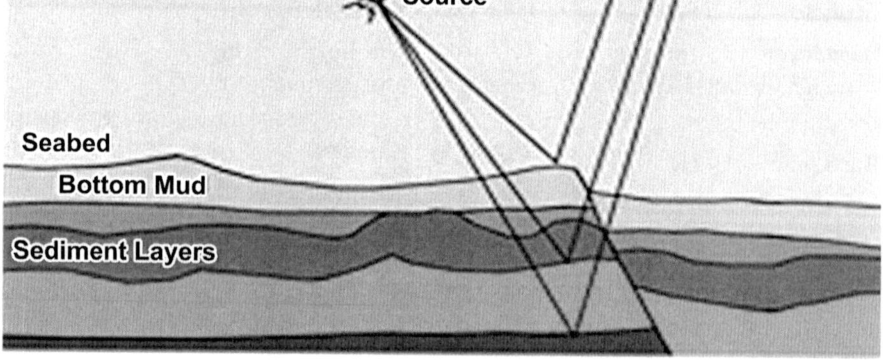

Battleships Grid

	1	2	3	4	5	6	7	8	9	10
A										
B										
C										
D										
E										
F										
G								X		
H								X		
I								X		
J										

versa) superimposed on a map, used to find and identify features. Alphanumeric grids are commonly used as a reference system on local street maps.

Baud rate In communications, the number of electrical cycles, or signals, transmitted per second. At lower transfer speeds the baud rate equals the data transfer rate measured in bps, or bits per second. Baud rate and bps are still sometimes used interchangeably, though inaccurately, since current standards allow for the encoding of multiple bits into a single cycle.

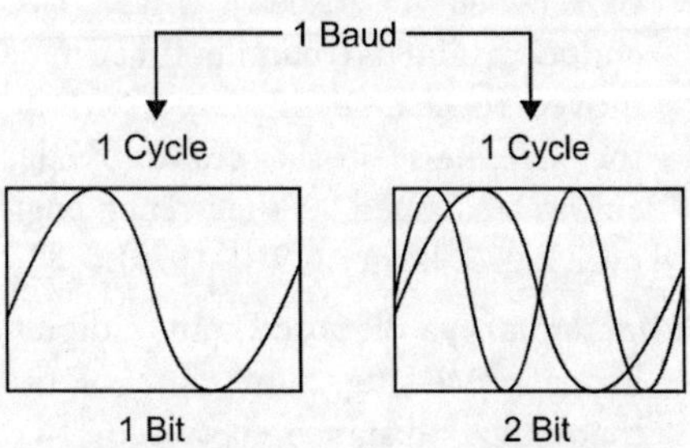

Bayes' theorem A theorem developed by English mathematician Thomas Bayes (1702–1761) about conditional probability. It states that the probability of a given event, given the original data and some new data, is proportional to the probability of the event given the original data only, and the probability of the new data given the original data and the event.

$$P(h|D) = \frac{P(D|h)P(h)}{P(D)}$$

$P(h|D)$ = posterior probability of h
$P(h)$ = prior probability of h
$P(D|h)$ = probability of observing D given that h holds
$P(D)$ = probability of observing D

Bayesian statistics A statistical approach to measuring likelihood. Bayesian estimates are based on the synthesis of a prior distribution and current sample data. Classical approaches to statistics estimate the probability of an event by averaging all possible data. The Bayesian approach, in contrast, weights probability according to actual data from a particular situation. It also factors in data from sources outside the statistical investigation, such as past experience, expert opinion, or prior belief. This outside information is described by a distribution that includes all possible values for the parameter.

Beach (sea beach) Zone of unconsolidated material that extends landward from the low water line to the place where there is marked change in material or physiographic form, or to the line of permanent vegetation (usually the effective lint of storm waves).

Beam A focused pulse of energy. The antenna beam of a side-looking radar (SLAR) is directed perpendicular to the flight path and illuminates a swath parallel to the platform ground track. Due to the motion of the satellite,

Bayesian Statistics

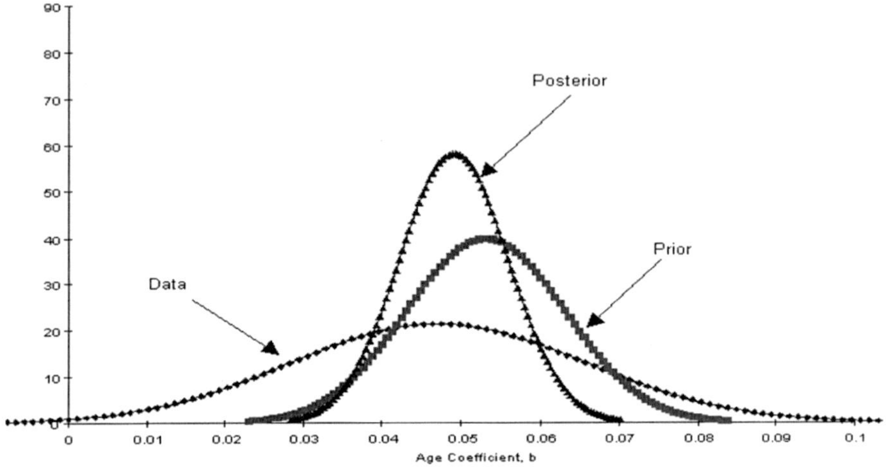

each target element is illuminated by the beam for a period of time, known as the integration time.

Beam mode The SAR operating configuration defined by the swath width and resolution.

Beam position The area within the total possible swath that is actually illuminated while being governed by the characteristics of a specific beam mode.

Bearing The compass direction from position to a destination.

Bearing method In Survey Analyst for field measurements, one of two methods for computing the coordinate geometry traverses. The bearing method uses compass directions for the orientation of each course.

Behaviour The actions or characteristics exhibited by an object in a database, as defined by a set of rule.

Bench mark Relatively permanent material object, natural or artificial, bearing a marked point whose elevation above or below an adopted datum is known.

Best available map The most suitable data source for a map.

Best route The route of least impedance between two or more locations, taking into account connectivity and travel restrictions such as one-way streets and rush-hour traffic.

Beta nought B° A radar brightness coefficient. The reflectivity per unit area in slant range which is dimensionless.

Bézier curve A curved line whose shape is derived mathematically rather than by a series of connected vertices. In graphics programs, a Bézier curve usually has two endpoints and two handles that can be moved to change the direction and the steepness of the curve. Bézier curves are named for the French engineer Pierre Bézier (1910–1999).

Bhattacharyya distance In digital image processing, a measure of the theoretical distance between two normal distributions of spectral classes, which acts as an upper limit on the probability of error in a Bayesian estimate of correct classification. Bhattacharyya distance is named for the Indian mathematician Anil Kumar Bhattacharyya (1915–1996).

Big endian A computer hardware architecture in which, within a multibyte numeric representation, the most significant byte has the lowest address and the remaining bytes are encoded in decreasing order of significance.

Bilinear interpolation A resampling method that uses a weighted average of the four nearest cells to determine a new cell value.

Billboarding A method for displaying graphics associated with features in a three-dimensional map display by posting them vertically as two-dimensional symbols and orienting them to always face the user.

Bin One of a series of equal intervals in a range of data, most commonly employed to describe the divisions in a histogram.

Binary Numerical system using the base 2. Based upon the integer two. Binary Code is composed of a combination of entities that can assume one of two possible conditions (0 or 1).

Binary file A file that contains data encoded as a sequence of bits (ones and zeros) instead of plain text. A binary file, such as a DLL or an executable file contains information that can be directly loaded or executed by a computer.

Binary large object A large block of data, such as an image, a sound file, or geometry, stored in a database. The database cannot read the BLOB's structure and only references it by its size and location.

Binding In computer programming, the process by which a program discovers an object's methods and properties.

Bingo grid A grid of numbered rows and lettered columns (or vice versa) superimposed on a map, used to find and identify features. Alphanumeric grids are commonly used as a reference system on local street maps.

Binomial distribution A distribution describing the probability of obtaining exactly K successes in N independent trials, where each trial results in either a success or a failure.

Biogeography The study of the geographical distribution of living things.

Biomass The total amount of organic matter in a defined area; usually refers to vegetation.

Bit A single digital unit of information. A bit can have one of two values, 1 and 0, that can represent on and off, yes and no, or true and false.

Bit depth The range of values that a particular raster format can store, based on the formula 2^n. An 8-bit depth dataset can store 256 unique values.

Bit error rate The fraction of a sequence of message bits that are in error (a bit error rate of 10–6 means that there is an average of one error per million bits).

Bitmap An image format in which one or more bits represent each pixel on the screen. The number of bits per pixel determines the shades of gray or number of colours that a bitmap can represent.

Black The signal sent to the human brain when human eyes are unable to detect optical photons.

Blackbody An ideal substance that absorbs all the radiant energy incident on it and emits radiant energy at the maximum possible rate per unit area at each wavelength for any given temperature. No actual substance is a true blackbody, although some substances, such as lampblack, approach its properties.

Blind digitizing A method of manual digitizing in which the operator has no graphic display on hand with which to see the digitized coordinates as they are captured.

Blind spot The point of the optic nerve to the retina where no radiation is detected by the eye.

BLOB Acronym for binary large object. A large block of data, such as an image, a sound file, or geometry, stored in a database. The database cannot read the BLOB's structure and only references it by its size and location.

Block In ArcGIS, a group of records in a compressed file geodatabase feature class or table that are stored together. The arrangement of compressed data into blocks helps optimize query performance.

Block attribute In CAD, a collection of objects that can be associated to form a single object.

Block group A unit of US census geography that is a combination of census blocks. A block group is the smallest unit for which the US Census Bureau reports a full range of demographic statistics. There are about 700 residents per block group. A block group is a subdivision of a census tract.

Block kriging A kriging method in which the average expected value in an area around an unsampled point is generated rather than the estimated exact value of an unsampled point. Block kriging is commonly used to provide better variance estimates and smooth interpolated results.

Blocking In ArcGIS, a geocoding indexing process that reduces the number of potential matches that need to be checked.

Blunder In surveying, a defective measurement that can be detected by a statistical test.

Bookmark In ArcWeb Services, a shortcut that saves the current map view, including extents, active layers, and styles for future use.

Boolean expression An expression, named for the English mathematician George Boole (1815–1864), that results in a true or false (logical) condition. For example, in the Boolean expression 'HEIGHT > 70 AND DIAMETER = 100', all locations where the height is greater than 70 and the diameter is equal to 100 would be given a value of 1, or true, and all locations where this criteria is not met would be given a value of 0, or false.

Boolean operation A GIS operation that uses Boolean operators to combine input datasets into a single output dataset.

Boolean operator A logical operator used in the formulation of a Boolean expression. Common Boolean operators include AND, which specifies a combination of conditions (A and B must be true); OR, which specifies a list of alternative conditions (A or B must be true); NOT, which negates a condition (A but not B must be true);

and XOR (exclusive or), which makes conditions mutually exclusive (A or B may be true but not both A and B).

Border arcs The arcs that create the boundary line of a polygon coverage.

Boundary A line separating adjacent political entities, such as countries or districts; adjacent tracts of privately-owned land, such as parcels; or adjacent geographic zones, such as ecosystems. A boundary is a line that may or may not follow physical features, such as rivers, mountains, or walls.

Boundary effect A problem created during spatial analysis, caused by arbitrary or discrete boundaries being imposed on spatial data representing non-discrete or unbounded spatial phenomena. Boundary problems include edge effects, in which patterns of interaction or interdependency across the borders of the bounded region are ignored or distorted, and shape effects, in which the shape imposed on the bounded area affects the perceived interactions between phenomena.

Boundary feature weight One of two types of feature weights that allow control of how labels are placed relative to polygon features in ArcMap. Higher feature weights prevent labels from being placed over features. A high boundary feature weight keeps labels off of the edge of a polygon, but does not prevent the label from being placed within the boundary.

Boundary line A division between adjacent political entities, tracts of private land, or geographic zones. Boundary lines may be imaginary lines, physical features that follow those lines, or the graphical representation of those lines on a map. Boundary lines between privately owned land parcels are usually called property lines.

Boundary monument Material object placed on or near a boundary line to preserve and identify the location of the boundary line on the ground.

Boundary network In Survey Analyst - Cadastral Editor, an irregular mesh of parcel boundaries, connection lines and control points representing a cadastral fabric. A boundary network represents parcels implicitly joined together, and is used by least-squares adjustment to distribute error from fixed control points based on the precision of boundary dimensions (bearings and distances).

Boundary survey Survey made to establish or to re-establish a boundary line on the ground, or to obtain data for constructing a map or plate showing a boundary line.

Bounding rectangle The rectangle, aligned with the coordinate axes and placed on a map display, that encompasses a geographic feature or group of features or an area of interest. It is defined by minimum and maximum coordinates in the x and y directions and is used to represent, in a general way, the location of a geographic area.

Bowditch rule A widely used rule for adjusting a traverse that assumes the precision in angles or directions and is equivalent to the precision in distances. This rule distributes the closure error over the whole traverse by changing the northings and eastings of each traverse point in proportion to the distance from the beginning of the traverse. More specifically, a correction factor is computed for each point as the sum of the distances along the traverse from the first point to the point in question, divided by the total length of the traverse. The correction factor at each point is multiplied by the overall closure error to get the amount of error correction distributed to the point's coordinates. The compass rule is also known as the Bowditch rule, named for the American mathematician and navigator Nathaniel Bowditch (1773–1838).

Bragg scattering Bragg Scattering is the enhanced backscatter due to coherent combinations of signals reflected from a rough surface having features, with periodic distribution in the direction of wave propagation, and whose spacing is equal to half of the wavelength as projected onto the surface.

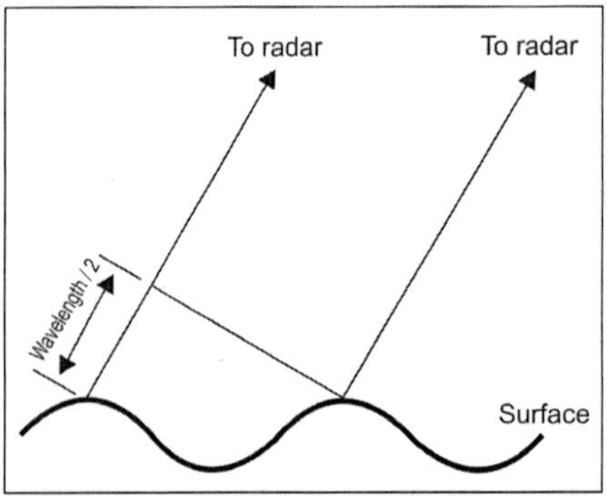

Break In ArcGIS Network Analyst, an object used in vehicle routing problem (VRP) analysis. A break can be used to model a specified period of rest along a route within a vehicle routing problem (VRP) instance.

Breakline A line in a TIN that represents a distinct interruption in the slope of a surface, such as a ridge, road, or stream. No triangle in a TIN may cross a breakline (in other words, breaklines are enforced as triangle edges). Z-values along a breakline can be constant or variable.

Bridging a method of establishing horizontal control points on a base map using aerial photographs.

Brightness Magnitude of the response produced in the eye by light. The attribute of visual perception in accordance with which an area appears to emit more or less light. Property of an image in which the strength of the radar reflectivity is expressed as being proportional to a digital number (digital image file) or to a grey scale (photographic image), which for a photographic positive shows bright as white. The attribute of visual perception in accordance with which an area appears to emit more or less light. Brightness may be a result of variations in tone, texture, or in the case of radar imagery, radar artefacts. The topography and surface roughness of the terrain will affect the image brightness. Where the local incidence angle is large, the image will be dark. Conversely, the image will be brighter where the local incidence angle is small.

Brightness theme In 3D Analyst and Spatial Analyst for ArcView 3.x, a grid theme whose cell values are used to vary the brightness of another grid theme. The cell values in one grid can be visually plotted against those in another. Most commonly, hillshade grids are used as brightness themes for elevation grids. The effect is to display the elevation surface in relief.

Browse data A reduced data volume representation of an image scene that can be viewed to determine general ground area coverage and cloud coverage. Browse data typically consist of three spectral bands.

Browse graphic An image associated with data to provide a general idea of what the service looks like. In ArcWeb Services, the browse graphic appears in the Content Library and Service Library and is defined in the metadata.

Brute force radar See real-aperture radar.

B-tree A tree data structure used for indexing data within a database or file system implementation. In a B-tree structure, data is sorted into a set of hierarchical nodes, normally using only three or four levels. The limited number of levels makes effective searches possible, because most of the nodes in the tree do not have to be accessed during a search.

Buffer A zone around a map feature measured in units of distance or time. A buffer is useful for proximity analysis.

Bug In computing, a flaw or error in a software program or hardware component that prevents it from performing the way it should.

Build In ArcGIS, the process of creating a network system. For geometric networks, this includes establishing connectivity, creating network features, and creating logical network tables. For network datasets, this includes establishing connectivity, creating network elements, and assigning network attribute values.

Build parcel In Survey Analyst - Cadastral Editor, a cadastral fabric editing command that creates a parcel from construction lines and legal records.

Builder A Web tool for creating and modifying content and publishing services. Users can access Builder through the ArcWeb Services Web-site.

Bus A set of conductors that provide communications links between the various functional components of a computer, such as memory and peripheral devices.

Business table In a geodatabase, the spatially-enabled DBMS table that holds the main attribute values of a dataset. A business table with a spatial column is a feature class, and a business table with a raster column is a raster dataset or a raster catalog. In the database, the business table name is the dataset name.

Bust A discrepancy between existing coordinates and computed coordinates that occurs when the final point of a closed traverse has known coordinates and the final course of a traverse computes different coordinates for the same survey point.

Button A command that executes a function, macro, or custom code when clicked.

Byte A group of eight bits of digital data.

C

C A common, flexible, programming language.

C4I In defense, an abbreviation used to signify that a computer program or system supports command, control, communication, computers, and information.

C/A code The standard (Clear/Acquisition) GPS code also known as the 'civilian acquisition code' or S-code, is a sequence of 1,023 pseudo-random binary bi-phase modulation of the GPS carrier wave at a chipping rate of 1.023 MH_Z and a code repetition period of 1 millisecond.

CAD Acronym for *computer-aided design*. A computer-based system for the design, drafting, and display of graphical information. Also known as computer-aided drafting, such systems are most commonly used to support engineering, planning, and illustrating activities.

CAD dataset The feature representation of a CAD file in a geodatabase-enforced schema. A CAD feature dataset is comprised of five read-only feature classes: points, polylines, polygons, multipatch and annotation. ArcGIS supported formats include DWG (AutoCAD), DXF (AutoDesk Drawing Exchange Format), and DGN (the default Microstation file format).

CAD drawing The digital equivalent of a drawing, figure, or schematic created using a CAD system.

CAD drawing dataset The pictorial representation of an entire CAD file that can be viewed in any ArcGIS application with a display. The CAD drawing dataset is a vector data source of a mixed feature type in which the symbology is set to mimic that of the originating CAD application. The graphic properties of a CAD drawing dataset's objects can be identified, but the dataset is not usable for feature class-based queries or analysis.

CAD feature class A read-only member of a CAD feature dataset, comprised of one of the following: polylines, points, polygons, multipatch, or annotation. The feature attribute table of a CAD feature class is a virtual table comprised of select CAD graphic properties and any existing field attribute values.

CAD feature dataset The feature representation of a CAD file in a geodatabase-enforced schema. A CAD feature dataset is comprised of five read-only feature classes: points, polylines, polygons, multipatch and annotation. ArcGIS supported formats include DWG (AutoCAD), DXF (AutoDesk Drawing Exchange Format), and DGN (the default Microstation file format).

CAD file The digital equivalent of a drawing, figure, or schematic created using a CAD system. CAD files are the data source for CAD drawing datasets, feature datasets and feature classes. ArcGIS software-supported formats include DWG (AutoCAD), DXF (AutoDesk Drawing Exchange Format), and DGN (the default Microstation file format). A CAD file is represented in ArcCatalog with a CAD feature dataset and a CAD drawing dataset.

CAD layer A layer that references a set of CAD data. CAD data is vector data of a mixed feature type. CAD layers may be of two types: CAD drawing dataset layers, in which one map layer represents the entire CAD file, and CAD feature layers, in which data is organized by geometry type.

CAD staging geodatabase A normalized, fixed set of feature classes and data tables of a predefined schema from a collection of input CAD drawings.

Cadastral fabric In Survey Analyst - Cadastral Editor, a network of connected parcels. Parcels are represented by parcel line features, parcel point features, and parcel polygon features, referred to in aggregate as parcel features. Parcel topology in the cadastral fabric is stored explicitly through shared or common parcel point features.

Cadastral fabric accuracy levels In Survey Analyst - Cadastral Editor, a number assigned to a parcel line that determines how much the line influences the coordinates that result from a least-squares adjustment. A line with a high accuracy level will have more influence than a line with a lower accuracy level. The highest accuracy level in the cadastral fabric is 1.

Cadastral fabric history In Survey Analyst - Cadastral Editor, the record of changes to the legal and system state of the cadastral fabric.

Cadastral fabric job In Survey Analyst - Cadastral Editor, a collection of parcels that have been extracted from the cadastral fabric for editing and least-squares adjustment. Parcels in the cadastral fabric are always edited in cadastral fabric jobs.

Cadastral fabric layer In Survey Analyst - Cadastral Editor, the map representation of a cadastral fabric for which display properties may be set; a layer in ArcMap that represents the cadastral fabric.

Cadastral fabric line point In Survey Analyst - Cadastral Editor, a point that allows a parcel corner to lie on an adjacent parcel boundary line without splitting the boundary line. Line points are constrained to lie on their parcel lines.

Cadastral fabric parcel line In Survey Analyst - Cadastral Editor, a parcel line feature in the cadastral fabric that contains dimension information. Represents parcel boundaries. Parcel lines connect up to form parcel polygons and always connect two point features in the cadastral fabric.

Cadastral fabric parcel line category In Survey Analyst - Cadastral Editor, a line type that defines how a parcel line in the cadastral fabric will be managed by the cadastral fabric editor. For example, connection lines are managed differently than boundary lines.

Cadastral fabric parcel point In Survey Analyst - Cadastral Editor, a point represents a parcel corner or the end of a connection line. A parcel point always has computed x- and y-coordinates.

Cadastral fabric sub layer In Survey Analyst - Cadastral Editor, an internal feature class that forms part of the cadastral fabric. Cadastral Fabric sublayers include lines, points, polygons, line points, and control points.

Cadastral fabric topology In Survey Analyst - Cadastral Editor, the topological relationships explicit in the cadastral fabric data model.

Cadastral survey Survey relating to land boundaries, made to create units suitable for title transfer or to define the limitations of title. Derived from 'cadastre' meaning a register of land quantities, values and ownership used for levying taxes; the term may properly be applied to surveys of a similar nature outside the public lands, such surveys are more commonly called 'land surveys' or 'property surveys'.

Cadastre An official record of the dimensions and value of land parcels, used to record ownership and assist in calculating taxes.

Calibration The act or process of comparing certain specific measurements in an instrument with a standard. In remote sensing the activities involved in adjusting an instrument to be intrinsically accurate, either before or after launch, i.e. instrumental calibration. It is also process of collecting instrument characterization information like scale, offset, nonlinearity, operational and environmental effects; using either laboratory standards, field standards or modelling, which is used to interpret instrument measurements, i.e. data calibration.

Calibration Data In remote sensing, measurements pertaining to the spectral or geometric characteristics of a sensor or radiation source. Calibration data are obtained by using a fixed energy source such as a calibration lamp, a temperature plate, or a geometric test pattern.

Calibration Parameter File (CPF) The Calibration Parameter File (CPF) supplies the radiometric and geometric correction parameters required during Level 1 processing to create superior products of uniform consistency across the Landsat 7 system. They fall into one of three major categories: geometric parameters, radiometric parameters, or artifact removal parameters. Each CPF is stamped with applicability dates; then sent to the LP DAAC for storage and eventual bundling with outbound Level 0 products. The Image Assessment System (IAS) updates and distributes the CPF at least every 90 days.

Callout line A line on a map extending between a feature's geographic position and its corresponding symbol or label, used in areas where there is not enough room to display a symbol or label in its correct location.

Calorie Amount of heat required to raise the temperature of 1g of water by 1 EC.

Camera An optical instrument using photographic film to reproduce a scene.

Camera station In aerial photography, each point in the flight path at which the camera exposes the film.

Camouflage detection photographs Another term for IR colour photograph.

Candidate A record returned as a potential match for an address in the geocoding process.

Candidate key In a relational database, any key that can be used as the primary key in a table.

Capacity In location-allocation, the maximum number of people or units that a centre can service, contain, or have assigned to it.

Caption In ArcGIS, the text for a command that appears with the 'Text Only' and 'Image and Text' display types. As part of the user interface, captions are customizable by the user.

Cardinal direction One of the four compass directions on the earth's surface: north, south, east, or west.

Cardinal point One of the four compass directions on the earth's surface: north, south, east, or west.

Cardinal point effect In radar, very bright signatures caused by optimally oriented corner reflectors, such as buildings.

Cardinality The correspondence or equivalency between sets; how sets relate to each other. For example, if one row in a table is related to three rows in another table, the cardinality is one to many.

Carrier A radio wave having at least one characteristic (e.g. frequency, amplitude, phase) that can be varied from a known reference value by modulation.

Carrier frequency The frequency of an unmodulated output of a radio transmitter. The GPS L1 carrier frequency is 1,575.42 MHz and the GPS L2 carrier frequency is 1,227.6 MHz.

Carrier phase GPS measurements based on the L1 or L2 carrier signal between a satellite and receiver.

Carrier smoothing A signal processing method by which the carrier wave is used to increase the accuracy of C/A code positioning by smoothing the variability in individual positions in a sequence of positions.

Carrier wave Simple radio sine wave characterized by frequency, amplitude and wavelength, any of which can be modulated from a known reference value.

Carrier-aided tracking A signal processing strategy that uses the GPS carrier signal to achieve an exact lock on the pseudo random code. More accurate than standard approach.

Carrying contour A single line representing multiple coincident contour lines, used to show vertical topographic features such as cliffs, cuts, and fills.

Cartesian coordinate system A two-dimensional, planar coordinate system in which horizontal distance is measured along an x-axis and vertical distance is measured along a y-axis. Each point on the plane is defined by an x,y coordinate. Relative measures of distance, area, and direction are constant throughout the Cartesian coordinate plane. The Cartesian coordinate system is named for the French mathematician and philosopher René Descartes (1596–1650).

Cartogram A diagram or abstract map in which geographical areas are distorted proportionally to the value of an attribute.

Cartographer One who practices the art and science of expressing graphically, usually through maps, the natural and social features of the earth.

Cartographic generalization The abstraction, reduction, and simplification of features so that a map is clear and uncluttered at a given scale.

Cartography Science and art of making maps and charts. The term may be taken broadly as comprising all the steps needed to produce a map: planning, aerial photography, field surveys, photogrammetric, editing, colour separation, and multicolour printing. Mapmakers, however, tend to limit use of the term to the map-finishing operations, in which the master manuscript is edited and colour separation plates are prepared for lithographic printing.

Cartouche An ornamental frame on a map, usually around the map's title. Cartouches are rarely used on modern maps.

Cascading style sheets A standard for defining the layout or presentation of an HTML or XML document. Style information includes font size, background colour, text alignment, and margins. Multiple style sheets may be applied to 'cascade' over previous style settings, adding to or overriding them. The World Wide Web Consortium maintains the CSS standard.

CASE Acronym for computer-aided software engineering. Any software that assists with the development and maintenance of software, especially the analysis and design. Complex tasks that often require many lines of code are simplified with CASE user interfaces and code generators.

Catalog tree In ArcCatalog, a hierarchical view of folder connections which provide access to GIS data stored on local disks or shared on a network that allows users to manage connections to databases and GIS servers.

Catchment A basin-like terrestrial region consisting of all the land that drains water into a common terminus.

Categorical data A data that is non-numerical. Example, nominal or an ordinal scale.

Categorical raster A raster that typically represents phenomena that have clear boundaries with attributes that are descriptions, classes, or categories. Generally, integers are used for the cell values. In a raster of land cover, for example, the value 1 might represent forest land, the value 2 urban land, and so on. It is assumed that the phenomena that each value represents fill the entire area of the cell. Rasters representing land use, political boundaries or ownership are examples of discrete rasters.

Category In ArcWeb Services, a collection of related data files, layers, data sources, or classes that show the primary use for data and help with display and navigation. Administrative, cultural, and elevation are examples of categories within ArcWeb Services.

Cathode ray tube (CRT) A vacuum tube with a phosphorescent screen on which images are displayed by an electron beam.

CATID Acronym for Component Category ID. A unique string assigned to locally related COM classes to group them together. A CATID is a type of Globally Unique IDentifier (GUID).

C-Band A nominal frequency range, from 8 to 4 Ghz (3.75 to 7.5 cm wavelength) within the microwave (radar) portion of the electromagnetic spectrum. C-band has been the frequency of choice for several experimental aircraft SAR systems as well as a series of single-band satellite SAR systems, including the European ERS–1, ERS–2 and Envisat SAR systems and Canada's RADARSAT SAR. The corresponding wavelength for these systems is on the order of 5.6 cm, which has been found useful in sea ice surveillance as well as in other applications. Imaging radars equipped with C-band are generally not hindered by atmospheric effects and are capable of 'seeing' through tropical clouds and rain showers. Its penetration capability with regard to vegetation canopies or soils is limited and is restricted to the top layers. C-band is also used in range instrumentation radars.

CBSA A geographic region containing at least one urban area with a population of at least 10,000, defined by the US Office of Management and Budget for use by federal statistical agencies, including the US Census Bureau. A core-based statistical area can be a metropolitan statistical area or a micropolitan statistical area.

CCD Charge-coupled detector. CCD is a device for the movement of electrical charge, usually from within the device to an area where the charge can be manipulated into a digital value (Digital Number) value.

CCT Computer-compatible tape.

CD An optical disk, slightly less than 5 inches in diameter, used to store up to approximately 650 megabytes of data.

Celestial sphere the sky, considered as the inside of a sphere of infinitely large radius that surrounds the earth, on which all celestial bodies except the earth are imagined to be projected.

Cell The smallest unit of information in raster data, usually square in shape. In a map or GIS dataset, each cell represents a portion of the earth, such as a square metre or square mile, and usually has an attribute value associated with it, such as soil type or vegetation class.

Cell assemblies The linked receptors, retinal neurons, and neural cells in the visual cortex of the brain which enable interaction between perception and past experience.

Cell selection The process of selecting raster cells either interactively or by using a SQL query.

Cell size The dimensions on the ground of a single cell in a raster, measured in map units. Cell size is often used synonymously with pixel size.

Cellular automaton A mathematical construction consisting of a row or grid of cells in which each cell has an initial value—from a known and limited number of possible values—and all cells are simultaneously evaluated and updated according to their internal states and the values of their odelling. The simplest cellular automaton is a row in which each cell has one of two values, such as red or green. In this case, there are eight possible value combinations for a cell and its odelling. (If a green cell with two red odelling is notated RGR, then the eight combinations are RRR, RRG, RGR, GRR, RGG, GRG, GGR, GGG.) A set of rules determines whether or not a cell changes value when it is evaluated. A sample rule might be, 'A green cell becomes red if it has a red odelling on both sides.' Successive updates, or generations, of a cellular automaton may produce complex patterns. Cellular automata are of interest in spatial odelling and are often used to model land-cover change.

Census block The smallest geographic entity for which the US Census Bureau tabulates decennial census data. Many blocks correspond to city blocks bounded by streets, but blocks in rural areas may include several square miles and have some boundaries that are not streets. The Census Bureau established blocks covering the entire nation for the first time in 1990. Previous censuses dating back to 1940 had blocks established only for part of the nation.

Census geography Any one of various types of precisely defined geographic areas used by the US Census Bureau to collect and aggregate data. The largest unit of area is the entire United States, while the smallest is a census block.

Census tract A small, statistical subdivision of a county that usually includes approximately 4,000 inhabitants but may include from 2,500 to 8,000 inhabitants. A census tract is designed to encompass a population with relatively uniform economic status, living conditions, and some demographic characteristics. Tract boundaries normally follow physical features but may also follow administrative boundaries or other nonphysical features. A census tract is a combination of census block groups.

Central lines the central parallel and the central meridian. Together, they define the centre or the origin of a map projection.

Central meridian The line of longitude that defines the centre and often the x-origin of a projected coordinate system. In planar rectangular coordinate systems of limited extent, such as state plane, grid north coincides with true north at the central meridian.

Centre point The optical centre of a photograph.

Centreline vectorization The generation of vector features along the centre of connected cells. It is typically used for vectorizing scanned parcel and survey maps.

Centroid The geometric centre of a feature. Of a line, it is the midpoint; of a polygon, the centre of area; of a three-dimensional figure, the centre of volume.

CGI Acronym for Common Gateway Interface. A standard for scripts that run external programs from a World Wide Web server. CGI typically specifies how to pass arguments to the

Centroid

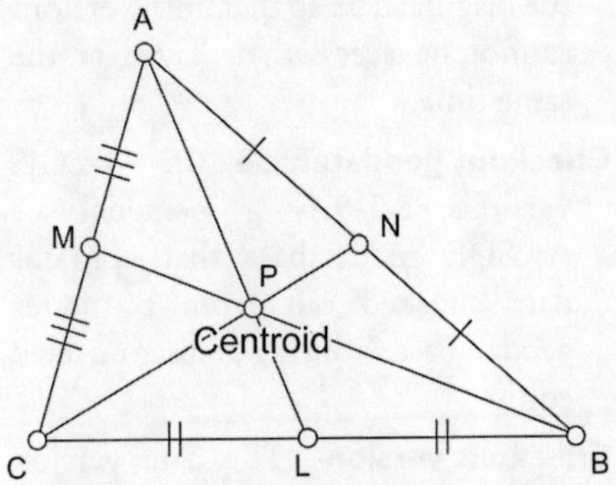

program via HTTP requests; defines a set of environmental variables made available to the program; and generates output, usually in HTML format, that is passed back to the browser. CGI scripts are frequently designed to access information in a database and format the results as HTML, convert information retrieved from an interactive Web page into a database, send datasets, and so on.

Chain Unit of length equal to 66 feet, used especially in the US public land surveys. The original measuring instrument (Gunter's chain) was literally a chain consisting of 100 iron links, each 7.92 inches long. Steel-ribbon tapes began to supersede chains around 1900, but surveying tapes are often still called 'chains' and measuring with a tape is often called 'chaining.' The chain is a convenient unit in cadastral surveys because 10 square chains equal 1 acre.

Chain code A method of drawing a polygon as a series of straight line segments defined as a set of directional codes, with each code following the last like links in a chain.

Change detection A process that measures how the attributes of a particular area have changed between two or more time periods. Change detection often involves comparing aerial photographs or satellite imagery of the area taken at different times. The process is most frequently associated with environmental monitoring, natural resource management, or measuring urban development.

Change-detection images A difference image prepared by digitally comparing images acquired at different times. The grey tones or colours of each pixel record the amount of difference between the corresponding pixels of the original images.

Channel A one-way communication link. The individual data streams from a multispectral device. A range of wavelength intervals selected from the electromagnetic spectrum.

Character A letter, digit, or special graphic symbol treated as a single unit of data and usually stored as one byte.

Chart Special-purpose map designed for navigation or to present specific data or information. The term 'chart' is applied chiefly to maps made primarily for nautical and aeronautical navigation, and to maps of the heavens, although the term is sometimes used to describe other special-purpose maps.

Chart, aeronautical Charts designed to meet requirements of aerial navigating, produced in several series, each on a specified map projection and differing in scale, format and content, for use as dictated by type of aircraft and whether the flight is to be conducted under visual or instrument flight rules.

Chart, bathymetric See: map, bathymetric

Chart, nautical Representation of a portion of the navigable waters of the earth and adjacent coastal areas on a specified map projection and designed specifically to meet requirements for marine navigation. Included on most nautical charts are depths of water, characteristics of the bottom, elevations of selected topographic features, general configurations and characteristics of the coast, the shoreline (usually the mean high water line), dangers, obstructions and aids to navigation limited tidal data, and information about magnetic variation in the charted area.

Check plot A plot of digitizing map e.g. *Chirp* - Typical phase coding or modulation applied to the range pulse of imaging radar designed to achieve a large time-bandwidth product. The resulting phase is quadratic in time, which has a linear derivative. Such coding is often called linear frequency modulation, or linear FM.

Check-in In disconnected editing, the procedure that transfers a copy of data into a master geodatabase, overwriting the original copy of that data and re-enabling it so it can be accessed and saved from that location. In checkout/check-in replication, check-in is the procedure that synchronizes the data in the parent replica with that in the child replica..

Checkout A procedure in disconnected editing that records the duplication of data from one geodatabase to another and disables the original data so that both versions cannot be accessed or saved at the same time.

Checkout geodatabase In ArcGIS versions 8.3–9.1, a personal or ArcSDE geodatabase that contains data checked out from a master geodatabase during disconnected editing.

Checkout version The data version created in a checkout geodatabase when data is checked out to that database during disconnected editing. The checkout version is created as a copy of the synchronization version. Only the edits made to this checkout version can be checked back in to the master geodatabase.

Checkout/check-in replication A type of geodatabase replication that involves copying data to a destination geodatabase, editing that data in the destination, and then merging the changes with the source geodatabase. In ArcGIS and ArcSDE, the destination can be a file, personal, or ArcSDE geodatabase, while the source must be an ArcSDE geodatabase. Once the data is merged (synchronized), checkout/check-in replication is completed.

Child replica In geodatabase editing, data that has been copied to a destination geodatabase during the replication process.

Chirp compression The echo signal is correlated with a suitable reference function. This correlation is performed in the frequency domain after suitable Fast Fourrier Transform from the time domain. The

Chirp Frequency (1)

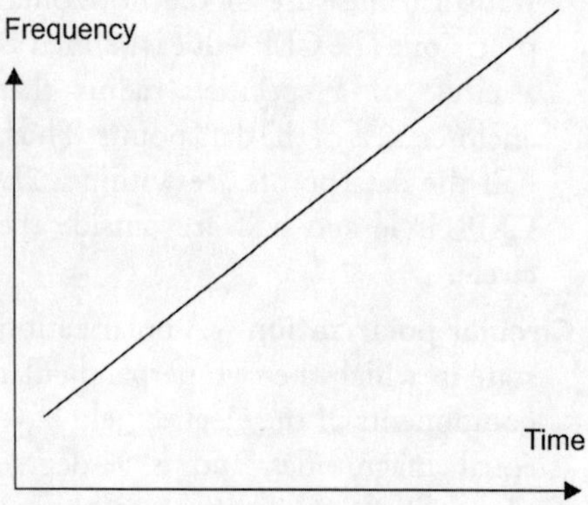

Chirp Frequency (2)

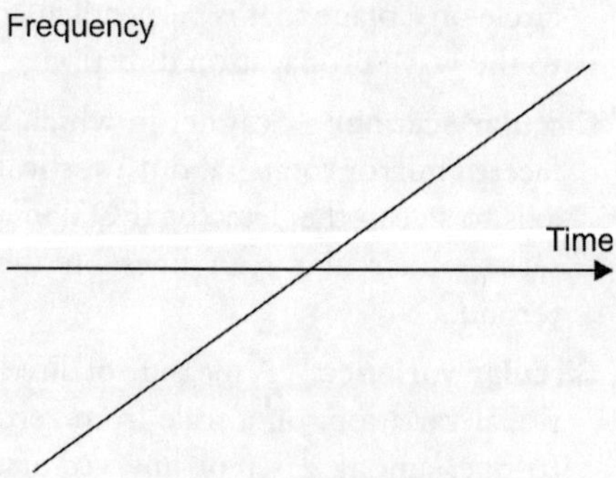

Chirp Frequency (3)

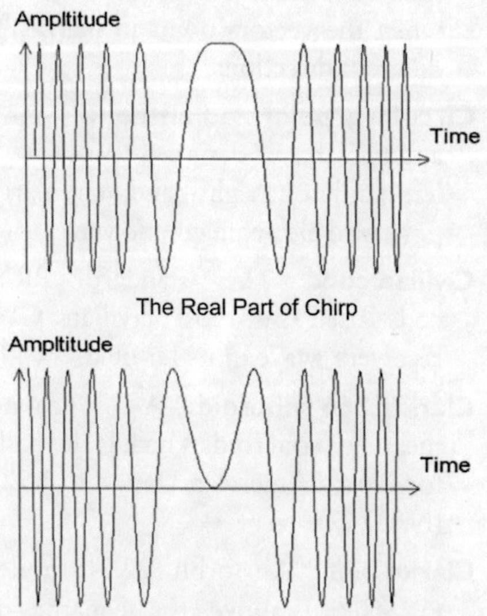

reference function of interest should represent the chirp signal which illuminates the target. This chirp varies from one swath to another due to the phase and amplitude settings introduced to produce beam steering. The beam steering (and shaping if necessary) is achieved by changing the excitations - phase and/or amplitude - to the many phase centres of the planar array antenna. The reference function should reflect these differences. In ASAR the phase of the excitation to any one row is constant along its length. Thus, any row can be considered to be homogeneous. The characteristics of such a row are determined by exciting just that row - and hence deriving a reference function for it. A total reference function can then be derived by combining 32 such reference functions - using the commanded excitation phases.

Chi-square statistic A statistic used to assess how well a model fits the data. It compares categorized data with a multinomial model that predicts the relative frequency of outcomes in each category to see to what extent they agree.

Chlorosis Yellowing of plant leaves resulting from an imbalance in the iron metabolism caused by excess concentrations of copper, zinc, manganese, or other elements in the plant.

Chord A straight line that joins two points on a curve.

Choropleth map A thematic map in which areas are distinctly coloured or shaded to represent classed values of a particular phenomenon.

Chroma The saturation, purity, or intensity of a colour.

Chromatic aberration colour distortion in an image produced by a lens, caused by the inability of the lens to bring the various colours of light to focus at a single point.

Chromatic vision The perception by the human eye of changes in hue. The normal ability to see colour.

Chronometer An extremely accurate clock that remains accurate through all conditions of temperature and pressure. The chronometer was developed in the eighteenth century for determining longitude at sea, but its scientific and navigational use has been made obsolete by the invention of quartz and atomic clocks.

CHUM Acronym for Chart Updating Manual. A document containing updates to aeronautical information, used by the US military to update their current published products with the latest information.

Circle A two-dimensional geometric shape for which the distance from the centre to any point on the edge is equal; the closed curve defining such a shape.

Circular arc A curved line that is a section of a circle, with two vertices, one situated at each endpoint.

Circular error probable (CEP) A statistical measure of the horizontal precision. The CEP value is defined as a circle of a specified radius that encloses 50% of the data points. Thus, half the data points are within a 2D CEP circle and half are outside the circle.

Circular polarization A polarization state in which the two perpendicular components of the electric field have equal magnitudes and a 90-degree phase difference. In this case, the tip of the electric field vector traces a circle on a plane that is perpendicular to the wave propagation direction.

Circular scanner Scanner in which a faceted mirror rotates about a vertical axis to sweep the detector IFOV in a series of circular scan lines on the terrain.

Circular variance A measure of directional variation, on a scale from zero to one, among a set of line vectors. Circular variance approaches zero when all vectors point in roughly the same direction and approaches one when the vectors point in markedly different directions.

Circularly polarized antenna An antenna that is designed to radiate a left-hand or right-hand circularly polarized electromagnetic wave.

Civilian code The standard PRN code used by most civilian GPS receivers' wave in its far field.

Clark 1,866 spheroid A ground measured spheroid, which is the basis for North American Datum of 1,927 (NAD 27).

Clarke belt An orbit 22,245 miles (35,800km) above the equator in which a satellite travels at the same speed that the earth rotates. The Clarke Belt was named after the writer and scientist Arthur C. Clarke. It is also referred to as a geostationary orbit.

Clarke ellipsoid of 1866 A reference ellipsoid having a semi-major axis of approximately 6,378,206.4 metres and a flattening of 1/294.9786982. It is the basis for the North American Datum of 1927 (NAD27) and other datums. The Clarke ellipsoid of 1,866 is also known as the Clarke spheroid of 1866.

Class A set of entities grouped together on the basis of shared attribute values.

Class identifier A COM term referring to the globally unique number that is used by the system registry and the COM framework to identify a particular coclass.

Class intervals A set of categories for classification that divide the range of all values so that each piece of data is contained within a non-overlapping category.

Classification Process of assigning individual pixels of an image to categories, generally on the basis of spectral reflectance characteristics. In the computer, when image pixels are the same colour, or nearly the same colour, an image 'classification' computer program can recognize this and group such pixels together. Such a grouping is called a 'class' and the process of doing the grouping is called 'classification'. The remote sensing researcher then has the challenge of identifying just what each 'class' represents in the real environment (pine trees? pavement? shallow water? dry grass?).

Classification table An ASCII file in the geocoding rule base that identifies and classifies keywords that may appear in an address, such as street types and directions. Classification tables have a .cls file extension.

CLDC Acronym for Connected Limited Device Configuration. A framework for developing J2ME applications for devices with very limited resources, such as wireless devices.

Clean data Data that is free from error.

Cleaning Improving the appearance of scanned or digitized data by correcting overshoots and undershoots, closing polygons, performing coordinate editing, and so on.

Clearinghouse A repository structure, physical or virtual, that collects, stores, and disseminates information, metadata, and data. A clearinghouse provides widespread access to information and is generally thought of as reaching or existing outside organizational boundaries.

Client An application, computer, or device in a client/server model that makes requests to a server.

Client sample A set of live and downloadable code samples designed to show users how to access an ArcWeb service. The live samples allow users to interact with an ArcWeb service using a browser. The downloadable code samples are available in several programming languages.

Client/server architecture A software system with a central processor (server) that accepts requests from one or more user applications, computers, or devices (clients). Although client/server architecture can exist on one computer, it is more relevant to (and is typically thought of as relating to) network systems that distribute applications over computers in different locations.

Client-side address locator An address locator that is created and used on the same computer.

Clinometer An instrument for measuring angles of slope (or tilt), elevation or inclination of an object with respect to gravity. Clinometers measure both inclines (positive slopes, as seen by an observer looking upwards) and declines (negative slopes, as seen by an observer looking downward).

Clip A command in GIS environment.

Clock bias The difference between the clock's indicated time and true universal time.

Cloning In object-oriented programming, the process of creating a new instance of a class with the same state as an existing instance.

Close coupling A high or complex degree of interconnections between the components within a program or between programs, that requires substantial overlap between methods, ontologies, class definitions, and so on.

Closed loop traverse In surveying, a traverse that starts and ends with the same survey point.

Closest facility analysis In ArcGIS Network Analyst, a type of network analysis for finding the closest locations (facilities) from sites (incidents), based on the impedance chosen—for example, finding hospitals near a car accident. When finding the closest facilities, users can specify how many to find and whether the direction of travel is toward or away from the site (incident). Users can also specify a cut-off threshold beyond which ArcGIS Network Analyst will not search for a facility—for example, finding hospitals within 6 miles of a car accident.

Closure error A discrepancy between existing coordinates and computed coordinates that occurs when the final point of a closed traverse has known coordinates and the final course of a traverse computes different coordinates for the same survey point.

Closure report The summary of the difference between the endpoint coordinate of a traverse and the calculated endpoint.

Cloud If there are clouds in the area and an optical satellite like SPOT, IRS or Landsat passes overhead, the satellite image will show the cloud but the features below the cloud won't be seen. A cloud shadow, the same shape as the cloud, will be seen nearby.

Clustering Multivariate statistical technique which separates image data into groups such that the between-group variance of the specified number of groups is maximized. A mathematical procedure for organizing multispectral data into spectrally homogeneous groups.

Clutter Unwanted echoes in a radar return. See also Backscatter.

CMOS chip A digital image-plane chip, similar to CCD chips. The acronym stands for Complementary

Metal Oxide Semiconductor. Much rarer in use than CCD chips. The differences between CCD and CMOS chips are primarily in the way the data from the individual detectors are transmitted from the data chip.

CMYK A colour model that combines the printing inks cyan, magenta, yellow, and black to create a range of colors. Most commercial printing uses this colour model.

CNT A component of the PLTS knowledge base that contains SQL statements and custom code for feature validation extended beyond standard geodatabase domains. PLTS utilizes condition tables for enhanced validation during both database production and quality control.

Coarse acquisition (C/A) A Code spread spectrum direct sequence code that is used primarily by commercial GPS receivers to determine the range to the transmitting GPS satellite.

Coastal zone colour scanner (CZCS) A satellite-carried multi-spectral scanner designed to measure chlorophyll concentrations in the oceans.

Coclass A template for an object that can be instantiated in memory.

Code phase GPS measurements based on the C/A code.

Coded value domain A type of attribute domain that defines a set of permissible values for an attribute in a geodatabase. A coded value domain, consists of a code and its equivalent value. For example, for a road feature class, the numbers 1, 2, and 3 might correspond to three types of road surface: gravel, asphalt, and concrete. Codes are stored in a geodatabase, and corresponding values appear in an attribute table.

Code-phase GPS GPS measurements calculated using the PRN code transmitted by a GPS satellite.

Cognitive map A person's perception of a place. A mental map may include the physical characteristics of a place, such as boundaries of a neighbourhood, or the attributes of a place, such as a neighbourhood's perceived unsafe areas. A mental map is primarily a psychological construct, although it may also be rendered as an actual map.

COGO Acronym for coordinate geometry. A method for calculating coordinate points from surveyed bearings, distances, and angles.

COGO composite measurement In Survey Analyst for field measurements, measurements that comprise a set of interdependent COGO simple measurements.

Coherence Coherence is the fixed relationship between waves in a beam of electromagnetic (EM) radiation. Two wave trains of EM radiation are coherent when they are in phase. That is, they vibrate in unison. In terms of the application to things like radar, the term coherence is also used to describe systems that preserve the phase of the received signal.

Coherence in an Interferogram In an interferogram, coherence is a measure of correlation. It ranges from 0.0, where there is no useful information in the interferogram; to 1.0, where there is no noise in the interferogram (a perfect interferogram). Both extremes are rarely seen – most images lie somewhere in between. Coherence is affected by, (in approximate order): Local slope (steep slopes lead to low coherence). Properties of the surface being imaged (vegetated or moving surfaces have low coherence). Time lag between the passes in an interferogram (long lags lead to low coherence). The baseline (large baselines lead to low coherence). Technical details of the generation of the interferogram (poor co-registration or resampling leads to low coherence). Coherence can serve as a measure of the quality of an interferogram; tells you more about the surface type (vegetated vs. rock); or tells you when a tiny, otherwise invisible change has occurred in the image, and it is only visible in the phase image of an interferogram. High coherence makes for attractive, not-noisy interferograms. Here is a coherent phase image of a mountain.

Coherent Phase Image of a Mountain

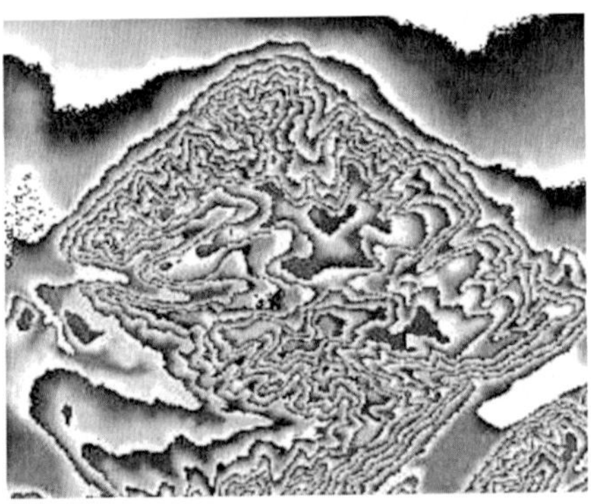

Low coherence makes unattractive, noisy interferograms, as shown below. Often, these interferograms are difficult to phase unwrap.

Incoherent Phase Image of a Mountain

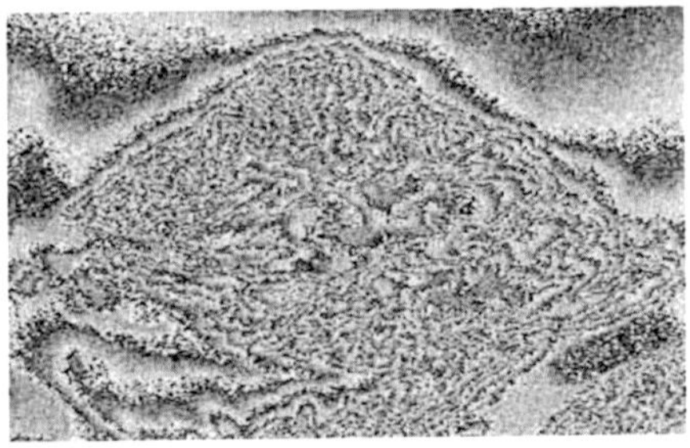

Coherence is the magnitude of an interferogram's pixels, divided by the product of the magnitudes of the original image's pixels. It is usually calculated on a small window of pixels at a time, from the complex interferogram and images. Interferometric SAR (InSAR) techniques make use of the coherence of the radar signal and the fact that the signal phase is equal to twice the path length between the sensor and the earth's surface. The phase difference between measurements generated from two SAR images with the sensor separated by a baseline, allows measurement of the slant range difference to fractions of a radar wavelength. The slant range difference can be geometrically related to the terrain height. By subtracting two signals, you generate an interference pattern. By subtracting the phases of two co-registered SAR images, you generate an interferogram. There are two basic SAR interferometry methods. In the first, two antennas are placed on the same platform and simultaneously acquire images of the scene from two different angles. The relative phase difference may then be used to construct a Digital Elevation Model (DEM). In the second, a pair of images from the same sensor are taken at different times. This is now a well-proven concept and has been demonstrated with several space borne SAR sensors, including RADARSAT. For this repeat-pass interferometry, the scenes are acquired at different times, so there is a time difference as well as viewing geometry to consider. The passes must have rather similar geometry in order to allow extraction of the relative phase difference. This usually requires that the satellite be on an exact repeat orbit.

Coherent noise A spurious, periodic pattern of noise within an image, generally of electronic origin. It is a systematic period noise in image or calibration data which can be characterized in terms of frequency, phase and amplitude. It is often quantified using Fourier analysis. Coherent noise is usually electronic, coming from sources such as power supplies, transmitters and clocks. It is desirable to remove, isolate or characterize sources of instrumental coherent noise before launch. It is possible to characterize a single independent coherent noise most easily on dark background data such as night scenes. It is also visible in dark radiometrically homogeneous region of a scene such as water. If the coherent noise is characterized well enough and is stable with time then it can be significantly reduced with notch filtering using inverse Fourier Transforms as a part of radiometric processing to produce the user.

Coherent radiation Electromagnetic radiation whose waves are equal in length and are in phase, so that waves at different points in space act in unison, as in laser and synthetic aperture radar.

Coherent reflector Simple or complex surface, such as a corner reflector, from which reflected wave components are coherent with respect to each other, and thus, combine to yield larger effective power than

Coherent Noise

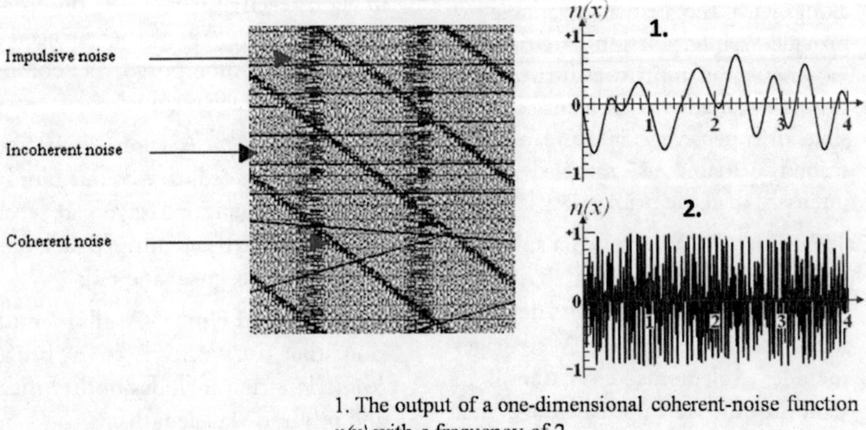

1. The output of a one-dimensional coherent-noise function *n(x)* with a frequency of 2
2. The output of a non-coherent-noise function *n(x)*

Source: 1. http://www.pgs.com
2. http://libnoise.sourceforge.net/coherentnoise/index.html

would be observed from a diffused scattering surface of the same area.

Coincident Occupying the same space. Coincident features or parts of features occupy the same space in the same plane.

Coincident geometry In a geodata base, how the coordinates of coincident features are stored. For example, if two lines are coincident, they will both be drawn in ArcMap, with one line lying precisely on top of the other. For two adjacent polygons, the coordinates for the shared boundary will be stored with each polygon and the boundary will be drawn twice.

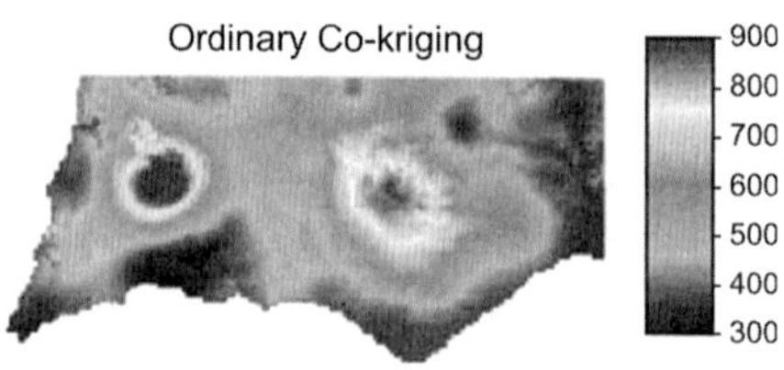

Co-kriging A form of kriging in which the distribution of a second, highly correlated variable (covariate) is used along with the primary variable to provide interpolation estimates. Co-kriging can improve estimates if the primary variable is difficult, impossible, or expensive to measure, and the second variable is sampled more intensely than the primary variable.

ColdFusion A Macromedia software product that integrates databases and Web pages using a server and development tools. ColdFusion Web pages include elements written in ColdFusion Markup Language that simplify integration with databases.

ColdFusion connector An ArcIMS Application Server Connector. In this environment, a request that includes ColdFusion tags is first executed on the ColdFusion Server. The ArcIMS custom ColdFusion tags are then passed to the ArcIMS Application Server for processing.

Color palette In ArcWeb Services, a tool for selecting colours and transparency. The Colour Palette lets a user select an object's fill and outline colour by choosing from a colour spectrum or by defining colours numerically. A user can also define the transparency of the colour for an object.

Colour A concept created by the human brain to help interpret spectral information received by the human eyes. Applied to *photons* in a *band* as well as a *monochromatic* beam.

Colour composite A colour image produced by assigning a colour to a particular spectral band.

Colour composite image Colour image prepared by projecting individual black-and-white multispectral images, each through a different colour filter. When the projected images are superposed, a colour composite image results.

Colour infrared A false colour film different from ordinary colour film in that the three sensitized layers are sensitive to green, red and infrared radiation instead of blue, green and red.

Colour infrared film A film emulsion that is sensitive to a broad *bandwidth* that includes both optical and infrared wavelengths.

Colour infrared photograph A type of film used in aerial photography that is exposed to green, red, and

Colour Composite Image

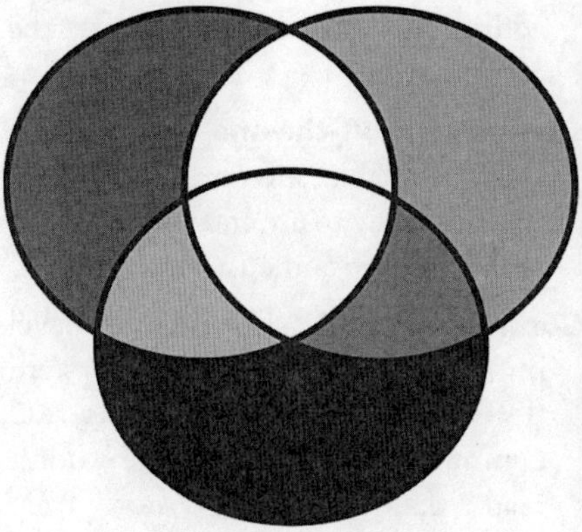

near-infrared wavelengths, which are depicted as blue, green and red in the photograph. It is also called false colour photograph.

Colour map A set of values that are associated with specific colours. Colour maps are most commonly used to display a raster dataset consistently on many different platforms.

Colour model Any system that organizes colours according to their properties for printing or display. Examples include RGB (red, green, blue), CMYK (cyan, magenta, yellow, black), HSB (hue, saturation, brightness), HSV (hue, saturation, value), HLS (hue, lightness, saturation), and CIE-L*a*b (Commission Internationale de l'Eclairage-luminance, a, b).

Colour ratio composite image Colour composite image prepared by combining individual ratio images for a scene using a different colour for each ratio image.

Colour separation Process of preparing a separate drawing, engraving, or negative for each colour required in the printing production of a map or chart.

Column An item in an attribute table.

COM Acronym for Component Object Model. A binary standard that enables software components to interoperate in a networked environment regardless of the language in which they were developed. Developed by Microsoft, COM technology provides the underlying services of interface negotiation, life-cycle management (determining when an object can be removed from a system), licensing, and event handling. The ArcGIS system is created using COM objects.

COM contract The COM requirement that interfaces, once published, cannot be altered.

COM interface A grouping of logically related virtual functions, implemented by a server object, allowing a client to interact with the server object. Interfaces form the basis of COM's communication between objects and the basis of the COM contract.

Combinatorial operator A kind of mathematical operator that interprets input with Boolean values. Combinatorial operators assign a different number to each unique combination of input values.

Combo box A user interface tool that combines the features of a text box and a drop-down list. For example, the Location combo box in ArcCatalog allows the selection of an item in the Catalog tree by typing its path or choosing its path from a drop-down list.

Command An instruction to a computer program, usually one word or concatenated words or letters, given by the user from a control device, such as a keyboard, or read from a file by a command interpreter.

Command bar A toolbar, menu bar, menu, or shortcut menu in an ArcGIS application.

Command line A string of text that acts as a command, typed at an interface prompt.

Command line interface A format of the input and output of a program in which the user enters commands by means of strings of text typed on a keyboard, as opposed to selecting commands from graphical prompts such as icons or dialog boxes.

Command line window In geoprocessing, a window that provides a command line for running tools and a message window for viewing the status messages created when running those tools.

Command prompt window A window accessible from the Windows Start menu in which MS-DOS commands are typed.

Comment In Survey Analyst for field measurements, a field that provides additional information about the computation.

Commercial off-the-shelf Commercially available software or systems that are ready to use and which do not require significant customization.

Committed read The isolation level in a database management system (DBMS) in which transactions read committed data only — they don't read data that has not been committed.

Common gateway interface A standard for scripts that run external programs from a World Wide Web server. CGI typically specifies how to pass arguments to the program via HTTP requests; defines a set of environmental variables made available to the program; and generates output, usually in HTML format, that is passed back to the browser. CGI scripts are frequently designed to access information in a database and format the results as HTML, convert information retrieved from an interactive Web page into a database, send datasets, and so on.

Common language runtime The execution engine for .NET Framework applications, providing services such as code loading and execution and memory management.

Compact disc An optical disk, slightly less than 5 inches in diameter, used to store up to approximately 650 megabytes of data.

Compaction A process that rearranges and consolidates the data in a file so that it occupies a single, contiguous space, allowing the data in each file to be accessed more efficiently.

Comparison threshold The degree of uncertainty that can be tolerated in the spelling of a keyword used in a search, including phonetic errors and the random insertion, deletion, replacement, or transposition of characters.

Compass An instrument used to find the direction of north from one's current location, consisting of a case with compass points marked around its edge and a floating magnetic needle that pivots to point to magnetic north.

Compass north The direction from a point on the earth's surface following a great circle toward the magnetic north pole, indicated by the north-seeking end of a compass.

Compass point An indication of direction. One of the 32 divisions into which the circle around the needle of a compass is divided, each equal to 11.25 degrees.

Compass rose A diagram of compass points drawn on a map or chart, subdivided clockwise from 0 to 360 degrees with 0 indicating true north. On older maps and charts a compass rose was a decorated diagram of cardinal points, divided into 16 or 32 points.

Compass rule A widely used rule for adjusting a traverse that assumes the precision in angles or directions is equivalent to the precision in distances. This rule distributes the closure error over the whole traverse by changing the northings and eastings of each traverse point in proportion to the distance from the beginning of the traverse. More specifically, a correction factor is computed for each point as the sum of the distances along the traverse from the first point to the point in question, divided by the total length of the traverse. The correction factor at each point is multiplied by the overall closure error to get the amount of error correction distributed to the point's coordinates. The compass rule is also known as the Bowditch rule, named for the American mathematician and navigator Nathaniel Bowditch (1773–1838).

Compilation Preparation of a new or revised map or chart, or portion thereof, from existing maps, aerial photographs, field surveys, and other sources.

Compiler A program used in software development that translates the lines of a programmer's code from one programming language to another, usually from a high-level language to the ones and zeros of machine language.

Complementary colours Two primary colours of light (one additive and the other subtractive) that produce white light when added together. Red and cyan are complimentary colours.

Complex dynamic event In ArcGIS Tracking Analyst, a type of complex temporal event that includes two components and involves a moving object, such as an airplane. The moving object's geographical location changes through time, so its additional attributes are stored in an input table.

Complex edge feature In a geodatabase, a linear network feature that corresponds to one or more network elements in the logical network.

Complex junction feature In ArcGIS 8.3 and previous versions, a junction feature in a geodatabase that corresponds to more than one network element in the logical network. For example, the state of the junction determines whether features can be connected or disconnected. This is not a concept that can be modelled in ArcGIS 9.0 and later versions.

Complex market area An area calculated by finding the outermost customers of a store along several vectors and connecting them. Complex market areas are more accurate than simple market areas because they respond to physical and cultural barriers. They are sometimes called amoebas because of their irregular shapes.

Complex number For radar systems, a complex number implies that the representation of a signal, or data file, needs both magnitude and phase measures. In the digital SAR context, a complex number is often represented by an equivalent pair of numbers, the real in-phase component (I) and the imaginary quadrature component (Q). For coherent systems such as SAR, the role of complex numbers is an essential part of the signal, since signal phase is used in the processor to obtain high-resolution.

Complex Number

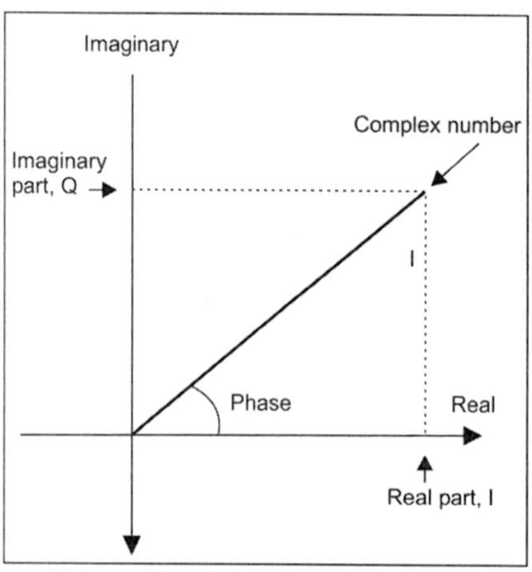

Complex stationary event A type of complex temporal event in ArcGIS Tracking Analyst that includes two components and involves a stationary object, such as a traffic sensor. The sensor's geographical location will not change, so its location information is stored in the input feature class.

Complex temporal event An event in ArcGIS Tracking Analyst that contains two components: one with persistent object information, and one with observations of the object through time. The merger of the temporal observations with the temporal object creates a complex event record or message. There are two types of complex temporal events: dynamic and stationary.

Component In COM, a binary unit of code that can be used to create COM objects.

Component category A section of the registry that can be used to categorize classes by their functionality. Component categories are used extensively in ArcGIS to allow extensibility of the system.

Component category manager An ArcGIS utility program (Categories.exe) that can be used to view and manipulate component category information.

Component object model A binary standard that enables software components to interoperate in a networked environment regardless of the language in which they were developed. Developed by Microsoft, COM technology provides the underlying services of interface negotiation, life-cycle management (determining when an object can be removed from a system), licensing, and event handling. The ArcGIS system is created using COM objects.

Composite image We can make a 'composite' image by selecting the most appropriate parts of other images. For instance, we could take only the cloud-free parts of many images to make a 'composite' image of all of Canada showing no clouds at all. It would not be a realistic scene, since we always have some clouds, but it would show all of Canada without allowing cloud cover to mask parts of it.

Composite measurement In Survey Analyst for field measurement, a set of simple measurements that are related and applied as a group.

Composite relationship A link or association between objects where the lifetime of one object controls the lifetime of its related objects. For example, the association between a highway and its shield markers is a composite relationship, since the shield markers should not exist without the highway.

Composition A UML term used to describe a form of association in which the lifetime of the whole controls the lifetime of the parts. In a composition, the instances of two classes depend on each other. The whole controls the location and lifetime of its parts. For example, in ArcMap, a map is composed of layers. If you move a map on a layout, the layers move, and if you delete the map, its layers get deleted; therefore, the lifetime of these objects depend on one another.

Compound element Within metadata, a group of data elements (including other compound elements) that together describe a characteristic of a spatial dataset in more detail than can be described by an individual data element.

Compound key A primary key that requires two or more fields to be unique.

Compression The process of reducing the size of a file or database. Compression improves data handling, storage, and database performance. Examples of compression methods include quadtrees, run-length encoding, and wavelets.

Compromise projection A projection that does not have equal area, conformal, or equidistant characteristics. The compromise projection is an attempt at balance between these characteristics, and is often used in thematic mapping.

Computation In Survey Analyst for field measurements, a process that requires a set of input parameters to apply a set of rules, and an algorithm to calculate output parameters. The input parameters are typically coordinates and measurements. The output parameters are usually coordinates.

Computation name In Survey Analyst for field measurements, a unique identifier that calls or retrieves a specific type of computation, defined on the General tab of the Survey Explorer.

Computation network In Survey Analyst for field measurements, a sequence of computation dependencies—the output points of some computations are used as the input for one or more others.

Computation network analysis window In Survey Analyst for field measurements, a window that displays information about the computation network, such as breaks in the sequence, computation states, and computation network cycles.

Computation network cycles In Survey Analyst for field measurements, a problem that occurs when a point's coordinates are used as both input and output within the same computation network. Cycles must be fixed before the whole network can be validated and brought to a state in which all computations are valid.

Computation state In Survey Analyst for field measurements, the condition of a computation. A computation may be in four different states: valid, out-of-date, incorrect, or incomplete.

Computation tool In Survey Analyst for field measurements, a tool that interacts with the map to add measurement values to computation pages.

Computational geometry A branch of mathematics that uses algorithms to solve geometry problems. Computational geometry is used in many GIS operations, including proximity analysis, feature generalization, and automated text placement.

Computer-aided design A computer-based system for the design, drafting, and display of graphical information. Also known as computer-aided drafting, such systems are most commonly used to support engineering, planning, and illustrating activities.

Computer-aided software engineering Any software that assists with the development and maintenance of software, especially the analysis and design. Complex tasks that often require many lines of code are simplified with CASE user interfaces and code generators.

Computer-assisted learning Instruction or training that uses computer-based media instead of hard-copy materials. Computer-assisted learning is generally designed to use the strengths of computer-based media such as the ability to navigate in a nonlinear fashion through the use of hyperlinks.

Computer-compatible tape (CCT) The magnetic tape on which the digital data for Landsat MSS and TM images are distributed.

Concatenate To join two or more character strings together, end to end; for example, to combine the two strings 'spatial' and 'analysis' into the single string 'spatial analysis.'

Concatenate events In linear referencing, a command that combines event records in tables containing events on the same route with the same value for specified fields. Only events in situations where the to-measure of one event matches the from-measure of the next event are combined. The concatenate events command is available for line event tables only.

Concatenated key In a relational database table, a primary key made by combining two or more keys that together form a unique identifier.

Concentric rings A method of defining the rings in an analysis so that the values inside the rings are cumulative. For example, if you had an analysis with three concentric rings and 10 households in each, the total number of households for ring 1 would be 10, the total for ring 2 would be 20 (ring 1 + ring 2), and the total for ring 3 would be 30 (ring 1 + ring 2 + ring 3).

Concurrency The ability of a DBMS to support simultaneous access by more than one user.

Concurrency management A database management process for maintaining the consistency of data while supporting simultaneous editing by more than one user. A typical technique involves locking portions of the database to prevent data corruption caused by multiple users simultaneously editing data.

Concurrent use Floating software products that are administered by a license manager. A central license manager (installed anywhere on a network) allows users to install the floating products on any number of machines. The number of seats or licenses purchased determines the number of users who can run the applications simultaneously.

Condition table A component of the PLTS knowledge base that contains SQL statements and custom code for feature validation extended beyond standard geodatabase domains. PLTS utilizes condition tables for enhanced validation during both database production and quality control.

Conduction Transfer of electromagnetic energy through a solid material by molecular interaction.

Cones Receptors in the retina which are sensitive to colour. There are cones sensitive to the red, green, and blue components of light.

Confidence level In a statistical test, the risk, expressed as a percentage, that the null hypothesis will be incorrectly rejected because of sampling error when the null hypothesis is true. For example, a confidence level of 95 per cent means that if the same test was performed 100 times on 100 different samples, the null hypothesis would be incorrectly rejected five times.

Configuration file In ArcIMS, the file that contains the core site information. ArcIMS configuration files contain all the basic information about the content to be delivered, such as location of the data and layer symbology. Typically, a configuration file contains data that defines map content and has a file extension of .axl, but it can also be used to deliver metadata or route data (as .axl files) and to serve maps created in ArcMap (.mxd or .pmf files). Regardless of their type, configuration files contain content that the service registers to the ArcIMS spatial server and Web server for processing.

Configuration keyword In ArcSDE, a name for a group of parameters that defines how geodatabase objects are stored.

Conflation A set of procedures that aligns the features of two geographic data layers and then transfers the attributes of one to the other.

Conflict In database editing, a state of incompatibility that occurs when multiple users simultaneously edit a version or reconcile two versions. Conflicts occur when the same feature or topologically related features are edited in two versions, and it is unclear which representation of the database is valid.

Conflict resolution The process of solving uncertainty within a database that occurs when two versions of the same data are edited at the same time. Conflicts can occur when multiple users simultaneously edit the same feature or topologically related features, or reconcile two versions of a dataset. Resolving a conflict requires that the users make a decision about the feature's correct representation and identify it in the Conflict Resolution dialog box.

Conformal projection One of map projection that preserves local shapes.

Conformality The characteristic of a map projection that preserves the shape of any small geographic area.

Conic projection One type of map projection that uses a cone as the projection surface.

Conjoint boundary A boundary common to two features. For example, in a parcel database, adjacent parcels share a boundary. Another example is a parcel that shares a boundary on one side with a river. The segment of the river that coincides with the parcel boundary shares the same coordinates as the parcel boundary.

Connection In ArcCatalog, a mechanism used to access remote file systems and shared databases.

Connection line A cadastral fabric line with bearing and distance data attached to it, commonly used to tie parcels across roads, tie in control points, or tie the point of survey commencement to the point of beginning for a particular parcel. Connection lines do not necessarily indicate parcel boundaries.

Connectivity A topological relationship that stipulates that arcs connect to each other at nodes.

Connectivity analysis Any method of solving network problems such as traversability, rate of flow, or capacity, using network connectivity.

Connectivity group In network datasets, a logical grouping of point features, line features, or both, that controls how network elements are connected. Connectivity groups are defined when a network dataset is built. A network dataset may have multiple connectivity groups.

Connectivity policy In a network dataset, a property of network sources that defines how network elements connect to each other within a connectivity group. There are two types of edge-edge connectivity policies (end-point connectivity and any-vertex connectivity) and two types of edge-junction connectivity policies (honour and override).

Connectivity rule In geometric networks, a rule that constrains the type and number of network features that can be connected to one another. There are two types of connectivity rules: edge-edge and edge-junction.

Connector A visual representation of the relationship between elements in a model. Connectors join elements together to create processes. Typical processes connect an input data element, a tool element, and a derived data element.

Conservation of confusion Principle, for imagery derived from a given SAR, that the amount of information in the data is a constant. One expression of this rule is that the product of the range and the azimuth resolution divided by the number of statistically independent looks is a constant, which serves as a figure of merit of the system. (In this context, information is related to the statistical degrees of freedom in the data ensemble, and not necessarily to knowledge about objects in the scene.)

Conservation of coordinates Principle, for Synthetic Aperture Radar (SAR) imagery, that image position is not changed by pitch, roll, or yaw rotations of the radar, since range is determined by the speed of light, and azimuth is determined by the along-track radar velocity.

Conservation of energy (Radar) Principle, assuming that all available data is used for each case, that the average value of the estimated reflectivity from a scene is a constant for a given SAR and processor, independent of the number of looks used, and independent of any time varying non-coherence in the scene, such as from a moving surface of water, or in the radar/processor combination.

Constant azimuth A complex curve on the earth's surface that crosses every meridian at the same oblique angle. A rhumb line path follows a single compass bearing; it is a straight line on a Mercator projection, or a logarithmic spiral on a polar projection. A rhumb line is not the shortest distance between two points on a sphere.

Constellation Refers to either the specific set of satellites used in calculating positions or all the satellites visible to a GPS receiver at one time.

Constrained adjustment In Survey Analyst for field measurements, one of two phases involved when performing a least squares adjustment for a measurement network. In this phase, the emphasis is on testing the reference points as well as computing final coordinates.

Constraints Limits imposed on a model to maintain data integrity. For example, in a water network model, an 8-inch pipe cannot connect to a 4-inch pipe.

Construct features In ArcMap, an edit command that takes selected features from one or more feature classes and creates new features in a target feature class. The Construct Features tool uses the input geometries of the selected features to construct polygons or lines following polygon boundaries, depending on the geometry of the target feature class.

Construction line In Survey Analyst - Cadastral Editor, a line with a bearing and distance that is used to create geometry for a new cadastral fabric parcel.

Contact print A reproduction from a photographic negative in direct contact with photosensitive paper.

Container account The operating system account that run as server object container processes, which is specified by the GIS server post-installation utility. Objects running in a server container process have the same access rights to system resources as the container account.

Container process A process in which one or more server objects are running. Container processes run on SOC machines, and are started and shut down by the SOM. They are visible in the Windows Task Manager as ArcSOC.exe.

Containment A spatial relationship in which a point, line, or polygon feature or set of features is enclosed completely within a polygon.

Content In ArcWeb Services, data that may include data files, layers, or services.

Content finder web service A SOAP ArcWeb service that allows users to search the metadata information for available content in the ArcWeb system.

Content Library A Web-based tool for viewing and maintaining information about data files, layers, and services used in ArcWeb Services. Users can access Content Library through the ArcWeb Services Web site.

Content standard for digital geospatial metadata A publication authored by the FGDC that specifies the information content of metadata for digital geospatial datasets. The purpose of the standard is to provide a common set of terminology and definitions for concepts related to the metadata. All US government agencies (federal, state, and local) that receive federal funds to create metadata must follow this standard.

Conterminous Having the same or coincident boundaries.

Context The known environment of a particular feature on an image.

Context menu A list menu that pops up when the right mouse button is clicked in Windows applications. Some keyboards also have an application key that opens shortcut menus.

Contiguity In a coverage, the topological identification of adjacent polygons by recording the left and right polygon for each arc.

Contiguous Of polygons: adjacent; having a common boundary; sharing an edge.

Continuous data Data such as elevation or temperature that varies without discrete steps. Since computers store data discretely, continuous data is usually represented by TINs, rasters, or contour lines, so that any location has either a specified value or one that can be derived.

Continuous features spatial features that exist between observations.

Continuous raster A raster in which cell values vary continuously to form a surface. In a continuous raster, the phenomena represented have no clear boundaries. Values exist on a scale relative to each other. It is assumed that the value assigned to each cell is what is found at the centre of the cell. Rasters representing elevation, precipitation, chemical concentrations, suitability models, or distance from a road are examples of continuous rasters.

Continuous tone Image not broken into dots by photographic screen; contains unbroken gradient tones from black to white, and may be either in negative or positive form. Aerial photographs are examples of continuous-tone prints. Contrasted with halftone (screened) and line copy.

Continuous tone image A photograph that has not been screened and so displays all the varying tones from dark to light.

Continuum spectrum *A spectrum* shaped like a gentle broad hill, rising to a peak that is at a wavelength determined by the temperature of the opaque, radiating body that creates it.

Contour Imaginary line on ground, all points of which are at the same elevation above or below a specific datum.

Contour interval The difference in elevation between adjacent contour lines.

Contour line A line on a map that connects points of equal elevation based on a vertical datum, usually sea level.

Contour tagging Assigning elevation values to contour lines.

Contrast The ratio between the energy emitted or reflected by an object and its immediate surroundings.

Contrast enhancement Image-processing procedure that improves the contrast ratio of images. The original narrow range of digital values is expanded to utilize the full range of available digital values.

Contrast ratio On an image, the ratio of reflectances between the brightest and darkest parts of an image.

Contrast stretching Expanding a measured range of digital numbers in an image to a larger range, to improve the contrast of the image and its component parts.

Control A basic element of a software application's GUI. Examples of controls include menus, buttons, tools, check boxes, slider bars, text input boxes, and combo boxes.

Control mapping Points of established position or elevation, or both, which are used to fix references in positioning and correlating map features. Fundamental control is provided by stations in the national networks of triangulation and traverse (horizontal control) and levelling (vertical control). Usually it is necessary to extend geodetic surveys, based on fundamental stations, over the area to be mapped, to provide a suitable density and distribution of control points. Supplemental control points are those needed to relate the aerial photographs used for mapping with the system of ground control. These points must be positively photo identified; that is, the points must be positively correlated with their images on the photographs.

Control point Also called a control station. A monumental point to which coordinates have been, or are being assigned by the use of surveying observations. The National Geodetic Survey maintains a nation-wide set of control points.

Control segment A worldwide network of GPS monitoring and control stations that ensure the accuracy of satellite positions and their clocks.

Control station Point on the ground whose position (horizontal or vertical) is known and can be used as a base for additional survey work.

Convection Transfer of heat through the physical movement of heated matter.

Coordinate geometry A method for calculating coordinate points from surveyed bearings, distances, and angles.

Coordinate geometry traverse In Survey Analyst, a process of computing a sequence of survey point locations starting from an initial known point. Each new survey point is defined by a traverse course and is used as the take-off point for the next point in the sequence. A traverse course can be defined using various combinations of directions, distances, angles, and circular arc parameters. The coordinate geometry traverse is primarily used to define coordinates based on values taken from subdivision plans.

Coordinate system A reference framework consisting of a set of points, lines, and/or surfaces, and a set of rules, used to define the positions of points in space in either two or three dimensions. The Cartesian coordinate system and the geographic coordinate system used on the earth's surface are common examples of coordinate systems.

Coordinate transformation The process of converting the coordinates in a map or image from one coordinate system to another, typically through rotation and scaling.

Coordinated universal time The official time keeping system of the world's nations since 1972. It refers to local time throughout the world from prime meridian, and is based on atomic clocks, but is periodically and artificially adjusted so that it always remain within 0.9 seconds of universal time. The adjustment is made by the addition of leap seconds to the course of atomic time. Coordinated universal time is abbreviated 'UTC.'

Coordinates Linear and (or) angular quantities that designate the position of a point in relation to a given reference frame.

Coordinates, origin of Points in a system of coordinates which serves as a zero point in computing the system's elements or in prescribing its use.

Co-polarization maxima The antenna polarization state for which maximum backscattered power is received from a particular target. For co-polarization, the transmit and receive antennas are the same.

Co-polarization nulls The antenna polarization state for which zero backscattered power is received from a particular target. For co-polarization, the transmit and receive antennas are the same. Co-polarization nulls may not correspond to the maximum cross polarization received power.

Co-polarization signature The received signature when the transmit and receive antennas have the same polarization properties.

Core-based statistical area A geographic region containing at least one urban area with a population of at least 10,000, defined by the US Office of Management and Budget for use by federal statistical agencies, including the US Census Bureau. A core-based statistical area can be a metropolitan statistical area or a micropolitan statistical area.

Corner reflector Cavity formed by two or three smooth planar surfaces intersecting at right angles. Electromagnetic waves entering a corner reflector are reflected directly back toward the source.

Correlation An association between data or variables that change or occur together. For example, a positive correlation exists between housing costs and distance from the beach; generally, the closer a home is to the beach, the more it costs. Correlation does not imply causation. For example, there is a statistical correlation between ice cream sales and crime rates, but neither causes the other. The correlation coefficient is an index number between –1 and 1 indicating the strength of the association between two variables.

Corridor A buffer drawn around a line.

Corridor analysis A form of spatial analysis usually applied to environmental and land-use data in order to find the best locations for building roads, pipelines, and other linear transportation features.

COSMIC Computer Software Management and Information Centre, University of Georgia. This facility distributes computer programs developed by US government-funded projects.

Cost A function of time, distance, or any other factor that incurs difficulty or an outlay of resources.

Cost grid A raster dataset that identifies the cost of travelling through each cell in the raster. A cost raster can be used to calculate the cumulative cost of travelling from every cell in the raster to a source or a set of sources.

Cost raster A raster dataset that identifies the cost of travelling through each cell in the raster. A cost raster can be used to calculate the cumulative cost of travelling from every cell in the raster to a source or a set of sources.

Cost-benefit analysis An appraisal that attempts to compare the benefits (including social benefits) expected from a project with the costs (sometimes including social costs) incurred by the project over its lifetime. Generally cost-benefit analyses are used to compare alternative proposals, or to make a case for the implementation of a particular plan or system.

Cost-distance analysis The calculation of the least cumulative cost from each cell to specified source locations over a cost raster.

Cost-weighted allocation An ArcGIS Spatial Analyst function that identifies the nearest source from each cell in a cost-weighted distance grid. Each cell is assigned to its nearest source cell, in terms of accumulated travel cost.

Cost-weighted direction An ArcGIS Spatial Analyst function that provides a road map from the cost weighted distance grid, identifying the route to take from any cell, along the least-cost path, back to the nearest source.

Cost-weighted distance An ArcGIS Spatial Analyst function that uses a cost grid to assign a value—the least accumulative cost of getting back to the source—to each cell of an output grid.

COTS Acronym for commercial off-the-shelf. Commercially available software or systems that are ready to use and which do not require significant customization.

County The primary legal subdivision of all US states except Alaska and Louisiana. The US Census Bureau uses counties or equivalent entities (boroughs in Alaska, parishes in Louisiana, the District of Columbia in its entirety, and municipios in Puerto Rico) as statistical subdivisions.

County subdivision A statistical division of a county recognized by the US Census Bureau for data presentation. County subdivisions can include census county divisions, census subareas, minor civil divisions, and unorganized territories.

Course made good (CMG) The bearing from your starting point to your present position.

Covariance A statistical measure of the linear relationship between two variables. Covariance measures the degree to which two variables move together relative to their individual mean returns.

Coverage The term used by ESRI for a vector based digital map with topology.

Coverage feature class In ArcInfo, a classification describing the format of geographic features and supporting data in a coverage. Feature classes include point, arc, node, route, route system, section, polygon, and region. One or more coverage features are used to model geographic features; for example, arcs and nodes can be used to model linear features, such as street centrelines. The tick, annotation, link, and boundary feature classes provide supporting data for coverage data management and viewing.

Coverage units The units of the coordinate system in which a coverage is stored (for example, feet, metres, inches).

Cracking In ArcGIS, a part of the topology validation process in which vertices are created at the intersection of feature edges.

Crandall rule A special-case, least-squares-based method for adjusting the closure error in a traverse. The Crandall rule is most frequently used in a closed traverse that represents a parcel from a subdivision plan to ensure that tangency between courses remains intact as, for example, when applied to a tangent curve. It assumes that course directions and angles have no error and, therefore, all error corrections are applied only to the distances. This method uses a least-squares adjustment to distribute the closure error, and applies infinite weight to the angles or direction measurements to ensure that they are not adjusted. In some circumstances the results of this adjustment method may be unexpected, or the adjustment may not be possible, and an alternative method is required. The Crandall rule was developed by C.L. Crandall around 1901.

Creation date In Survey Analyst for field measurements, an attribute of the computation that records the date of origin.

Creation time The time it takes to initialize an instance of a server object when server objects are created in the GIS server, either as a result of the server starting or in response to a request for a server object by a client.

Credit A charge to access an ArcWeb service. Users use credits each time they request an ArcWeb service to perform a task, such as to find a route.

Critical value The specific cut-off point that determines acceptance or rejection of a hypothesis. Critical values are determined by the choice of a level of significance (á).

Crop marks Also known as trim marks, are the marks that indicate the edge of the page of a finished, printed map. Crop marks are used as a reference for trimming excess paper after printing.

Cross correlation Statistical correlation between spatial random variables of different types, attributes, names, and so on, where the correlation depends on the distance or direction that separates the locations.

Cross covariance The statistical tendency of variables of different types, attributes, names, and so on, to vary in ways that are related to each other. Positive cross covariance occurs when both variables tend to be above their respective means together, and negative cross covariance occurs if one variable tends to be above its mean when the other variable is below its mean.

Cross polarization maxima The antenna polarization state for which maximum cross-polarized backscattered power is received from a particular target. Co-polarization nulls may not correspond to the maximum cross polarization received power.

Cross polarization nulls The antenna polarization state for which zero cross-polarized backscattered power is received from a particular target. For co-polarization, the transmit and receive antennas are the same. Note that for cross polarization nulls the co-polarization power is maximum.

Cross polarization signature The received signature when the transmit and receive antennas have orthogonal polarizations.

Cross tabulation In a GIS, comparing attributes in different coverages or map layers according to location.

Cross validation A procedure for testing the quality of a predicted data distribution. In cross validation, a piece of data whose value is known independently is removed from the dataset and the rest of the data is used to predict its value. Full cross validation is done by removing, in turn, each piece of data from the dataset and using the rest of the data to predict its value.

Cross variogram A function of the distance and direction separating two locations, used to quantify cross correlation. The cross variogram is defined as the variance of the difference between two variables of different types or attributes at two locations. The cross variogram generally increases with distance, and is described by nugget, sill, and range parameters.

Cross-polarized Describes a radar pulse in which the polarization direction of the return is normal to the polarization direction of the transmission. Cross-polarized images may be HV (horizontal transmit, vertical return) or VH (vertical transmit, horizontal return).

Cross-polarized waves (or orthogonal waves) Each wave in a pair of cross-polarized waves is completely polarized. However, an antenna optimized to receive the co-polarization maximum of one wave will receive no power from the other wave. Note that, in general, an arbitrary wave may be treated as the sum of two cross-polarized waves.

Cross-reference databaseA database containing tables with information defining the mapping between a data source schema and an output geodatabase schema. Cross- reference databases are used by the PLTS data loader to batch populate a geodatabase from a variety of sources.

Cross-tile indexing A technique for indexing features that cross tile boundaries in a map library by storing them as one or more features in each tile, instead of storing them each as a single feature.

Cross-track error (XTE) The distance you are off a desired course in either direction.

Cross-track scanner Scanner in which a faceted mirror rotates about a horizontal axis to sweep the detector IFOV in a series of parallel scan lines oriented normal to the flight direction.

CRT Cathode ray tube.

CSDGM A publication authored by the FGDC that specifies the information content of metadata for digital geospatial datasets. The purpose of the standard is to provide a common set of terminology and definitions for concepts related to the metadata. All US government agencies (federal, state, and local) that receive federal funds to create metadata must follow this standard.

CSS Acronym for Cascading Style Sheets. A standard for defining the layout or presentation of an HTML or XML document. Style information includes font size, background colour, text alignment, and margins. Multiple style sheets may be applied to 'cascade' over previous style settings, adding to or overriding them. The World Wide Web Consortium maintains the CSS standard.

Cubic convolution A technique for resampling raster data in which the average of the nearest 16 cells is used to calculate the new cell value.

Cull In ArcScene and ArcGlobe, to selectively choose not to draw one side of an areal feature.

Cultural feature A human-made feature represented on a map, such as a building, road, tower, or bridge.

Cultural geography The field of geography concerning the spatial distribution and patterns created by human cultures and their effects on the earth.

Culture Features constructed by man that are under, on, or above the ground which are delineated on a map. These include roads, trails, buildings, canals, sewer systems and boundary lines. In a broad sense, the term also applies to all names, other identification and legends on a map.

Curb approach In network analysis, a network location property that models a path for approaching a stop from a specific side, based on edge direction. For example, a school bus must approach a school from its door side so that students exiting the bus will not have to cross the street. There are three types of curb approaches: left, right, or both.

Current cadastral fabric In Survey Analyst - Cadastral Editor, the most up-to-date legal state of the cadastral fabric.

Current coordinate In Survey Analyst for field measurements, the single coordinate for a survey point that is the best representation for its location within each project. A current coordinate is required when the same project computes or imports more than one coordinate for a particular survey point.

Current task During editing in ArcMap, a setting in the Task drop-down list that determines which operation the sketch construction tools will perform. Examples of edit tasks include creating new features and modifying existing features. The edit task is set by clicking a task in the Task drop-down list.

Current workspace A user-specified path to a container for file-based geographic data, set in the Environment Settings dialog box. Data from the current workspace can be accessed from any tool dialog box (including script and model dialog boxes), or at the command line, simply by typing its name.

Curve fitting Converting short connected straight lines into smooth curves to represent features such as rivers, shorelines, and contour lines. The curves that result pass through or close to the existing points.

Custom behaviour A set of methods, functions or operations associated with a database object that has been specifically created or overridden by a programmer.

Custom feature In geodatabases, a feature with specialized behaviour instantiated in a class by a programmer.

Custom functionality A modification to or enhancement of standard software functionality to meet a specific user's needs.

Custom group layer A layer created in ArcMap using the ArcGIS Image Server extension. Custom group layers display the contents of an image service definition, including the footprint, boundary, seamline and preview raster layer.

Custom object An object with custom behaviour provided by a developer.

Custom tool In geoprocessing, a tool created by a user and added to a toolset and/or toolbox. Custom tools may only be added to custom toolsets and/or toolboxes.

Custom toolset In geoprocessing, a subset of a toolbox created by a user to hold custom tools or a group of frequently used tools.

Customer market analysis A type of market analysis that focuses on data about customers, rather than about a store or stores. An example is desire line analysis.

Customer profiling A process that establishes common demographic characteristics for a set of customers within a geographic area.

Customer prospecting A type of market analysis that locates regions with appropriate demographic characteristics for targeting new customers.

Cut off The digital number in the histogram of a digital image which is set to zero during contrast stretching. Usually, this is a value below which atmospheric scattering makes a major contribution.

Cut/fill An ArcGIS Spatial Analyst and 3D Analyst function that summarizes areas and volumes of change between two surfaces.

Cycle One complete oscillation of a wave.

Cycle slip A discontinuity of an integer number of cycles in the measured carrier beat phase resulting from a temporary loss-of-lock in the carrier tracking loop of a GPS receiver.

Cylindrical projection one of the types of map projection that uses a cylinder as the projection surface.

CZCS Coastal Zone colour scanner.

D

D A technique for increasing the accuracy of GPS measurements by comparing the readings to two receivers—one roving and the other a fixed base station—and a known location.

Daemon In programming, a process that continuously runs in the background, without being explicitly implemented, and performs a specified operation at predefined times or in response to certain events. Daemon is a UNIX term; the Windows equivalent is a system agent or service.

Dangle The endpoint of a dangling arc.

Dangle length In ArcInfo coverages, the minimum length allowed for dangling arcs by the clean process, which removes dangling arcs shorter than the dangle tolerance.

Dangle tolerance A tolerance used in ARC/INFO that specifies the minimum length for dangling arcs on the output coverage.

Dangling arc An arc having the same polygon on both its left and right sides and having at least one node that does not connect to any other arc. It often occurs where a polygon does not close properly, where arcs do not connect properly (an undershoot), or where an arc was digitized past its intersection with another arc (an overshoot). A dangling arc is not always an error; for example, it can represent a cul-de-sac in a street network.

Dangling node A node at the end of an arc that is not connected to other arcs or it is the endpoint of a dangling arc.

Dasymetric mapping A technique in which attribute data that is organized by a large or arbitrary area unit is more accurately distributed within that unit by the overlay of geographic boundaries that exclude, restrict, or confine the attribute in question. For example, a population attribute organized by census tract might be more accurately distributed by the overlay of water bodies, vacant land and other land-use boundaries within which it is reasonable to infer that people do not live.

Data Any collection of related facts arranged in a particular format; often, the basic elements of information that are produced, stored, or processed by a computer.

Data capture Any operation that converts GIS data into computer-readable form. Geographic data can be captured by being downloaded directly into a GIS from sources such as remote sensing or GPS data, or it can be digitized, scanned, or keyed in manually from paper maps or photographs.

Data change message In ArcGIS, a message containing any inserts, updates, and deletes applied to a replica. These messages are used to synchronize replicas in a disconnected environment.

Data collection system (DCS) On Landsats 1 and 2, the system that acquired information from seismometers, flood gauges, and other measuring devices. These data were relayed to ground receiving stations.

Data continuity A NASA requirement to ensure that Landsat 7 data are compatible to those obtained by earlier Landsat satellites.

Data conversion The process of translating data from one format to another which are compatible with the GIS package.

Data cube A term used to describe any set of data that has three dimensions, for example temperatures at various depths (z) in a specified area (x,y) of ocean. Often used to describe the intensity of radiation in hyper spectral data measured on the Earth, where the spatial coordinates are two dimensions (x, y) and the wavelength (λ) for the channel is the third dimension. Compare with hyperdimensional data cube.

Data definition language A set of SQL statements that can be used either interactively or within a programming language to create a new database, set permissions on it, and define its attributes.

Data dictionary A catalog or table containing information about the datasets stored in a database. In a GIS, a data dictionary might contain the full names of attributes, meanings of codes, scale of source data, accuracy of locations, and map projections used.

Data element The smallest unit of information used to describe a particular characteristic of a spatial dataset. A data element is a logically primitive description that cannot be further subdivided.

Data exploration Data-centred query and analysis.

Data features Spatial features that do not exist between observations, form separate entities and are individually distinguishable.

Data file A file that holds text, graphics, or numbers.

Data format The structure used to store a computer file or record.

Data frame A map element that defines a geographic extent, a page extent, a coordinate system, and other display properties for one or more layers in ArcMap. A dataset can be represented in one or more data frames. In data view, only one data frame is displayed at a time; in layout view, all the map's data frames are displayed at the same time. Many cartography texts use the term 'map body' to refer to what ESRI calls a data frame.

Data integration The process of sharing and combining data between two organizations or systems.

Data integrity The degree to which the data in a database is accurate and consistent according to data model and data type.

Data logger Also called a data recorder. A hand held, lightweight data entry computer. It can be used to store additional data obtained by GPS receiver.

Data manager web service A SOAP ArcWeb service that allows users to upload their own point, line, or polygon data and attribute information to use in custom map and spatial query data sources.

Data message (1) A 1,500-bit message included in the GPS signal which reports the satellite's location, clock corrections, and health. (2) Information in a satellite's GPS signal that reports its orbital position, operating health, and clock corrections.

Data model In GIS, a mathematical construct for representing geographic objects or surfaces as data. For example, the vector data model represents geography as collections of points, lines and polygons; the raster data model represents geography as cell matrixes that store numeric values; and the TIN data model represents geography as sets of contiguous, non-overlapping triangles.

Data product In ArcWeb Services, a collection of layers that has a single credit cost. Every data layer has an associated data product.

Data recorder A lightweight, handheld field computer used to store data collected by a GPS receiver.

Data repository A repository structure, physical or virtual, that collects, stores, and disseminates information, metadata, and data. A clearinghouse provides widespread access to information and is generally thought of as reaching or existing outside organizational boundaries.

Data set A logically meaningful grouping or collection of similar or related data.

Data sharing Making data available and accessible to organizations or individuals other than the creator of the data.

Data snooping In Survey Analyst, the process of testing each measurement using the W-test.

Data source Any data. Data sources may include coverages, shapefiles, rasters, or feature classes.

Data structure The organization of data within a specific computer system that allows the data to be stored and manipulated effectively; a representation of a data model in computer form.

Data synchronization The process of applying data changes from a replica to the relative replica in a replica pair. Data changes include row or feature inserts, updates, and deletes. Synchronization can be performed in one or both directions between a replica pair.

Data transfer The process of moving data from one system to another or from one point on a network to another.

Data type The attribute of a variable, field, or column in a table that determines the kind of data it can store. Common data types include character, integer, decimal, single, double, and string.

Data view An all-purpose view in ArcMap and ArcReader for exploring, displaying, and querying geographic data. This view hides all map elements, such as titles, north arrows, and scale bars.

Database One or more structured sets of persistent data, managed and stored as a unit and generally associated with software to update and query the data. A simple database might be a single file with many records, each of which references the same set of fields. A GIS database includes data about the spatial locations and shapes of geographic features recorded as points, lines, areas, pixels, grid cells, or TINs, as well as their attributes.

Database administrator The person who manages a database. Database administration includes user setup, security, backup and recovery procedures for data, and optimization of physical data storage for best performance.

Database connection A link to a database from a software application. Database connections have two states: connected to or disconnected from the database. Deletion of a database connection only deletes the connection itself, not the database or its contents. Creation of a database connection requires selection of a data provider for data retrieval.

Database design The development of the conceptual, logical, and physical structures of a database in order to meet user requirements.

Database generalization The abstraction, reduction, and simplification of features and feature classes for deriving a simpler model of reality or decreasing stored data volumes.

Database management system (DBMS) (1) A set of computer programs for managing an integrated and shared database for such tasks as data input, search, retrieval, manipulation, and output. (2) A set of software applications used to create and maintain databases according to a schema. Database management systems provide tools for adding, storing, changing, deleting, and retrieving data.

Database role The database permissions category assigned to a group of users who perform the same types of tasks and, therefore, require the same level of database access. Database roles are used by database administrators to simplify the administration of user privileges.

Database statistics Mathematically calculated data that describes the state of a database and which the database management system (DBMS) uses to optimize query response.

Database support The proprietary database platforms supported by a program or component.

Data-driven ring analysis A type of market analysis primarily used to look at competing sites or to select potential new locations.

Dataflow The route of data passage through a system.

Dataset Any collection of related data, usually grouped or stored together.

Dataset precision The mathematical exactness or detail with which a value is stored within a dataset, based on the number of significant digits that can be stored for each coordinate.

Datum A set of reference points on the Earth's surface against which position measurements are made, and an associated model of the shape of the earth (reference ellipsoid) to define a geographic coordinate system. In surveying, a reference system for computing or correlating the results of surveys. There are tow principal types of datums: vertical and horizontal. A vertical datum is a level surface to which heights are referred. In the United States, the generally adopted vertical datum for levelling operations is the National Geodetic Vertical Datum of 1929. The horizontal datum is used as a reference for position. The North American Datum of 1927 is defined by the latitude and longitude of an initial point (Meade's Ranch in Kansas), the direction of a line between this point and a specified second point, and two dimensions that define the spheroid. The new North American Datum of 1983 is based on a newly defined spheroid (GRS80); it is an Earth-centred datum having no initial point or initial direction.

Datum level A surface to which heights, elevations, or depths are referenced.

Datum plane A surface to which heights, elevations, or depths are referenced.

Datum points In Survey Analyst for field measurements, survey points that are not defined by computations, but that are the input data for the initial computations in a sequence of computations.

Datum shift A systematic conversion of the latitude-longitude values for a set of points from one geographic coordinate system to equivalent values in another geographic coordinate system. Depending on the geographic coordinate systems involved, the transformation can be accomplished in various ways. Typically, equations are used to model the position and orientation of the 'from' and 'to' geographic coordinate systems in three-dimensional coordinate space; the transformation parameters may include translation, rotation, and scaling. Other methods, including one used in transformations between NAD 1927 and NAD 1983, use files in which the differences between the two geographic coordinate systems are given for a set of coordinates; the values of other points are interpolated from these.

Datum transformation A systematic conversion of the latitude-longitude values for a set of points from one geographic coordinate system to equivalent values in another geographic coordinate system. Depending on the geographic coordinate systems involved, the transformation can be accomplished in various ways. Typically, equations are used to model the position and orientation of the 'from' and 'to' geographic coordinate systems in three-dimensional coordinate space; the transformation parameters may include translation, rotation, and scaling. Other methods, including one used in transformations between NAD 1927 and NAD 1983, use files in which the differences between the two geographic coordinate systems are given for a set of coordinates; the values of other points are interpolated from these.

DBF file A database file format. In the ArcWeb Services Web site, users can upload data into Builder as either a DBF file or a shapefile (SHP).

DCOM Acronym for Distributed Component Object Model. Extends COM to support communication among objects on different computers on a network.

DDL A set of SQL statements that can be used either interactively or within a programming language to create a new database, set permissions on it, and define its attributes.

Dead reckoning A navigation method of last resort that uses the most recently recorded position of a ship or aircraft, along with its speed and drift, to calculate a new position.

Debug To test for, detect, and correct errors in a computer program or component.

Decimal degrees Values of latitude and longitude expressed in decimal format rather than in degrees, minutes, and seconds.

Decimal degrees (DD) system A measurement for longitude and latitude values such as 78.5°.

Decision support system A computer program that includes data presentation and modelling tools that help people understand problems and find solutions.

Declination (1) In a spherical coordinate system, the angle between the equatorial plane and a line to a point somewhere on the sphere. (2) In astronomy, the angular distance of a celestial body above (north, plus) or below (south, minus) the celestial Equator. Magnetic declination is the angular difference between magnetic north and true (geographic) north at the point of observation; it is not constant but varies with time because of the 'wandering' of the magnetic north pole.

Deeply stateful application An application that uses the GIS server to maintain application state by changing the state of a server object or its related objects. Deeply stateful applications require non-pooled server objects.

Default cadastral fabric In Survey Analyst-Cadastral Editor, the most up-to-date version of the cadastral fabric in the database that corresponds to the default version in the database. The default cadastral fabric does not represent the current legal state of the cadastral fabric.

Default interface When a Component Object Model (COM) is created, the interface that is returned automatically if no other interface is specified. Most ArcObjects classes specify are known as the default interface.

Default junction type In geometric networks, the user-established junction type that automatically connects two edges in the absence of a current user choice. An edge may also have a default end junction type, used for the free ends of new edges.

Default style The original style that is used to symbolize a layer class in ArcWeb Services. Symbols can be overwritten in a map data source.

Defined study area A study area with a defined boundary, such as a city.

Definition query In ArcMap, a request that examines feature or tabular attributes based on user-selected criteria and displays only those features or records that satisfy the criteria.

Deflection The creation of a segment at an angle relative to an existing segment.

Degree A unit of angular measure represented by the symbol E. The earth is divided into 360 degrees of longitude and 180 degrees of latitude.

Degree minute second (DMS) system A measuring system for longitude and latitude values such as 42°30'00'. One degree equals 60 minutes and one minute equals 60 seconds.

Degree of polarization An electromagnetic wave can have a polarized and a non-polarized component. The degree of polarization is given by the ratio of the power in the polarized part of an electromagnetic wave to the total power in the electromagnetic wave.

Degree slope One method for representing the measurement of an inclined surface. The steepness of a slope may be measured from 0 to 90 degrees.

Degrees-minutes-seconds The unit of measure for describing latitude and longitude. A degree is 1/360th of a circle. A degree is further divided into 60 minutes, and a minute is divided into 60 seconds.

Delaunay triangles The components of Delaunay triangulation. Delaunay triangles cannot exist alone; they must exist as part of a set or collection that is typically referred to as a triangulated irregular network (TIN). A circle circumscribed through the three nodes of a Delaunay triangle will not contain any other points from the collection in its interior.

Delaunay triangulation (1) A technique for creating a mesh of contiguous, non-overlapping triangles from a dataset of points. Each triangle's circumscribing circle contains no points from the dataset in its interior. Delaunay triangulation is named for the Russian mathematician Boris Nikolaevich Delaunay. (2) An iteractive process of connecting points with their two nearest neighbours to form triangles as equiangular as possible in triangulated irregular network (TIN).

Deletes table The geodatabase system table, created when a feature class or table is registered as versioned, that maintains information on all rows that have been deleted or updated.

Delimiter A character, such as a space or comma, that separates words or values.

Delta database A file that contains data edits that can be exchanged between geodatabases or between geodatabases and other data stores. The edits can come from a checkout geodatabase, modified rows between source and target versions, or a custom application. Supported delta file formats are XML (delta XML file) and delta database (.mdb file).

Delta file A file that contains data edits that can be exchanged between geodatabases or between geodatabases and other data stores. The edits can come from a checkout geodatabase, modified rows between source and target versions, or a custom application. Supported delta file formats are XML (delta XML file) and delta database (.mdb file).

Delta table One of two geodatabase system tables—the adds and deletes tables—created for a feature class or table when it is registered as versioned. These tables record changes made to a version during editing.

DEM Acronym for digital elevation model. The representation of continuous elevation values over a topographic surface by a regular array of z-values, referenced to a common datum. DEMs are typically used to represent terrain relief.

Demographics The statistical characteristics (such as age, birth rate, and income) of a human population.

Demography The statistical study of human populations, especially their locations, distribution, economic statistics, and vital statistics.

Densify To add vertices to a line at specified distances without altering the line's shape.

Densitometer (1) An instrument for measuring the opacity of translucent materials such as photographic negatives and optical filters. (2) Optical device for measuring the density of photographic transparencies.

Density In spatial measurements, the quantity per unit area or length.

Density slicing (1) A technique normally applied to a single-band monochrome image for highlighting areas that appear to be uniform in tone, but are not. Greyscale values (0–255) are converted into a series of intervals, or slices, and different colours are assigned to each slice. Density slicing is often used to highlight variations in vegetation. (2) Process of converting the continuous grey tones of an image into a series of density intervals, or slices, each corresponding to a specific digital range. The density slices are then displayed either as uniform grey tones or as colours.

Density, of images Measure of the opacity, or darkness, of a negative or positive transparency.

Density, of materials (r) Ratio of mass to volume of a material, typically expressed as grams per cubic centimetre.

Dependent variable The variable representing the process being predicted or modelled, such as crime, foreclosure, or rainfall. The dependent variable is a function of the independent variables. Regression can be used to predict the dependent variable, using known (observed) values to build (calibrate) the regression model. In the regression equation, the dependent variable appears on the left side of the equal sign.

Deployment The installation of a component or application on a target machine.

Depolarization The polarization state of an electromagnetic wave can change when the wave scatters from a target. Depolarization is a measure of the change in the degree of polarization of a partially polarized wave upon scattering. For example, a target may scatter a wave with a greater degree of polarization than the incident wave, in which case the depolarization is negative. Depolarization is also used to indicate spatial or temporal variation of the degree of polarization for a completely polarized wave.

Depolarized Refers to a change in polarization of a transmitted radar pulse as a result of various interactions with the terrain surface.

Depot In ArcGIS Network Analyst, a network location used to represent a starting, stopping, or renewal location for routes in vehicle routing problem (VRP) analysis. Users can specify multiple depots. Depots are used as locations for loading/unloading vehicles within the fleet.

Depot visit In ArcGIS Network Analyst, an object used to represent a single visit to a specific depot in vehicle routing problem (VRP) analysis. A depot visit may occur at the start of a route, the end of the route, or as a renewal midway along a route.

Depression angle (y) In radar, the angle between the imaginary horizontal plane passing through the antenna and the line connecting the antenna and the target.

Depression contour On a topographic map, concentric contour lines drawn with hachures to indicate a closed depression or basin. Concentric contour lines drawn without hachure marks indicate a hill.

Depth contour A line on a map connecting points of equal depth below a hydrographic datum.

Depth curve A line on a map connecting points of equal depth below a hydrographic datum.

Derived data In ModelBuilder, data created by running a geoprocessing operation on existing data. Derived data from one process can serve as input data for another process.

Derived value In ModelBuilder, data created by running a geoprocessing operation on existing data. Derived data from one process can serve as input data for another process.

Descending node (1) Direction a satellite is travelling relative to the Equator. A descending node implies a southbound Equatorial crossing. (2) The point at which a satellite travelling north to south crosses the equator.

Descriptive data Tabular or textual data describing the geographic characteristics of features.

Descriptor A type of attribute for network elements that cannot be apportioned. The value of a descriptor stays the same through the length of an edge element in a network dataset. Descriptors describe characteristics of the element; for example, the number of lanes for a particular road in a road network.

Designated market area A television market as defined by Nielsen Media Research. Most DMAs correspond to whole counties, but there are a few exceptions where counties are split into different DMAs.

Desire-line analysis A type of market analysis that draws lines from a set of geocoded points (usually customers) to a single, central point (usually a store). Desire lines can be weighted.

Desktop GIS Mapping software that is installed onto and runs on a personal computer and allows users to display, query, update, and analyse data about geographic locations and the information linked to those locations.

Destination The secondary object in a relationship class, such as a table containing attributes associated with features in a related table.

Destination table In ArcView 3.x, one of the two tables involved in a join operation. The destination table must be the active table; the attributes of the source (inactive) table are appended to it.

Detail page One of two types of pages in the Survey Analyst Survey Explorer. The Detail page displays a detailed view of individual survey objects.

Detectability Measure of the smallest object that can be discerned on an image.

Detection Processing stage at which the strength of the signal is determined for each pixel value. Detection removes phase information from the data file. The preferred detection scheme uses a magnitude squared method, which is energy conserving, and has units of voltage squared per pixel.

Detector Component of a remote sensing system that converts electromagnetic radiation into a recorded signal.

Determinate flow direction A conclusively definitive line or course in which something is issuing or moving in a stream. For an edge feature, this occurs when the flow direction can be ascertained from the connectivity of a network, the locations of sources and sinks, and the enabled or disabled states of features.

Deterministic model In spatial modelling, a type of model or a part of a model in which the outcome is completely and exactly known based on known input; the fixed or non-random components of a spatial model. The spline and inverse distance weighted interpolation methods are deterministic since they have no random components. The kriging and cokriging interpolation methods may have a deterministic component, often called the trend.

Detrending The process of removing the trend from a spatial model by subtracting the trend surface (usually polynomial functions of the spatial x- and y-coordinates) from the original data values. The resulting detrended values are called residuals.

Developable surface A geometric shape such as a cone, cylinder, or plane that can be flattened without being distorted. Many map projections are classified in terms of these shapes.

Developer product Products that can be used on one machine, similar to single use products. Each developer product requires a unique registration number used to generate the authorization file. The Software Authorization Wizard is then used to enter the authorization file and unlock the software for use.

Developer sample A sample contained in the ArcGIS Developer Help system.

Developing Chemical processing of an exposed photographic emulsion to produce an image.

Development environment A software product used to write, compile, and debug components or applications.

Device context Represents a surface that can be drawn to, for example, a screen, bitmap, or printer. In ArcGIS, the Display abstract class is used to abstract a device context.

Device coordinates The coordinates shown on a digitizer or display, as opposed to those of a recognized datum or coordinate system.

DGIWG Acronym for Digital Geographic Information Working

Group. A group established in 1983 to develop standards for spatial data exchange among nations participating in the North Atlantic Treaty Organization (NATO). The goals of the group are interoperability and burden sharing among nations, and its membership has recently expanded beyond NATO nations. While DGIWG is not an official NATO body, its work on standards has been recognized by the NATO Geographic Conference (NGC).

DGPS A technique for increasing the accuracy of GPS measurements by comparing the readings to two receivers – one roving and the other a fixed base station – and a known location.

DHTML An extension to HTML that allows Web designers to make elements on a Web page interactive, rather than changeable only when the page is loaded.

Dialog box In geoprocessing, a form consisting of a tool's parameters.

Diazo process Rapid method for copying documents in which the image is developed by exposure to ammonia.

Dielectric constant Electrical property of matter that influences radar returns. Also referred to as complex dielectric constant.

Difference image Image prepared by subtracting the digital values of pixels in one image from those in a second image to produce a third set of pixels. This third set is used to form the difference image.

Differential correction A technique for increasing the accuracy of GPS measurements by comparing the readings to two receivers – one roving and the other a fixed base station – and a known location.

Differential GPS (DGPS) A technique to improve GPS accuracy that uses pseudo range errors measured at a known location to improve the measurements made by other GPS receivers within the same general geographic area.

Differential positioning Precise measurement of the relative positions of two receivers tracking the same GPS signals.

Diffuse reflector Surface that reflects incident radiation nearly equally in all directions.

Diffusion The spread of an innovation or technology use among a group of people or organizations.

DIGEST Acronym for Digital Geographic Information Exchange Standard. A standard for spatial data transfer among nations, data producers, and data users. The Digital Geographic Information Working Group (DGIWG) developed the standard to support interoperability within and between nations and share the burden of digital data production. The standard addresses the exchange of raster, matrix, and vector data (and associated text) and a range of levels of topological structures.

Digital (1) A means for encoding information in a communications signal by using bits (binary digits). Conversion of information into bits of data for transmission through wire, fibre optic cable, satellite or over air techniques. Method allows simultaneous transmission of voice, data or video. (2) Represented in discrete, quantified units rather than continuously. Computers process and store information in digital form.

Digital analysis If you have a digital satellite image, then it is useful to analyse it digitally. Special computer programs are available for this. Such programs can stretch and distort a digital image to make it fit a map; they can enhance it to make it show some features more clearly, they can classify the image into categories which contain similar features, and much more.

Digital data Information that is made up of numbers is digital data. Telephone numbers are digital data, so are the percentage scores on your last test. So are digital images from satellites. The opposite of digital is ANALOGUE.

Digital display A form of data display in which values are shown as arrays of numbers.

Digital elevation model (DEM) The representation of continuous elevation values over a topographic surface by a regular array of z-values, referenced to a common datum. DEMs are typically used to represent terrain relief.

Digital geographic information exchange standard A standard for spatial data transfer among nations, data producers, and data users. The Digital Geographic Information Working Group (DGIWG) developed the standard to support interoperability within and between nations and share the burden of digital data production. The standard addresses the exchange of raster, matrix and vector data (and associated text), and a range of levels of topological structures.

Digital geographic information working group A group established in 1983 to develop standards for spatial data exchange among nations participating in the North Atlantic Treaty Organization (NATO). The goals of the group are interoperability and burden sharing among nations, and its membership has recently expanded beyond NATO nations. While DGIWG is not an official NATO body, its work on standards has been recognized by the NATO Geographic Conference (NGC).

Digital image An image stored in binary form binary codes from 0 to 255, or as one byte and divided into a matrix of pixels. Each pixel consists of a digital value of one or more bits, defined by the bit depth. The digital value may represent, but is not limited to, energy, brightness, colour, intensity, sound, elevation or a classified value derived through image processing. A digital image is stored as a raster and may contain one or more bands.

Digital image processing (DIP) Any technique that changes the digital values of an image for the sake of analysis or enhanced display, such as density slicing or low- and high-pass filtering.

Data files containing vector representations of cartographic information derived from USGS maps and related sources. DLGs include information from the USGS planimetric map base categories such as transportation, hydrography, contours, and public land survey boundaries.

Digital line graph (DLGs) Digital representations of point, line and area features from USGS quads including contour lines, spot elevations, hydrography, boundaries transportation and US Public Land Survey System.

Digital nautical chart A nautical database developed from existing hard-copy charts, digital data, bathymetric survey information, imagery and various raster data. DNCs are used by the US military and its allies for marine navigation.

Digital number (DN) Value assigned to a pixel in a digital image.

Digital orthophoto quadrangle A computer-generated, uniform-scale image created from an aerial photograph. Digital orthophoto quadrangles are true photographic maps in which the effects of tilt and relief are removed by a mathematical process called transformation or rectification. The uniform scale of a DOQ allows accurate measurement of distances.

Digital orthophoto quarter quadrangle A digital orthophoto quadrangle (DOQ) divided into four quadrants.

Digital raster graphic A raster image of a scanned USGS standard series topographic map, usually including the original border information, referred to as the map collar, map surround or marginalia. Source maps are georeferenced to the surface of the earth, fit to the universal transverse Mercator (UTM) projection and scanned at a minimum resolution of 250 dpi. The accuracy and datum of a DRG matches the accuracy and datum of the source map.

Digital terrain elevation data A format for elevation data, usually tiled in 1-degree cells, produced by the National Geospatial-Intelligence Agency and US allies for military applications.

Digital terrain model (DTM) A topographic model of the bare earth-terrain relief that can be manipulated by computer programs. It describes only the terrain heights without artificial objects.

Digitization Process of converting an analogue display into a digital display.

Digitizer Device for scanning an image and converting it into numerical format. It is connected to a computer, consisting of a tablet and a handheld puck, that converts positions on the tablet surface as they are traced by an operator to digital x,y coordinates, yielding vector data consisting of points, lines, and polygons.

Digitizing The process of converting the geographic features on an analog map into digital format using a digitizing tablet, or digitizer, which is connected to a computer. Features on a paper map are traced with a digitizer puck, a device similar to a mouse, and the x,y coordinates of these features are automatically recorded and stored as spatial data.

Digitizing mode A way of using a digitizing tablet in which locations on the tablet are mapped to specific locations on the screen. Moving the digitizer puck on the tablet surface causes the screen pointer to move to precisely the same position on the screen.

Digitizing table A table with a built in electronic mesh that can sense the position of the cursor and can transmit its x, y coordinates to the connected computer.

Dijkstra's algorithm An algorithm that examines the connectivity of a network to find the shortest path between two points. Dijkstra's algorithm is named after the Dutch computer scientist Edsger Dijkstra (1930–2002).

Dike Bank of earth or stone used to form a barrier, frequently and confusingly interchanged with levee. A dike restrains water within an area that normally is flooded.

Dilution of precision (DOP) A n indicator of satellite geometry for a constellation of satellites used to determine a position. Positions with a lower DOP value generally constitute better measurement results than those with higher DOP. Factors determining the total GDOP (geometric DOP) for a set of satellites include PDOP (positional DOP), HDOP (horizontal DOP), VDOP (vertical DOP), and TDOP (time DOP).

DIME Acronym for Dual Independent Map Encoding. A data storage format for geographic data developed by the US Census Bureau in the 1960s. DIME-encoded data was stored in Geographic Base Files (GBF). The Census Bureau replaced the DIME format with Topologically Integrated Geocoding and Referencing (TIGER) in 1990.

Dimension A length of a certain distance and bearing.

Dimension construction method One of a number of procedures that dictate what type of dimension feature is created and the number of points required to complete the feature's geometry. Construction methods include simple aligned, aligned, linear, rotated linear, free aligned and free linear.

Dimension feature In ArcMap, a special kind of geodatabase annotation that shows specific lengths or distances on a map. A dimension feature may indicate the length of a side of a building or land parcel, or it may indicate the distance between two features, such as a fire hydrant and the corner of a building.

Dimension feature class A geodata base feature class that stores dimension features.

Dimension style Description of a dimension feature's symbology, what parts of it are drawn, and how it is drawn. Every time a new dimension feature is created, it is assigned a particular style according to its shared characteristics. A collection of dimension styles is associated with a dimension feature class. Styles for a dimension feature class are created, copied and managed using ArcCatalog or the editing capabilities in ArcMap. Styles are then assigned to individual dimension features.

Dimensioning toolbar A toolbar in ArcMap that facilitates the creation of dimension features.

Direct connect An ArcSDE connection that results in a two-tier configuration, in which the ArcSDE client application communicates directly with the database.

Direct translation use of a translator or algorithm in a GIS package to directly convert spatial data from one format to another.

Directed link In ArcInfo Workstation, an arc between two nodes with one direction specified.

Directed network flow A network state in which edges have an associated direction of flow. In a directed network flow, the resource that traverses a network's components cannot choose a direction to take, as in hydrologic and utility systems.

Direction In an ArcGIS vertical coordinate system, an identifier that indicates whether z-values are positive up or positive down. Heights or elevations are usually positive up, against the force of gravity (indicated by +1). Depths are usually positive down, with the force of gravity (indicated by −1).

Direction field In Survey Analyst for field measurements, a field in the computation page that allows the entry of bearings or azimuths between an input survey point and a computed survey point.

Directional filter In image processing, an edge-detection filter that enhances those linear features in an image that are oriented in a particular direction.

Directional influences Natural or physical processes that affect a measured trait or attribute so that the magnitude of the effects on the attribute vary in different directions.

Directory An area of a computer disk that holds a set of data files, other directories, or both. Operating systems use directories to organize data. Directories are arranged in a tree structure, in which each branch is a subdirectory of its parent branch. The location of a directory is specified with a path, for example, C:\gisprojects\shrinkinglemurhabitatgrids.

Directory Service An OpenLS ArcWeb service used to find points of interest and other entities around a given location.

Dirty areas Regions surrounding features that have been altered after the initial topology validation process and require additional topology validation to be performed to find any errors.

Disabled feature In geometric networks, an object or shape representing a geographic object through which flow is impossible.

Disconnected editing The process of copying data to another geodatabase, editing that data, then merging the changes with the data in the source or master geodatabase.

Discrete data Data that represents phenomena with distinct boundaries. Property lines and streets are examples of discrete data.

Discrete digitizing A method of digitizing in which points are placed individually to define a feature's shape.

Discrete feature A feature that has definite feature boundaries.

Discrete raster A raster that typically represents phenomena that have clear boundaries with attributes that are descriptions, classes, or categories. Generally, integers are used for the cell values. In a raster of land cover, for example, the value 1 might represent forestland, the value 2 urban land, and so on. It is assumed that the phenomena that each value represents fill the entire area of the cell. Rasters representing land use, political boundaries or ownership are examples of discrete rasters.

Disk A storage medium for recording digital information, consisting of a round, flat, spinning plate coated with a magnetic material.

Disk cache In ArcGlobe, the folder on a computer's disk drive where ArcGlobe stores layer cache files.

DispID binding A type of early binding in which DispID properties and methods are accessed at compile time, so there is no need to call them at run time. DispID binding is used for components that have type libraries but do not support vTable binding. MapObjects uses DispID binding.

Displacement Image displacement in a remote sensing image (away from the nadir point in optical imagery and toward the nadir point in radar imagery) due to sensor/target imaging geometries.

Displacement link In ArcGIS, a link created to define the source and destination coordinates for a spatial adjustment. Links are represented as arrows with the arrowhead pointing toward the destination location. Links can be created manually or loaded from a link file.

Displacement vector In Survey Analyst-Cadastral Editor, the difference in the coordinates of a point between successive least-squares adjustments. A set of displacement vectors is recorded for a point when the point is part of a least-squares adjustment. Adjustment vectors can be used to track spatial changes in the cadastral fabric over time, and to adjust associated feature classes to match the most up-to-date cadastral fabric location.

Display Often used to refer to subclasses of the Display abstract class. For example, 'when drawing to the display' means when drawing to any of the display coclasses; 'the display pipeline' refers to the sequence of calls made when drawing occurs.

Display projection The coordinate system used for displaying geographic data. Examples include the data frame in ArcMap and the view in ArcView GIS.

Display scale The scale at which data is rendered on a computer screen or on a printed map.

Display type The mode of command representation on a computer screen. The display type controls whether you see a command's image, its caption, or both when it appears on a toolbar or in a menu.

Display unit The unit of measure used to render dimensions of shapes, distance tolerances, and offsets on a computer screen or on a printed map. Although they are stored with consistent units in the dataset, users can choose the units in which coordinates and measurements are displayed – for example, feet, miles, metres, or kilometres.

Dissemination The spread of an innovation or technology use among a group of people or organizations.

Dissolve A geoprocessing command that removes boundaries between adjacent polygons that have the same value for a specified attribute.

Dissolve route events In linear referencing, a procedure that combines event records in tables where there are events on the same route that have the same value for specified fields. The Dissolve Route Events tool combines events when there is measure overlap, and is available for both line and point event tables.

Distance The measure of separation between two entities or locations that may or may not be connected, such as two points. Distance is differentiated from length, which implies a physical connection between entities or locations.

Distance decay A mathematical representation of the effect of distance on the accessibility of locations and the number of interactions between them, reflecting the notion that demand drops as distance increases. Distance decay can be expressed as a power function or as an exponential function.

Distance field In the Survey Analyst for field measurements Survey Explorer, a field for entering distance for use in a computation.

Distance unit The unit of measurement for distance, such as feet, miles, metres, and kilometres.

Distortion On a map or image, the misrepresentation of shape, area, distance, or direction of or between geographic features when compared to their true measurements on the curved surface of the earth.

Distortion circle A graphical representation of the spatial distortion at a particular map location. The indicatrix is the figure that results when a circle on the earth's surface is plotted to the corresponding point on a map. The shape, size, and orientation of an indicatrix at any given point depend on the map projection used. In conformal (shape-preserving) projections, the indicatrix is a circle; in nonconformal projections, it is an ellipse at most locations. As a visual aid, indicatrices convey a general impression of distortion; as mathematical tools, they can be used to quantify distortion of scale and angle precisely. The indicatrix is named for Nicolas Auguste Tissot, the French mathematician who developed it.

Distributed component object model Extends COM to support communication among objects on different computers on a network.

Distributed data Data spread over multiple platforms or a network by a process referred to as replication.

Distributed database A database with records that are dispersed between two or more physical locations. Data distribution allows two or more people to be working on the same data in separate locations.

Distribution The frequency or amount at which a thing or things occur within a given area.

Dithering (1) The approximation of shades of grey or colours in a computer image made by arranging pixels of black and white or other colours in alternate layers. The technique gives the appearance of a wider range of colour or shades than is actually present in the image. It is widely used to improve the appearance of images displayed on devices with limited colour palettes. (2) The introduction of digital noise. This is the process the Department of Defence (DoD) uses to add inaccuracy to GPS signals to induce Selective Availability.

Diurnal Daily, as in the revolution of the earth.

Diurnal arc The apparent path from rise to set made by a heavenly body across the sky.

DLG Acronym for digital line graph. Data files containing vector representations of cartographic information derived from USGS maps and related sources. DLGs include information from the USGS planimetric map base categories such as transportation, hydrography, contours, and public land survey boundaries.

DLL Acronym for dynamic-link library. A type of file that stores shared code to be used by multiple programs (a 'code library'). Programs access the shared code by linking to the DLL file when they run, a process referred to as dynamic linking. The DLL file must be registered for other programs to locate it.

DMA Acronym for Designated Market Area. A television market as defined by Nielsen Media Research. Most DMAs correspond to whole counties, but there are a few exceptions where counties are split into different DMAs.

DMS The unit of measure for describing latitude and longitude. A degree is 1/360th of a circle. A degree is further divided into 60 minutes, and a minute is divided into 60 seconds.

DNC Acronym for digital nautical chart. A nautical database developed from existing hard-copy charts, digital data, bathymetric survey information, imagery, and various raster data. DNCs are used by the US military and its allies for marine navigation.

DNS Acronym for domain name system. The Internet distributed system that stores IP addresses and domain names to assist with the routing of network traffic.

Dockable window A window that can exist in a floating state or be attached to the main application window.

Docking Moving a floating toolbar or window to a fixed location in the graphical user interface.

Document A component of an ArcView 3.x project. Each document type (view, table, chart, layout, script) has its own window and interface.

Documentation Supporting information for software data and tools. Documentation may be descriptive or instructional, and is published in a variety of formats, including user's guides and manuals, desktop help systems, embedded or context-sensitive help, tutorials, reports, and metadata.

Documentation editor In geoprocessing, the interface used to write documentation for tools, toolsets, toolboxes, and processes within a model.

Domain The range of valid values for a particular metadata element.

Domain name The unique name of a computer system on the Internet, such as 'esri.com'.

Domain name system The Internet distributed system that stores IP addresses and domain names to assist with the routing of network traffic.

Dongle A small hardware device that provides the unique number used in the generation of a license file. The key is plugged into either the parallel or

USB port on the License Manager Server. UNIX users of ArcGIS software do not need a hardware key. Hardware keys are also called dongles.

Donut rings A method of defining the rings in an analysis so that the values inside the rings are exclusive. For example, in an analysis with three donut rings and 10 households in each, the total number of households for each ring would be 10.

DOP Acronym for dilution of precision. See Dilution of precision.

Doppler frequency The Doppler frequency depends on the component of satellite velocity in the line-of-sight direction to the target. This direction changes with each satellite position along the flight path, so the Doppler frequency varies with azimuth time. For this reason, azimuth frequency is often referred to as Doppler frequency.

Doppler principle Describes the change in observed frequency that electromagnetic or other waves undergo as a result of the movement of the source of waves relative to the observer.

Doppler radar A radar system which differentiates between fixed and moving targets by detecting the change in frequency of the reflected wave caused by the doppler effects. The system can also measure target velocity with high accuracy.

Doppler shift The apparent change in frequency of sound or light waves caused by the relative motion between a source and an observer. As they approach one another, the frequency increases; as they draw apart, the frequency decreases. The Doppler shift is also known as the Doppler effect, and is named for Austrian physicist and mathematician Christian Andreas Doppler.

Doppler-aided GPS Signal processing that uses a measured Doppler shift to help the receiver track the GPS signal.

Doppler-aiding A signal processing strategy that uses a measured doppler shift to help the receiver smoothly track the GPS signal. This allows more precise velocity and position measurement.

DOQ Acronym for digital orthophoto quadrangle. A computer-generated, uniform-scale image created from an aerial photograph. Digital orthophoto quadrangles are true photographic maps in which the effects of tilt and relief are removed by a mathematical process called transformation or rectification. The uniform scale of a DOQ allows accurate measurement of distances.

DOQQ Acronym for digital orthophoto quarter quadrangle. A digital orthophoto quadrangle (DOQ) divided into four quadrants.

Dot density map A quantitative, thematic map on which dots of the same size are randomly placed in proportion to a numeric attribute associated with an area. Dot density maps convey the intensity of an attribute.

Dot distribution map A map that uses dots or other symbols to represent the presence, quantity, or value of a phenomenon or thing in a specific area. In a dot distribution map, the size of the dots is scaled in proportion to the intensity of the variable.

Dot screen A photographic film covered with uniformly sized, evenly spaced dots used to break up a solid colour, producing an apparently lighter colour.

Dots per inch A measure of the resolution of scanners, printers, and graphic displays. The more dots per inch, the more detail can be displayed in an image.

Double precision The level of coordinate exactness based on the possible number of significant digits that can be stored for each coordinate. Datasets can be stored in either single or double precision. Double- precision geometries store up to 15 significant digits per coordinate (typically 13 to 14 significant digits), retaining the accuracy of much less than 1 metre at a global extent.

Double-coordinate precision The level of coordinate exactness based on the possible number of significant digits that can be stored for each coordinate. Datasets can be stored in either single or double precision. Double-precision geometries store up to 15 significant digits per coordinate (typically 13 to 14 significant digits), retaining the accuracy of much less than 1 metre at a global extent.

Douglas-Peucker algorithm (1) A computer algorithm for line simplification. (2) An algorithm that simplifies complex lines by reducing the number of points used to represent them. The Douglas-Peucker algorithm was developed by the Canadian geographers David H. Douglas and Thomas K. Peucker.

Downlink A communications channel for receiving transmissions from a spacecraft.

Downstream In network tracing, the direction along a line or edge that is the same as the direction of flow.

Dpi Acronym for dots per inch. A measure of the resolution of scanners, printers, and graphic displays. The more dots per inch, the more detail can be displayed in an image.

Drafting A method of drawing with pencil or pen and ink, used in cartographic reproduction.

Drainage All map features associated with the movement and flow of water, such as rivers, streams and lakes.

Draped layer A layer in ArcGlobe that has been categorized to be draped on top of the globe surface.

Draping A perspective or panoramic rendering of a two-dimensional image superimposed onto a three-dimensional surface. For example, an aerial photograph might be draped over a digital elevation model (DEM) to create a realistic terrain visualization.

Drawing priority In 3D analysis, the order in which layers that occupy the same x,y,z positions are drawn in a scene. For example, if a road feature layer and an orthophoto are draped over the same terrain model, the roads and raster may appear patchy or broken up where they coincide. The drawing priority for the raster can be reduced so it will appear below the features. The drawing priority can only be changed for polygon features and surfaces.

DRG Acronym for digital raster graphic. A raster image of a scanned USGS standard series topographic map, usually including the original border information, referred to as the map collar, map surround, or marginalia. Source maps are georeferenced to the surface of the earth, fit to the universal transverse Mercator (UTM) projection, and scanned at a minimum resolution of 250 dpi. The accuracy and datum of a DRG matches the accuracy and datum of the source map.

Drift In geography, drift is a general general tendency of Surface Ocean water to flow in the direction of the prevailing winds. But in GIS (Interpolation method) drift is the general pattern of z-values throughout a kriging model. The drift, or structure, forms the model's basic shape.

Drive-time area A zone around a map feature measured in units of time needed for travel by car. For example, a store's 10-minute drive-time area defines the area in which drivers can reach the store in 10 minutes or less.

Drum scanner A type of scanner in which a hard-copy image or map is attached to a cylinder that spins while a sensor captures a digital image from the surface of the page.

DTD Acronym for document type definition. A set of rules that define the allowed structure and properties of XML documents.

DTED Acronym for digital terrain elevation data. A format for elevation data, usually tiled in 1-degree cells, produced by the National Geospatial-Intelligence Agency and US allies for military applications.

DTM Acronym for digital elevation model. The representation of continuous elevation values over a topographic surface by a regular array of z-values, referenced to a common datum. DEMs are typically used to represent terrain relief.

Dual independent map encoding A data storage format for geographic data developed by the US Census Bureau in the 1960s. DIME-encoded data was stored in Geographic Base Files (GBF). The Census Bureau replaced the DIME format with Topologically Integrated Geocoding and Referencing (TIGER) in 1990.

Duplicate labels Labels with identical content. Maplex for ArcGIS provides the option of labelling only a single feature out of a cluster of similar features. Identical labels will not be placed within the user-defined distance of a placed label.

Dwell time Time required for a detector IFOV to sweep across a ground resolution cell.

Dynamic binding A COM technique that an application uses for determining an object's properties and methods at run time, rather than when the code is compiled. Late binding is generally used by scripting languages.

Dynamic feature class A feature class consisting of points associated with address elements in an address data table that change, based on changes made to the address data table.

Dynamic HTML An extension to HTML that allows Web designers to make elements on a Web page interactive, rather than changeable only when the page is loaded.

Dynamic Range The range between the maximum and minimum amount of input radiant energy that an instrument can measure.

Dynamic segmentation (1) A data model that is built upon arcs of line coverage and allows the use of real world coordinates with linear measures. (2) The process of computing the map locations of linearly referenced data (for example, attributes stored in a table) at run time so they can be displayed on a map, queried, and analysed using a GIS. The dynamic segmentation process enables multiple sets of attributes to be associated with any portion of a line feature without segmenting the underlying feature. In the transportation field, examples of such linearly referenced data might include accident sites, road quality, and traffic volume.

Dynamic style In ArcWeb Services, a map style that changes based on the selected style sheet. Users can select from a set of defined dynamic styles for a map data source, but they cannot create dynamic styles.

Dynamic-link library A type of file that stores shared code to be used by multiple programs (a 'code library'). Programs access the shared code by linking to the DLL file when they run, a process referred to as dynamic linking. The DLL file must be registered for other programs to locate it.

E

Early binding A COM technique in application uses to access an object. In early binding, an object's properties and methods are discovered from an interface at compile time, instead of being checked at run time, as in late binding. This difference often gives early binding performance benefits over late binding. There are two types of early binding: DispID binding and vTable binding.

Earth observation Looking down at the Earth from aircraft and satellites using various sensors which make images that are afterwards used to study what is happening on or near the Earth's surface.

Earth-centred datum A horizontal geodetic datum based on an ellipsoid that has its origin at the earth's centre of mass. Examples are the World Geodetic System of 1984, the North American Datum of 1983, and the Geodetic Datum of Australia of 1994. The first uses the WGS84 ellipsoid; the latter two use the GRS80 ellipsoid. Geocentric datums are more compatible with satellite positioning systems, such as GPS, than are local datums.

Easting The distance east of the origin that a point in a Cartesian coordinate system lies, measured in that system's units.

Eccentricity A measure of how much an ellipse deviates from a circle, expressed as the ratio of the distance between the centre and one focus of an ellipsoid to the length of its semi-major axis. The square of the eccentricity (e^2) is commonly used with the semi-major axis a to define a spheroid.

Ecliptic The great circle formed by the intersection of the plane of the earth's orbit around the sun (or apparent orbit of the sun around the earth) and the celestial sphere.

Ecological fallacy The assumption that an individual from a specific group or area will exhibit a trait that is predominant in the group as a whole.

Economic geography The field of geography concerning the distribution and variation of economic factors by location, including how economic factors interact with geographic factors such as climate, land use, and geology.

EDC EROS Data Centre.

Edge (1) A boundary in an image between areas with different tones. (2) A line between two points that forms a boundary. In a geometric shape, an edge forms the boundary between two faces. In an image, edges separate areas of different tones or colours. In topology, an edge defines lines or polygon boundaries.

Edge connectivity policy In network datasets, a connectivity policy that defines how one edge may connect to another edge mid-span. There are two edge-edge connectivity policies:

endpoint connectivity and any-vertex connectivity.

Edge detection A digital image processing technique for isolating edges in a digital image by examining it for abrupt changes in pixel value.

Edge element A line between two points that forms a boundary. In a geometric shape, an edge forms the boundary between two faces. In an image, edges separate areas of different tones or colours. In topology, an edge defines lines or polygon boundaries.

Edge enhancement (1) A digital image processing technique for emphasizing the appearance of edges and lines in an image. (2) Image-processing technique that emphasizes the appearance of edges and lines.

Edge matching an operation in joining adjacent coverage.

Edge-edge rule In geometric networks, a connectivity rule that defines how one edge may connect to another edge through a junction.

Edge-junction cardinality In connectivity relationships for networks, the number of edges of one type that may be associated with junctions of another type. Edge-junction cardinality defines a range of permissible connections that may occur in a one-to-many relationship between a single junction and many edges.

Edge-junction rule In geometric networks, a connectivity rule that defines how an edge may connect to a junction.

Edgematching A spatial adjustment process that aligns features along the edge of an extent to the corresponding features in an adjacent extent.

Edit annotation tool A tool on the Annotation toolbar that is used to manipulate geodatabase annotation. Text can be interactively moved, scaled, and rotated. Shortcut menu options allow control of the position, orientation, symbology, content, size, and style of text.

Edit box An entity that displays text entered by a user or derived from another source for editing purposes.

Edit cache A setting used in ArcMap that allows temporary storage of geodatabase or ArcIMS feature service features from a given map extent in the desktop computer's RAM, which may result in performance improvements in ArcMap for editing, feature rendering, and labelling.

Edit mask The portion of a coverage where the geometry (or geographical features) has been altered, but where topology has not yet been restored.

Edit session In ArcMap, the environment in which spatial and attribute editing take place. After starting an edit session, a user can modify feature locations, geometry, or attributes. Modifications are not saved unless the user explicitly chooses to save them.

Edit sketch In ArcMap, a temporary sketch that is used to perform a variety of tasks. Creating an edit sketch is a standard way to edit feature geometry.

Edit task During editing in ArcMap, a setting in the Task drop-down list that determines which operation the sketch construction tools will perform. Examples of edit tasks include creating new features and modifying existing features. The edit task is set by clicking a task in the Task drop-down list.

Editdistance A tolerance used in ARC/INFO.

Editor toolbar In ArcMap, a set of tools that allows the creation and modification of features and their attributes.

EDMS Acronym for electronic document management system. A computer-based system for organizing, maintaining, and retrieving digital and hard-copy documents. An EDMS usually includes a check-in, check-out system for document tracking, versioning, and search-and-retrieval capabilities.

EDN Acronym for ESRI Developer Network. A subscription program used to acquire ESRI software for developers, which includes a special program for supporting the ESRI developer community. EDN is intended to promote collaboration and interaction among GIS developers and ESRI staff.

EJB The server-side component architecture for the J2EE platform. EJB enables development of distributed, transactional, secure, and portable Java applications.

Ektachrome A Kodak colour positive film.

Elastic transformation A procedure for adjusting the coordinates of all the data points in a dataset to allow a more accurate match between known locations and a few data points within the dataset. Rubber sheeting preserves the interconnectivity between points and objects through stretching, shrinking, or reorienting their interconnecting lines.

Electromagnetic Relating to the interplay between electric and magnetic fields.

Electromagnetic displacement Image distortion in the range-direction caused by Infra Red, microwave, and radio waves. By terrain features in the scene being above, or below, the reference elevation contour and, in fact, being closer to, or farther from, the radar than their planimetric position. The effect may be used to create radar stereo images.

Electromagnetic radiation Energy that moves through space at the speed of light as different wavelengths of time-varying electric and magnetic fields. Types of electromagnetic radiation include gamma, x, ultraviolet, visible, infrared, microwave and radio.

Electromagnetic spectrum The entire range of radiant energies or wave frequencies from the longest to the shortest wavelengths – the categorization of solar radiation. Satellite sensors collect this energy, but what the detectors capture is only a small portion of the entire electromagnetic spectrum. The spectrum is usually divided into seven sections: radio, microwave, infrared, visible, ultraviolet, x-ray and gamma-ray radiation.

Electronic atlas A mapping system that displays but does not allow for the spatial analysis of data.

Electronic distance measuring (EDM) device Instruments that measure the phase difference between transmitted and reflected or retransmitted electromagnetic waves of known frequency, or that measure the round-trip transit time of a pulsed signal, from which distance is computed.

Electronic document management system A computer-based system for organizing, maintaining, and retrieving digital and hard-copy documents. An EDMS usually includes a check-in, check-out system for document tracking, versioning and search-and-retrieval capabilities.

Electronic navigational chart A vector data product used for nautical navigation. ENC data is produced by nautical charting agencies throughout the world and uses the IHO (International Hydrographic Organization) S–57 standard for its database structure and attribution.

Element In geoprocessing in ArcGIS, a component of a model. Elements can be variables, such as input and derived data, or tools.

Elevation The vertical distance of a point or object above or below a reference surface or datum (generally mean sea level). Elevation generally refers to the vertical height of land.

Elevation guide A map element that displays a simplified representation of the terrain within a map's extent. Elevation guides are designed to provide a quick overview of topography, including the high and low points.

Elevation layer A layer in ArcGlobe that has been categorized to help define the geometry of the globe surface.

Elevation tints Hypsometric tint bands based on elevation ranges used in an elevation guide.

Ellipse A geometric shape described mathematically as the collection of points whose distances from two given points (the foci) add up to the same sum. An ellipse is shaped like a circle viewed obliquely.

Ellipsoid (1) A model that approximates the earth. Also called spheroid. (2) A three-dimensional, closed geometric shape, all planar sections of which are ellipses or circles. An ellipsoid has three independent axes, and is usually specified by the lengths a,b,c of the three semi-axes. If an ellipsoid is made by rotating an ellipse about one of its axes, then two axes of the ellipsoid are the same, and it is called an ellipsoid of revolution, or spheroid. If the lengths of all three of its axes are the same, it is a sphere.

Elliptical polarization A polarization state in which the two perpendicular components of the electric field have unequal magnitudes and a non-zero phase difference. In this case, the tip of the electric field vector traces an ellipse on a plane that is transverse to the wave propagation direction.

Elliptically polarized antenna An antenna that radiates elliptically-polarized electromagnetic waves in its far field.

Ellipticity A measure of how much an ellipse deviates from a circle, expressed as the ratio of the distance between the centre and one focus of an ellipsoid to the length of its semimajor axis. The square of the eccentricity (e^2) is commonly used with the semi-major axis a to define a spheroid.

Embedded feature class A multipoint feature class embedded into a terrain dataset. When a feature class is embedded, it is incorporated directly into the terrain pyramid and the terrain becomes the sole container of the data. Embedded feature classes can be used to reduce the amount of disk space required by mass point data such as lidar.

EMF Acronym for Enhanced Metafile. A spool file format used in printing by the Windows operating system.

Emission Process by which a body radiates electromagnetic energy. Emission is determined by kinetic temperature and emissivity.

Emissivity (e) Ratio of radiant flux from a body to that from a blackbody at the same kinetic temperature and emissivity.

Emit means the same as 'sent out' or 'given off'. The sun emits radiation, some of which we can feel as heat and some of which we can see as light. The radar sensor in RADARSAT emits a radar beam.

Emittance A term for the radiant flux of energy per unit area emitted by a body. (Now obsolete).

Emitted energy The radiated electromagnetic energy of the active sensor during emission which is determined by kinetic temperature and emissivity.

Empirical That property of a quantity that indicates that the quantity depends on data, observations, or experiment only; that is, it is not a model or part of a model. An empirical semivariogram is computed on data only, in contrast to a theoretical semivariogram model.

Emulsion Suspension of photosensitive silver halide grains in gelatin that constitutes the image-forming layer on photographic film.

Enabled feature In geometric networks, a network feature that allows flow to pass through it.

ENC Acronym for electronic navigational chart. A vector data product used for nautical navigation. ENC data is produced by nautical charting agencies throughout the world and uses the IHO (International Hydrographic Organization) S–57 standard for its database structure and attribution.

Enclosure A file describing the contents of an item included in metadata. Enclosing files in metadata works the same way as enclosing files in an e-mail message.

Encoding The recording or reformatting of data into a computer format. Data may be encoded to reduce storage, increase security, or to transfer it between systems using different file formats. In GIS, analog graphic data, such as paper maps and images, are encoded into computer formats by scanning or digitizing.

End hatch definition In linear referencing, a special type of hatch definition that draws hatch marks only at the low and high measure of a linear feature end offset. An adjustable value that dictates how far away from the end of a line an address location should be placed. Using an end offset prevents the point from being placed directly over the intersection of cross streets if the address happens to fall on the beginning or end of the street.

Endpoint connectivity In network datasets, a type of edge connectivity policy that states that an edge may only connect to another edge at its endpoints.

Energy flux Radiant flux.

Enhanced thematic mapper plus (ETM+) The sensor aboard Landsat 7 that picks up solar radiation reflected by or emitted from the Earth.

Enhancement In remote sensing, applying operations to raster data to improve appearance or usability by making specific features more detectable. Such operations can include contrast stretching, edge enhancement, filtering, smoothing, and sharpening.

Enterprise geodatabase A geodatabase managed in an RDBMS server by ArcSDE. Multiuser geodatabases can be very large and support multiple concurrent editors. They are supported on a variety of commercial RDBMS, including IBM DB2, IBM Informix, Oracle, Microsoft SQL Server, and PostgreSQL.

Enterprise GIS A geographic information system that is integrated through an entire organization so that a large number of users can manage, share, and use spatial data and related information to address a variety of needs, including data creation, modification, visualization, analysis, and dissemination.

Enterprise JavaBeans The server-side component architecture for the J2EE platform. EJB enables development of distributed, transactional, secure, and portable Java applications.

Envelope The rectangle surrounding one or more geographical features in coordinate space, determined by the minimum and maximum coordinates in the x and y directions, as well as the ranges of any z- or m-values that the features may have. An envelope can be used to filter data for analysis.

Environment settings Settings that can apply to all tools within the application, all tools within a model or script, or a particular process within a model or script. Environment settings include current workspace, output spatial reference, output spatial grids, cell size, and tile size. They are generally set before running tools.

Environment variable A variable maintained by the operating system and shared among programs. Environment variables function as placeholders for environment information, such as a drive, path or file name.

Environmental model An abstract representation of a complex environmental process, emphasizing relationships and patterns in natural systems. Environmental models allow decision makers to better understand the effects of natural systems or the impact of human activities on natural systems.

EOBrowser An ArcGIS utility application that can be used to investigate the contents of object libraries.

EOSAT The commercial company that took over operations of the Landsat system in 1985.

Ephemeris (1) A table of satellite orbital locations for specific time intervals. The ephemeris data help to characterize the conditions under which remotely sensed data are collected and are commonly used to correct the sensor data before analysis. (2) A table of the predicted positions of a satellite within its orbit for each day of the year, or for other regular intervals.

Ephemeris errors Errors which originate in the ephemeris data transmitted by a GPS satellite.

EPSG ID A coordinate system identification created by the European Petroleum Survey Group.

Equal competition area A trade area boundary set halfway between a store or service point and its neighbouring stores or service points.

Equal-area classification A data classification method that divides polygon features into groups so that the total area of the polygons in each group is approximately the same.

Equal-area projection A projection in which the whole of the map as well as each part has the same proportional area as the corresponding part of the earth. An equal-area projection may distort shape, angle, scale, or any combination thereof. No flat map can be both equal-area and conformal.

Equal-interval classification A data classification method that divides a set of attribute values into groups that contain an equal range of values.

Equator (1) An imaginary circle around a body that is everywhere equidistant from the poles, defining the boundary between the northern and southern hemispheres. (2) The parallel of reference that is equidistant from the poles and defines the origin of latitude values.

Equatorial aspect A planar (or azimuthal) projection with its central point located at the equator.

Equidistant projection A projection that maintains scale along one or more lines, or from one or two points to all other points on the map. Lines along which scale (distance) is correct are the same proportional length as the lines they reference on the globe. In the sinusoidal projections, for example, the central meridian and all parallels are their true lengths. An azimuthal equidistant projection centred on Chicago shows the correct distance between Chicago and any other point on the projection, but not between any other two points.

Equivalent projection A projection in which the whole of the map as well as each part has the same proportional area as the corresponding part of the earth. An equal-area projection may distort shape, angle, scale, or any combination thereof. No flat map can be both equal-area and conformal.

Erase In ArcInfo, a command that removes or deletes features from one coverage that overlap features in another coverage.

ERBSS Earth Radiation Budget Sensor System, carried by NOAA satellites.

EREP Earth Resources Experiment Package, carried on Skylab and consisting of cameras and multispectral scanner.

EROS the Greek god of love. Earth Resources Observation System.

EROS data centre (EDC) Earth Resources Observation System (EROS) Data Centre (EDC) is a national archive, production, distribution and research facility for remotely sensed data and other geographic information. Facility of the US Geological Survey at Sioux Falls, South Dakota, that archives, processes, and distributes images.

Erosion Group of natural processes including weathering, dissolution, abrasion, corrosion, and transportation that remove material from any part of the Earth's surface.

Error A measured, observed, calculated, or interpreted value that differs from the true value or the value that would be obtained by a perfect observer using perfect equipment and perfect methods under perfect conditions.

Error propagation In GIS data processing, the persistence of an error into new datasets calculated or created using datasets that originally contained errors. The study of error propagation is concerned with the effects of combined and accumulated errors throughout a series of data processing operations.

Error table A geodatabase table used by the GIS Data ReViewer to track error information through the quality control process. Defects are recorded, resolved and verified in the error table.

ERTS Earth Resource Technology Satellite, now called Landsat.

ESA European Space Agency, based in Paris. A consortium between several European states for the development of space science, including the launch of remote-sensing satellites.

ESRI adapters for IBM LAS A translator that allows users to access ArcWeb Services through the IBM WebSphere Everywhere Access (WEA) Location Aware Services (LAS).

ESRI developer network A subscription program used to acquire ESRI software for developers, which includes a special program for supporting the ESRI developer community. EDN is intended to promote collaboration and interaction among GIS developers and ESRI staff.

ESRI Grid An ESRI data format for storing raster data that defines geographic space as an array of equally sized square cells arranged in rows and columns. Each cell stores a numeric value that represents a geographic attribute (such as elevation) for that unit of space. When the grid is drawn as a map, cells are assigned colours according to their numeric values. Each grid cell is referenced by its x,y coordinate location.

ESRI maplex label engine In Maplex for ArcGIS, extended software that allows finer control of how labels are placed.

ESRI standard label engine In Arc Map, the software used to place labels.

EsriMapCatalog.ser A binary serialized file created in the ArcIMS 3.x AppServer directory to save site parameters. This file has now been replaced by ArcIMSSite.sez.

EsriMapCookies.ser Ephemeris errors are removed by differential correction

Estimated time enroute (ETE) The time left to your destination at your present speed.

Estimated time of arrival (ETA) The time of day of your arrival at your destination.

Estimation In spatial modelling, the process of forming a statistic from observed data to assign optimal parameters in a model or distribution.

Estuary That portion of a stream influenced by the tide of the body of water into which it flows; an arm of the sea at a river mouth.

ETC Earth-terrain camera.

ETM+ Format 1 The ETM+ Format 1 major frames contain all data (e.g. imaging and calibration) from and associated with Bands 1–6. The MSCD and PCD data are duplicated in both ETM+ formats.

ETM+ Format 2 The ETM+ Format 2 major frames contain all data (e.g. imaging and calibration) from and associated with Bands 6–8. The MSCD and PCD data are duplicated in both ETM+ formats.

Euclidean distance The straight-line distance between two points on a plane. Euclidean distance, or distance 'as the crow flies,' can be calculated using the Pythagorean theorem.

Euclidean distance analysis In ArcGIS Spatial Analyst, a description of each raster cell's relationship to the closest source.

European remote sensing (ERS) Satellite series (ERS–1 and ERS–2) launched by the European Space Agency (ESA) in July 1991 and April 1995. One instrument (AMI) includes a C-band SAR, VV polarization, and 23E incidence angle, and 30-metre resolution. Predecessors to the ENVISAT satellite.

Evaluator A function that determines attribute values for network elements in a network dataset. If a network source does not have an evaluator, the default evaluator for its element type is used.

Evaporative cooling Temperature drop caused by evaporation of water from a moist surface.

Event (1) A basic element of the dynamic segmentation model that associates attribute data such as pavement conditions, accidents and speed limits with routes. A binary serialized file created in the ArcIMS 3.x AppServer directory to store ArcIMS Folders and any submitted MapNotes and EditNotes. This file has now been replaced by ArcIMSFolders.sez. (2) A geographic location stored in tabular rather than spatial form. Event types include address events, route events, x,y events, and temporal events.

Event handling Watching for events that are broadcast by another class and taking action when they occur.

Event layer In ArcGIS, a layer created from an event table.

Event location A geographic location stored in tabular rather than spatial form. Event types include address events, route events, x,y events, and temporal events.

Event overlay In linear referencing, an operation that produces a route event table that is the logical intersection or union of two input route event tables. Event overlay is one way to perform line-on-line, line-on-point, and event point-on-point overlays.

Event table A data source containing location information in tabular format (called events) that is used to create a spatial dataset. For example, an event table might contain x,y coordinates or routes.

Event theme In ArcView 3.x, a spatial data theme created from an event table.

Event transform In linear referencing, an operation that produces a new table by copying and transforming events from one route reference to another. This allows the events to be used with a route reference having different route identifiers and/or measures.

Exception An error that is an acceptable violation of a topology rule. In ArcMap, for example, a cul-de-sac is a legitimate exception to the rule that prohibits dangles.

Executable file A binary file containing a program that can be run as a stand-alone application. In the Microsoft Windows program, executable files are designated with an .exe extension.

Exit state The condition of a tool upon closure. If a tool fails due to a programming bug or command failure, the exit state will be 'failed.'

Explanatory variable One or a set of variables used to model or predict the dependent variable. For example, a prediction of annual purchases for a proposed store (the dependent variable) might include independent variables representing the number of potential customers, distance to competition, store visibility, and local spending patterns. In the regression equation, independent variables appear on the right side of the equal sign and are often referred to as explanatory variables.

Explode An editing process that separates a multipart feature into its component features, which become independent features.

Exponent In mathematics, a number that indicates how many times a base value is multiplied by itself. Exponents are usually indicated with superscripts.

Export To move data from one computer system to another, and often, in the process, from one file format to another.

Exposure station In aerial photography, each point in the flight path at which the camera exposes the film.

Expression A sequence of operands and operators constructed according to the syntactic rules of a symbolic language that evaluates to a single number, string, or value.

Extended entity data In AutoCAD, extra optional data attached to an AutoCAD drawing element.

Extended postal code An enhanced ZIP Code that consists of the five-digit ZIP Code plus four additional digits that identify a specific geographic segment within the five-digit delivery area, such as a city block, office building, or other unit.

Extensible markup language Developed by the W3C, a standardized general purpose markup language for designing text formats that facilitates the interchange of data between computer applications. XML is a set of rules for creating standard information formats using customized tags and sharing both the format and the data across applications.

Extensible style language A set of standards for defining XML document presentation and transformation. An XSL style sheet may contain information about how to display tagged content in an XML document, such as font size, background colour, and text alignment. An XSL style sheet may also contain XSLT code that describes how to transform the tagged content in an XML document into an output document with another format. The W3C maintains the XSL standards.

Extensible style language transformations A language for transforming the tagged content in an XML document into an output document with another format. An XSL style sheet contains the XSLT code that defines each transformation to be applied. Transforming a document requires the original XML document, an XSL document containing XSLT code, and an XSLT processor (parser) to execute the transformations. The W3C maintains the XSLT standard.

Extension In ArcGIS, an optional software module that adds specialized tools and functionality to ArcGIS Desktop. ArcGIS Network Analyst, ArcGIS StreetMap, and ArcGIS Business Analyst are examples of ArcGIS extensions.

Extent The minimum bounding rectangle (xmin, ymin and xmax, ymax) defined by coordinate pairs of a data source. All coordinates for the data source fall within this boundary.

Extent rectangle A rectangle that is displayed in one data frame, showing the size and position of another data frame.

External polygon In coverages, the first record in a polygon attribute table, representing the area beyond the outer boundary of the coverage.

Extract server A private ArcIMS virtual server that clips data and saves it to a Zip file.

Extract wizard An ArcToolbox wizard that selects features from a coverage based on attribute values to create a new coverage.

Extraction guide A specification that defines parameters for feature extraction and attribution. Specifications typically include the size of features to be collected, density of feature collection, scale ranges, and attribute assignment.

Extrapolation Using known or observed data to infer or calculate values for unobserved times, locations or other variables outside a sampled area. In the absence of data, extrapolation is a common method for making predictions, but it is not always accurate. For example, based on observed economic indicators, an economist can make predictions about the state of the economy at a future time. These predictions may not be accurate because they cannot take into account seemingly random events such as natural disasters.

Extrusion The process of projecting features in a two-dimensional data source into a three-dimensional representation: points become vertical lines, lines become planes, and polygons become three-dimensional blocks. Uses of extrusion include showing the depth of well point features or the height of building-footprint polygons.

F

F A ratio of variances, calculated from a sample of data and used to provide information about a whole dataset. For example, statistic F may be used to provide estimates of variance, or differences, in a population based on observations from two or more random samples.

Fabric In Survey Analyst - Cadastral Editor, a network of connected parcels. Parcels are represented by parcel line features, parcel point features, and parcel polygon features, referred to in aggregate as parcel features. Parcel topology in the cadastral fabric is stored explicitly through shared or common parcel point features.

Face A planar surface of a geometric shape, bounded by edges.

Facility In ArcGIS Network Analyst, a network location used in closest facility and service area analyses.

False colour image (FCC) A colour image where parts of the non-visible EM spectrum are expressed as one or more of the red, green, and blue components, so that the colours produced by the Earth's surface do not correspond to normal visual experience. Also called a false-colour composite (FCC). The most commonly seen false-colour images display the very-near infrared as red, red as green, and green as blue.

False colour photograph Another term for IR colour photograph.

False easting The linear value added to all x-coordinates of a map projection so that none of the values in the geographic region being mapped are negative.

False northing (1) A value applied to the origin of the map projection to change the y-coordinate. (2) The linear value added to all y-coordinates of a map projection so that none of the values in the geographic region being mapped are negative.

Far range The portion of a radar image farthest from the aircraft or spacecraft flight path.

Feature A representation of a real-world object on a map.

Feature attribute table A database or tabular file containing information about a set of geographic features, usually arranged so that each row represents a feature and each column represents one feature attribute. In raster datasets, each row of an attribute table corresponds to a certain zone of cells having the same value. In a GIS, attribute tables are often joined or related to spatial data layers, and the attribute values they contain can be used to find, query, and symbolize features or raster cells.

Feature class In ArcGIS, a collection of geographic features with the same geometry type (such as point, line or polygon), the same attributes and the same spatial reference. Feature classes can be stored in geodatabases, shapefiles, coverages or other data formats. Feature classes allow homogeneous features to be grouped into a single unit for data storage purposes. For example, highways, primary roads and secondary roads can be grouped into a line feature class named 'roads'. In a geodatabase, feature classes can also store annotation and dimensions.

Feature code In Survey Analyst for field measurements, a block of text included in data collector file format, used to describe features surveyed in the field.

Feature data Data that represents geographic features as geometric shapes.

Feature dataset In ArcGIS, a collection of feature classes stored together that share the same spatial reference; that is, they share a coordinate system, and their features fall within a common geographic area. Feature classes with different geometry types may be stored in a feature dataset.

Feature displacement The movement of features that would otherwise overprint or conflict with other features. For example, if a river, a road and a railway run through a narrow valley, it is necessary, at some scales, to displace at least one of the features that represent them on the map to keep their symbols distinct.

Feature extraction In image processing, a method of pattern recognition in which patterns within an image are measured and then classified as features based on those measurements.

Feature layer A layer that references a set of feature data. Feature data represents geographic entities as points, lines and polygons.

Feature separation Process of preparing a separate drawing, engraving or negative for selected types of data in the preparation of a map or chart.

Feature service A service that uses the ArcIMS spatial server feature streaming capabilities. The server bundles data and streams it to the requesting client. Since more of the processing is done in the Java applet, requests are sent to an ArcIMS spatial server only when additional data is needed.

Feature streaming The process of delivering vector feature data defined for a service that uses the feature server component. On the client side, feature streaming allows you to access a published map and then add feature data for overlays, sharing, making EditNotes, and performing analysis. Feature streaming functionality minimizes the need for multiple server requests.

Feature table A table in an ArcSDE geodatabase that stores geometric shapes for each feature. Feature tables are used in geodatabases that store data as a binary data type, such as SQL Server geodatabases. They are related to the business table of a feature class through the feature ID. In the database, feature table names are prefaced with an 'f' and are stored in the schema of the user who owns the feature class.

Feature weight In ArcMap, a ranking system that indicates whether features from a given feature class may be covered by a label in cases where the label cannot be placed in free space. Feature classes with lower weights will tend to have labels placed over their features before feature classes with higher weights. Polygon feature classes have two types of weights: boundary weights and interior weights.

Feature-linked annotation Annotation that is stored in the geodatabase with links to features through a geodatabase relationship class. Feature-linked annotation reflects the current state of features in the geodatabase: it is automatically updated when features are moved, edited, or deleted.

Federal geographic data committee (FGDC) An organization established by the United States Federal Office of Management and Budget responsible for coordinating the development, use, sharing, and dissemination of surveying, mapping and related spatial data. The committee is comprised of representatives from federal and state government agencies, academia and the private sector. The FGDC defines spatial data metadata standards for the United States in its Content Standard for Digital Geospatial Metadata and manages the development of the National Spatial Data Infrastructure (NSDI).

Federal Radio navigation plan (FRP) Congressionally managed, joint Department of Defence (DoD) and Department of Transportation (DOT) effort to reduce the proliferation and overlap of federally funded radio navigation systems. The FRP is designed to delineate policies and plans for US government-provided radio navigation services.

FGDC clearinghouse An Internet server that hosts a collection of metadata and data maintained and stored on a computer server by a data provider. An NSDI Clearinghouse Node provides information about geographic data within the data provider's areas of responsibility. Nodes must host FGDC-compliant metadata and data and use a common access protocol.

FGDC standard A publication authored by the FGDC that specifies the information content of metadata for digital geospatial datasets. The purpose of the standard is to provide a common set of terminology and definitions for concepts related to the metadata. All US government agencies (federal, state, and local) that receive federal funds to create metadata must follow this standard.

FID In ArcGIS, a system-managed value that uniquely identifies a record or feature.

Fiducial mark An index mark rigidly connected with the camera body, which forms an image on the negative. This image is used to determine the position of the optical centre or principal point of the imagery. Also called collimating marks.

Field (1) A column in a table that describes an attribute of the main feature. Also called column or item. (2) A column in a table that stores the values for a single attribute.

Field mapping In geoprocessing, defining the field structure and content for an output dataset.

Field precision The number of digits that can be stored in a field in a table.

Field scale The number of decimal places for float or double-type geodatabase table fields.

Field view A philosophical view of geographic space in which space is completely filled by occurrences of phenomena, and in which phenomena are described by a range of values on a numeric scale. In this view, every spatial location is something, even if it is the zero value of a phenomenon.

File A collection of uniquely named information stored on a drive, disk, or tape. A file generally resides within a directory.

File geodatabase A geodatabase stored as a folder of files. A file geodatabase can be used simultaneously by several users, but only one user at a time can edit the same data.

File name The name that distinguishes a file from all other files in a particular directory. It can refer to the name of the file by itself (harold), the name plus the file extension (harold.shp), or the whole path of a file up to and including the file name extension (C:\mygisdata\shapefiles\harold.shp).

File name extension The abbreviation following the final period in a file name that indicates the file's format, such as .shp, .zip, or .tif. File name extensions are usually one to four letters long.

File transfer protocol A protocol that allows the transmission of files between computers over a network.

Fill The interior of a polygon; the area inside the perimeter.

Fill symbol A colour or pattern used to fill polygons on a map.

Fillet A segment of a circle used to connect two intersecting lines. Fillets are used to create smoothly curving connections between lines, such as pavement edges at street intersections or rounded corners on parcel features.

Film Light-sensitive photographic emulsion and its base.

Film speed Measure of the sensitivity of photographic film to light. Larger numbers indicate higher sensitivity.

Filter On a raster, an analysis boundary or processing window within which cell values affect calculations and outside which they do not. Filters are used mainly in cell-based analysis where the value of a centre cell is changed to the mean, the sum, or some other function of all cell values inside the filter. A filter moves systematically across a raster until each cell has been processed. Filters can be of various shapes and sizes, but are most commonly three-cell by three-cell squares.

Filter, digital Mathematical procedure for modifying values of numerical data.

Filter, optical A material that, by absorption or reflection, selectivity modifies the radiation transmitted through an optical system.

Find addresses An ArcWeb tool for ArcGIS that enables users to input a street address and receive a candidate list of matched locations for display on a map.

Find places An ArcWeb tool for ArcGIS that allows users to search for place-names and receive a candidate list of place-names and associated coordinates for display on a map.

Find points of interest An ArcWeb tool for ArcGIS that enables users to define spatial or attribute queries and display the results on a map.

Find route An ArcWeb tool for ArcGIS that lets users generate multipoint driving directions between user-defined locations. The Find Route tool determines the shortest or the quickest path between locations and creates a map of the travel route and a list of travel directions.

Find similar analysis A process that seeks out new market areas based on the characteristics of an existing market area.

First normal form The first level of guidelines for designing table and data structures in a relational database. The first normal form guideline recommends creating a unique key for every row in a database table, eliminating duplicate columns from a table and creating separate tables to contain related data. A database that follows these guidelines is said to be in first normal form.

Fitness for use The degree to which a dataset is suitable for a particular application or purpose, encompassing factors such as data quality, scale, interoperability, cost, data format, and so on.

Fix (1) A single position obtained by surveying, GPS, or astronomical measurements, usually given with altitude, time, date and latitude-longitude or grid position. (2) A single position with latitude, longitude (or grid position), altitude, time and date.

Fixed reference point In Survey Analyst for field measurements, a survey point used as an input to a computation. The coordinates are not updated by the computation.

Fixed use In software licensing, a software product that can be used on only one machine.

Fixed-time data In ArcGIS Tracking Analyst, stored temporal data that can be viewed in past, future and past-and-future time windows. This data is stored in a shapefile or as a feature class in a geodatabase.

Flag A marker that identifies or calls attention to something, indicating importance or the need for further attention.

Flash A browser-independent, vector graphic technology for creating interactive Web sites with video, graphics and animation. Browsers need plug-ins for Flash animations.

Flat file A database that contains all data in a large table.

Flatbed scanner A type of scanner with a flat, clear surface on which a map or image remains stationary while a sensor beam moves across it and captures a digital image.

Flattening A measure of how much an oblate spheroid differs from a sphere. The flattening equals the ratio of the semi-major axis minus the semi-minor axis to the semi-major axis.

Flex A set of tools that allows developers to create cross-platform, browser-independent Web applications using a standards-based programming language. Created by Macromedia in 2004 and now developed by Adobe, the Flex family of products includes Flex SDK, Flex Builder, Flex Data Services and Flex Charting.

Flight path Line on the ground directly beneath a remote sensing aircraft or space craft. Also called flight line.

Floating layer A layer in ArcGlobe that has been categorized to float independently above or below the globe surface.

Floating point A type of numeric field for storing real numbers with a decimal point. The decimal point can be in any position in the field and, thus, may 'float' from one location to another for different values stored in the field. For example, a floating-point field can store the numbers 23.632, 0.000087 and –96,432.15.

Floating use Floating software products that are administered by a license manager. A central license manager (installed anywhere on a network) allows users to instal the floating products on any number of machines. The number of seats or licenses purchased determines the number of users who can run the applications simultaneously.

Flood plain Belt of low flat ground bordering a stream channel that is flooded when run-off exceeds the capacity of the stream channel.

Flow direction The route or course followed by commodities proceeding through edge elements in a network.

Flow map A map that uses line symbols of variable thickness to show the proportion of traffic or flow within a network.

Fluorescence Emission of light from a substance following exposure to radiation from an external source.

F-number Representation of the speed of a lens determined by the focal length divided by diameter of the lens. Smaller numbers indicate faster lenses.

Focal analysis The computation of an output raster where the output value at each cell location is a function of the value at that cell location and the values of the cells within a specified neighbourhood around the cell.

Focal functions The computation of an output raster where the output value at each cell location is a function of the value at that cell location and the values of the cells within a specified neighbourhood around the cell.

Focal length In cameras, the distance from the optical centre of the lens to the plane at which the image of a very distant object is brought into focus.

Focus field In Survey Analyst for field measurements, the field that is the current target for data entry.

Folder A location on a disk containing a set of files, other folders, or both.

Folder connection In ArcCatalog, a top-level item in the Catalog tree that provides quick access to geographic data stored on local disks (including CD-ROMs) or shared on a network. Folder connections may provide access to individual folders.

Follow feature mode In ArcGIS, a method for placing geodatabase annotation relative to a line or polygon feature. For example, text next to a river may be dragged along it so that the text curves like the river.

Font A single typeface or a set of related patterns representing characters or symbols at one size.

Footprint In ArcGIS Image Server, the extent (often rectangular) of each raster dataset in the image service definition. The footprint is stored as a polygon shapefile within the image service definition.

Force concentration In MOLE, a component of a map display that shows where force units are located so that the map reader can see where forces are the strongest and weakest.

Force element In MOLE, a type of graphic that represents a military unit (such as Company A, 1st Battalion of the 135th Infantry), equipment, or installation (such as a hospital or radar site). Force elements are also known as force units or military units.

Foreground In a scene or display, the area that appears to be closest to an observer.

Foreign key An attribute or set of attributes in one table that match the primary key attributes in another table. Foreign keys and primary keys are used to join tables in a database.

Foreshortening A distortion in radar images causing the lengths of slopes facing the antenna to appear shorter on the image than on the ground. It is produced when radar wavefronts are steeper than the topographic slope.

Form lines Lines on a map that approximate the shape of terrain in lieu of actual contours. Form lines do not refer to a true datum and do not necessarily use regular intervals.

Format (Also called as size of an image) In computing, the structure and organization of digital information.

Formatted text tag Tags used with text in ArcGIS that allow formatting to be modified for a portion of a text string. This allows the creation of mixed-format text where, for example, one word in a sentence is underlined. Text formatting tags adhere to XML syntax rules and can be used most places where both a text string and a text symbol can be specified. The tags are most commonly used with labels, annotation, and graphic text.

Forward overlap The percentage of duplication by successive photographs along a flight line.

Fovea The region around that point on the retina intersected by the eye's optic axis, where receptors are most densely packed. It is the most sensitive part of the retina.

Fractal A geometric pattern that repeats itself, at least roughly, at ever smaller scales to produce self-similar, irregular shapes and surfaces that cannot be represented using classical geometry. If a fractal curve of infinite length serves as the boundary of a plane region, the region itself will be finite. Fractals can be used to model complex natural shapes such as clouds and coastlines.

Frame In MOLE, the geometric border of a graphic that indicates the affiliation, battle dimension, and status of the war fighting element that the MOLE graphic represents.

Framework The existing ArcObjects components that comprise the ArcGIS system.

Free network adjustment In Survey Analyst for field measurements, one of two phases involved when performing a least squares adjustment for a measurement network. The free network adjustment phase examines the overall geometry of the network by processing only the measurements and using the reference points only for position scale and orientation of the network. The emphasis is on testing the quality of the measurements rather than computing the coordinates.

Free representation In ArcGIS, the representation of a single feature that has been disconnected from its representation rule to create an independent rule whose structure can be changed, allowing full freedom of display. Geometric effects within the rule can be converted so that their results become static parts of the feature representation.

Freeze To fix a column in place in a table for better viewing of the table's contents. A frozen column will stay in place while the other columns scroll normally.

Frequency (υ) The number of wave oscillations per unit time or the number of wavelengths that pass a point per unit time. The rate at which the electric and magnetic fields in a *photon* alternate directions. Photons with higher frequencies have greater energies. Frequency is indicated in equations by the Greek letter υ (nu).

Frequency assignment Frequency at which a sensor (especially radar) operates.

Frequency modulation A technique in which the frequency of a signal is changed about a fundamental or carrier frequency. FM signals are used in many applications in remote sensing and in the radio broadcasting community. In SAR, the outgoing chirp signal has a linear FM of several megahertz. GPS signals are broadcast in FM.

Frequency rate The frequency rate is the coefficient of quadratic phase variation in the linear FM signal.

From-node Of an arc's two endpoints, the first one digitized. From- and to-nodes give an arc left and right sides and, therefore, direction.

F-statistic A ratio of variances, calculated from a sample of data and used to provide information about a whole dataset. For example, statistic F may be used to provide estimates of variance or differences in a population, based on observations from two or more random samples.

F-stop Focal length of a lens divided by the diameter of the len's adjustable diaphragm. Smaller numbers indicate larger openings, which admit more light to the film.

F-test A statistical test for determining the probability that the variances of two different samples are the same. The F-test uses a statistic known as statistic F to test statistical hypotheses about the variances of distributions from which samples have been drawn values in the geographic region being mapped are negative.

FTP Acronym for File Transfer Protocol. A protocol that allows the transmission of files between computers over a network.

Full cache In ArcGlobe, a layer cache saved on disk that contains complete levels of details for the entire layer.

Full WRS scene A full WRS scene with overlap is 375 ETM+ scans. A nominal scene without overlap is 335 ETM+ scans. Due to a movement of the bumpers, which resulted in a larger turn-around interval and an increase in the nominal number of minor frames per major frame (from 7,423 to 7,473), the nominal size of a scene without overlap may decrease. Further bumper wear changes may decrease the number of scans without overlap. The number of overlap scans will increase to compensate for the decrease in the non-overlap scene size.

Function An operation. In GIS, functions include data input, editing, and management; data query, analysis, and visualization; and output operations.

Functional surface In 3D Analyst, a surface that stores a single z-value for any given x,y location.

Fuzzy boundary A boundary that has a vague or indeterminate location, or that is a gradual transition between two zones.

Fuzzy classification Any method for classifying data that allows attributes to apply to objects by membership values, so that an object may be considered a partial member of a class. Class membership is usually defined on a continuous scale from zero to one, where zero is nonmembership and one is full membership. Fuzzy classification may also be applied to geographic objects themselves, so that an object's boundary is treated as a gradated area rather than an exact line. In GIS, fuzzy classification has been used in the analysis of soil, vegetation, and other phenomena that tend to change gradually in their physical composition and for which attributes are often partly qualitative in nature.

Fuzzy set In mathematics, a collection of elements that belong together based on specified criteria, so that elements with partial or uncertain degrees of membership may be included in the collection.

Fuzzy tolerance (1) A tolerance used in software that specifies the minimum distance between arc vertices on the output coverage. (2) The distance within which coordinates of nearby features are adjusted to coincide with each other when topology is being constructed or polygon overlay is performed. Nodes and vertices within the fuzzy tolerance are merged into a single coordinate location. Fuzzy tolerance is a very small distance, usually from 1/10,00,000 to 1/10,000 times the width of the coverage extent, and is generally used to correct inexact intersections.

G

Gain (1) A general term used to denote an increase in signal power in transmission from one point to another; usually expressed in decibels. (2) An increase or amplification. (3) A measure of amplification expressed in dB.

Gamma radiation Extremely high energy radiation, *wavelengths* shorter than 0.001 nm.

Gantt chart A project management graph that displays tasks on a schedule, often used to plan and track projects. The Gantt chart was developed by the American mechanical engineer and management consultant Henry Laurence Gantt.

Gateway An OpenLS ArcWeb service that determines the location of wireless devices.

Gateway service An OpenLS Arc Web service that determines the location of wireless devices.

Gaussian distribution A theoretical frequency distribution of a dataset in which the distribution of values can be graphically represented as a symmetrical bell curve. Normal distributions are typically characterized by a clustering of values near the mean, with few values departing radically from the mean. There are as many values on the left side of the curve as on the right, so the mean and median values for the distribution are the same. Sixty-eight per cent of the values are plus or minus one standard deviation from the mean; 95 per cent of the values are plus or minus two standard deviations; and 99 per cent of the values are plus or minus three standard deviations.

Gaussian noise Statistically random radio noise characterized by a wide frequency spectrum that is continuous and uniform over a specified frequency band.

Gauss-Kruger projection A projected coordinate system that uses the transverse Mercator projection to divide the world into standard zones 6 degrees wide. Used mainly in Europe and Asia, the Gauss-Krüger coordinate system is similar to the universal transverse Mercator coordinate system. The Gauss-Krüger projection is named for the German mathematician and scientist Karl Friedrich Gauss and the German geodesist and mathematician Johann Heinrich Louis Krüger.

Gazetteer A list of geographic place-names and their coordinates. Entries may include other information as well, such as area, population, or cultural statistics. Atlases often include gazetteers, which are used as indexes to their maps. Well-known digital gazetteers include the US Geological Survey Geographic Names Information System (GNIS) and the Alexandria Digital Library Gazetteer.

GBF/DIME Acronym for Geographic Base Files/Dual Independent Map Encoding. Vector geographic base files made for the 1970 and 1980 US censuses, containing address ranges, ZIP Codes, and the coordinates of street segments and intersections for most metropolitan areas in the United States. TIGER files replaced DIME files for the 1990 and subsequent censuses.

GCP Ground-control point.

GDB A database or file structure used primarily to store, query, and manipulate spatial data. Geodatabases store geometry, a spatial reference system, attributes, and behavioural rules for data. Various types of geographic datasets can be collected within a geodatabase, including feature classes, attribute tables, raster datasets, network datasets, topologies and many others. Geodatabases can be stored in IBM DB2, IBM Informix, Oracle, Microsoft Access, Microsoft SQL Server and PostgreSQL relational database management systems or in a system of files, such as a file geodatabase.

GDI Acronym for Graphical Device Interface. A standard for displaying and transmitting text and graphical objects output devices, such as monitors and printers. GDI generally refers to the Windows GDI API.

Gemini US programme of two-man earth-orbiting spacecraft in 1965 and 1966.

Generalization The abstraction, reduction and simplification of features for change of scale or resolution.

Genetic algorithm A search algorithm inspired by genetics and Darwin's theory of natural selection. The algorithm goes through an iterative process of applying genetic operators, such as reproduction, mutation and crossover, to a collection of data over several stages. At each stage the fitness of the results is evaluated and the best of the results population is retained, until the results present an optimal solution.

Geo TIFF GeoTIFF is a public domain metadata standard which allows georeferencing information to be embedded within a TIFF file. The potential additional information includes projections, coordinate systems, ellipsoids, datum and everything else necessary to establish the exact spatial reference for the file. The GeoTIFF format is fully compliant with TIFF 6.0, so software incapable of reading and interpreting the specialized metadata will still be able to open a GeoTIFF file.

Geocentric Measured from the earth or the earth's centre.

Geocentric coordinate system A three-dimensional, earth-centred reference system in which locations are identified by their x-, y- and z-values. The x-axis is in the equatorial plane and intersects the prime meridian (usually Greenwich). The y-axis is also in the equatorial plane; it lies at right angles to the x-axis and intersects the 90-degree meridian. The z-axis coincides with the polar axis and is positive toward the north pole. The origin is located at the centre of the sphere or spheroid.

Geocentric datum A horizontal geodetic datum based on an ellipsoid that has its origin at the earth's centre of mass. Examples are the World

Geodetic System of 1984, the North American Datum of 1983 and the Geodetic Datum of Australia of 1994. The first uses the WGS84 ellipsoid; the latter two use the GRS80 ellipsoid. Geocentric datums are more compatible with satellite positioning systems, such as GPS, then are local datums.

Geocentric latitude The angle between the equatorial plane and a line from a point on the surface to the centre of the sphere or spheroid. On a sphere, all lines of latitude are geocentric. Latitude generally refers to geodetic latitude.

Geocentric longitude The angle between the prime meridian and a line drawn from a point on the surface to the centre of a sphere or spheroid. For an ellipsoid of revolution (such as the earth), geocentric longitude is the same as geodetic longitude.

Geocloud The term 'cloud' represents the internet/web. Geocloud/Geocloud computing therefore, is the ability for consumers, developers and applications to access computing resources from the cloud. These geographically aware/spatial computing resources could be owned and operated by some third party provider – either for free or for a fee.

Geocode To assign a street address to a location.

Geocoded feature class A feature class created by batch geocoding.

GeocodeServer An ArcGIS Server software component that provides programmatic access to an address locator and performs single and batch address matching. It is designed for use in building Web services and Web applications using ArcGIS Server.

Geocoding (1) A GIS operation for converting street addresses into spatial data that can be displayed as features on a map, usually by referencing address information from a street segment data layer. (2) Process by which imagery is corrected for all source-dependent errors and geometrically transformed to the desired map projection, being resampled to a standard square pixel size.

Geocoding engine An entity in the geocoding framework that drives the geocoding process.

Geocoding index An index on reference data used by an address locator to search for matching records in the reference data. A geocoding index is either a file or a database table containing index attributes used by the address locator.

Geocoding platform A conceptual entity of the geocoding framework that combines the interaction of the ArcGIS interface with the input parameters set in the address locator and the processes of the geocoding engine.

Geocoding process The steps involved in translating an address entry, searching for the address in the reference data embedded in an address locator and delivering the best candidate or candidates. These steps include parsing the address, standardizing abbreviated values, assigning each address element to a category known as a match key, indexing the needed categories, searching the reference data, assigning a score to each potential candidate, filtering the list of candidates based on the minimum match score and delivering the best match. The process requires reference files, input address records, address locators and software.

Geocoding reference data Data that a geocoding service uses to determine the geometric representations for locations.

Geocoding rule base A collection of files that directs the geocoding engine in how to standardize address data and match it to the related location in the reference data. Each address locator style uses a specific rule base, designed for that style.

Geocoding service In ArcGIS 8.3 and previous versions, an object that defines the process for translating nonspatial descriptions of places, such as street addresses, into spatial data that can be displayed as features on a map. A geocoding service defines the path to the reference data source and the file of nonspatial data, algorithms for standardizing addresses and matching them to the reference data, and parameters for reading address data, matching address data to the reference data and creating output. In ArcGIS 9, a geocoding service is called an address locator.

Geocoding style A template on which a geocoding service is built. Each template is designed to accommodate a specific format of address and reference data and geocoding parameters. Geocoding style template files have a .lot file extension.

Geocomputation The application of computer technology to spatial problems, including problems of collecting, storing, visualizing and analysing spatial data, and of modelling spatial system dynamics.

Geodata Information describing the location and attributes of things, including their shapes and representation. Geographic data is the composite of spatial data and attribute data.

Geodatabase A database or file structure used primarily to store, query, and manipulate spatial data. Geodatabases store geometry, a spatial reference system, attributes, and behavioural rules for data. Various types of geographic datasets can be collected within a geodatabase, including feature classes, attribute tables, raster datasets, network datasets, topologies, and many others. Geodatabases can be stored in IBM DB2, IBM Informix, Oracle, Microsoft Access, Microsoft SQL Server and PostgreSQL relational database management systems or in a system of files, such as a file geodatabase.

Geodatabase data model The schema for the various geographic datasets and tables in an instance of a geodatabase. The schema defines the GIS objects, rules, and relationships used to add GIS behaviour and integrity to the datasets in a collection.

Geodatabase feature dataset In a geodatabase, a collection of feature classes stored together so they can participate in topological relationships with one another. All the feature classes in a feature dataset must share the same spatial reference; that is, they must have the same coordinate system and their features must fall within a common geographic area. Feature classes with different geometry types may be stored in a feature dataset. In ArcGIS, feature classes that participate in a geometric network must be placed in a feature dataset.

Geodatabase replication In ArcGIS, a method of distributing data across two or more geodatabases in order to synchronize data changes. An entire geodatabase or a subset of a geodatabase can be replicated. There are three types of geodatabase replication: two-way replication, one-way replication, and check-out replication.

GeoDataServer In ArcGIS, a coarse-grained object that represents a geodatabase. It allows software users to perform replication operations, data extraction and database query operations on a geodatabase over the WAN using ArcGIS Server. Application developers can also use the GeoDataServer with geodatabase connections made over the LAN.

Geodataset Any organized collection of data in a geodatabase with a common theme.

Geodesic The shortest distance between two points on the surface of a spheroid. Any two points along a meridian form a geodesic.

Geodesy (1) Science concerned with the measurement and mathematical description of the size and shape of the earth and its gravitational fields. Geodesy also includes the large-scale, extended surveys for determining positions and elevations of points, in which the size and shape of the earth must be taken into account. (2) The science of measuring and representing the shape and size of the earth and the study of its gravitational and magnetic fields.

Geodetic datum A datum that is the basis for calculating positions on the earth's surface or heights above or below the earth's surface.

Geodetic latitude The angle that a line drawn perpendicular to the surface through a point on a spheroid makes with the equatorial plane.

Geodetic longitude The angle between the plane of the meridian that passes through a point on the surface of the spheroid and the plane of a prime meridian, usually the Greenwich meridian.

Geodetic reference system of 1980 The standard measurements of the earth's shape and size adopted by the International Union of Geodesy and Geophysics in 1979.

Geodetic survey A global survey that takes the shape and size of the earth into account, used to precisely locate horizontal and vertical positions suitable for controlling other surveys (comprised of reference or control points).

Geodetic transformation A systematic conversion of the latitude-longitude values for a set of points from one geographic coordinate system to equivalent values in another geographic coordinate system. Depending on the geographic coordinate systems involved, the transformation can be accomplished in various ways. Typically, equations are used to model the position and orientation of the 'from' and 'to' geographic coordinate systems in three-dimensional coordinate space; the transformation parameters may include translation, rotation and scaling. Other methods, including one used in transformations between NAD 1927 and NAD 1983, use files in which the differences between the two geographic coordinate systems are given for a set of coordinates; the values of other points are interpolated from these.

Geofence A designated boundary around a geometry that, if crossed, initiates a notification. Geofences are often used in real-time route Web applications.

Geographic Of or relating to the earth.

Geographic constraint The minimum bounding rectangle (xmin, ymin and xmax, ymax) defined by coordinate pairs of a data source. All coordinates for the data source fall within this boundary.

Geographic coordinate data base (GCDB) A database that is being developed by the US Bureau of Land Management (BLM) to include longitude and latitude values and other descriptive information for section corners and monuments recorded in the PLSS.

Geographic coordinate system A reference system that uses latitude and longitude to define the locations of points on the surface of a sphere or spheroid. A geographic coordinate system definition includes a datum, prime meridian, and angular unit.

Geographic coordinates A measurement of a location on the earth's surface expressed in degrees of latitude and longitude.

Geographic data Information describing the location and attributes of things, including their shapes and representation. Geographic data is the composite of spatial data and attribute data.

Geographic database A database or file structure used primarily to store, query and manipulate spatial data. Geodatabases store geometry, a spatial reference system, attributes and behavioural rules for data. Various types of geographic datasets can be collected within a geodatabase, including feature classes, attribute tables, raster datasets, network datasets, topologies and many others. Geodatabases can be stored in IBM DB2, IBM Informix, Oracle, Microsoft Access, Microsoft SQL Server and PostgreSQL relational database management systems or in a system of files, such as a file geodatabase.

Geographic grid The location reference system for spatial features on the earth's surface.

Geographic information science The field of research that studies the theory and concepts that underpin GIS. It seeks to establish a theoretical basis for the technology and use of GIS, study how concepts from cognitive science and information science might apply to GIS, and investigate how GIS interacts with society.

Geographic information system (GIS) (1) A data-handling and analysis system based on sets of data distributed spatially in two dimensions. The data sets may be map oriented, when they comprise qualitative attributes of an area recorded as lines, points, and areas often in vector format, or image oriented, when the data are quantitative attributes referring to cells in a rectangular grid usually in raster format. It is also known as a geobased or geocoded information system. (2) An integrated collection of computer software and data used to view and manage information about geographic places, analyse spatial relationships and model spatial processes. A GIS provides a framework for gathering and organizing spatial data and related information so that it can be displayed and analysed.

Geographic north The direction from any point on the earth's surface to the geographic north pole.

Geographic primitive In MOLE, the most elementary part of a graphic. Icon, frame and fill are examples of components that make up MOLE graphics.

Geographic projection The coordinate system used for displaying geographic data. Examples include the data frame in ArcMap and the view in ArcView GIS.

Geographic transformation A systematic conversion of the latitude-longitude values for a set of points from one geographic coordinate system to equivalent values in another geographic coordinate system. Depending on the geographic coordinate systems involved, the transformation can be accomplished in various ways. Typically, equations are used to model the position and orientation of the 'from' and 'to' geographic coordinate systems in three-dimensional coordinate space; the transformation parameters may include translation, rotation and scaling. Other methods, including one used in transformations between NAD 1927 and NAD 1983, use files in which the differences between the two geographic coordinate systems are given for a set of coordinates; the values of other points are interpolated from these.

Geographic visualization Use of maps for data exploration and visual information processing.

Geographically referenced data Data that describe both the location and characteristics of spatial features on the earth's surface.

Geography The study of the earth's surface, encompassing the description and distribution of the various physical, biological, economic and cultural features found on the earth and the interaction between those features.

Geography level A division of statistical geographic data, such as country, province, postal code, tract or block group.

Geography markup language A n OpenGIS Implementation Specification designed to store and transport geographic information. GML is a profile (encoding) of XML.

Geoid (1) A hypothetical surface representing the form the earth's oceans would take if there were no land and the waters were free to respond to the earth's gravitational and centrifugal forces. The resulting geoid is irregular and varies from a perfect sphere by as much as 75 metres above and 100 metres below its surface. (2) Figure of the earth visualized as a mean sea level surface extended a continuously through the continents. It is a theoretically continuous surface that is perpendicular at every point to the direction of gravity (the plumbline).

Geoid height The height of the geoid above the ellipsoid.

Geoid-ellipsoid separation The distance from the surface of an ellipsoid to the surface of the geoid, measured along a line perpendicular to the ellipsoid. The separation is positive if the geoid lies above the ellipsoid, negative if it lies below.

Geolocation The process of creating geographic features from tabular data by matching the tabular data to a spatial location. An example of geolocation is creating point features from a table of x,y coordinate.

Geometric coincidence The distance within which features in a geometric network are deemed to be coincident and, therefore, connected.

Geometric correction (1) Image-processing procedure that corrects spatial distortions in an image. (2) The correction of errors in remotely sensed data, such as those caused by satellites or aircraft not staying at a constant altitude or by sensors deviating from the primary focus plane. Images are often compared to ground control points on accurate basemaps and resampled, so that exact locations and appropriate pixel values can be calculated.

Geometric dilution of precision An indicator of satellite geometry for a constellation of satellites used to determine a position. Positions with a lower DOP value generally constitute better measurement results than those with higher DOP. Factors determining the total GDOP (geometric DOP) for a set of satellites include PDOP (positional DOP), HDOP (horizontal DOP), VDOP (vertical DOP) and TDOP (time DOP).

Geometric effect In ArcGIS, a dynamic process that can be applied within a representation rule to dynamically alter the geometry of features before they are drawn. Geometric effects can act on a single symbol layer, or on all symbol layers in a representation rule, and can be chained together to create cumulative effects.

Geometric element One of the most basic parts or components of a geometric figure: that is, a surface, shape, point, line, angle or solid.

Geometric network Edge and junction features that represent a linear network, such as a utility or hydrologic system, in which the connectivity of features is based on their geometric coincidence. A geometric network does not contain information about the connectivity of features; this information is stored within a logical network. Geometric networks are typically used to model directed flow systems.

Geometric transformation (1) The process of converting a map or an image from one coordinate system to another by using a set of control points and transformation equations. (2) The process of rectifying a raster dataset to map coordinates or converting a raster dataset from one coordinate system to another.

Geometry The measures and properties of points, lines and surfaces. In a GIS, geometry is used to represent the spatial component of geographic features.

GeoMobility server An OpenLS platform for wireless Web services defined by the Open Geospatial Consortium.

Geomorphology The study of the nature and origin of landforms, including relationships to underlying structures and processes of formation.

Geoprocessing A GIS operation used to manipulate GIS data. A typical geoprocessing operation takes an input dataset, performs an operation on that dataset and returns the result of the operation as an output dataset. Common geoprocessing operations include geographic feature overlay, feature selection and analysis, topology processing, raster processing and data conversion. Geoprocessing allows for definition, management and analysis of information used to form decisions.

Geoprocessing server A computer in a network that is used to handle geoprocessing tasks. Geoprocessing servers may use UNIX or Windows platforms, and include a utility to schedule remote processing.

Geoprocessing settings A n y settings that affect working with or running tools. Geoprocessing settings include the state of the ArcToolbox window, the state of the Environment Settings dialog box, and variables that have been created at the command line. In ArcMap, geoprocessing settings are saved with a map document. In ArcCatalog, geoprocessing settings are persisted with the application.

Geoprocessing tool An ArcGIS tool that can create or modify spatial data, including analysis functions (overlay, buffer, slope), data management functions (add field, copy, rename) or data conversion functions.

GEOPUB30.DLL An ArcView Geocoding Windows dynamic link library (DLL) for use on ArcView 3.x. It exposes additional geocoding requests by making them public (rather than private), Avenue scripts-callable requests.

Georectification The digital alignment of a satellite or aerial image with a map of the same area. In georectification, a number of corresponding control points, such as street intersections, are marked on both the image and the map. These locations become reference points in the subsequent processing of the image.

Georeferencing Aligning geographic data to a known coordinate system so it can be viewed, queried and analysed with other geographic data. Georeferencing may involve shifting, rotating, scaling, skewing and in some cases warping, rubber sheeting or orthorectifying the data.

Georelational (data) model (1) A GIS package that sores spatial data and attribute data in two separate but related file systems, i.e. spatial data in binary files and attribute data in tables and links the tow data components by the Ids of map features. (2) A geographic data model that represents geographic features as an interrelated set of spatial and attribute data. The georelational model is the fundamental data model used in coverages.

GeoRSS Acronym for Geographically Encoded Objects for RSS feeds. Metadata for RSS documents that describes the location of Web content.

Geospatial data clearinghouse A community of digital spatial data providers that maintain NSDI Clearinghouse Nodes as part of the US National Spatial Data Infrastructure.

Geospatial technology A set of technological approaches, such as GIS, photogrammetry and remote sensing, for acquiring and manipulating geographic data.

Geospecific model A model used to represent a real-world feature. For example, a geospecific model for the White House would look exactly like the White House and be used to represent the White House on a map of Washington, D.C.

Geostationary (1) Positioned in an orbit above the earth's equator with an angular velocity the same as that of the earth and an inclination and eccentricity approaching zero. A geostationary satellite will orbit as fast as the earth rotates on its axis, so that it remains effectively stationary above a point on the equator. A geostationary satellite is geosynchronous, but geosynchronous satellites are not necessarily geostationary. (2) Refers to satellites travelling at the angular velocity at which the earth rotates; as a result, they remain above the same point on earth at all times. (3) Refers to satellites travelling at the angular velocity at which the earth rotates; as a result, they remain above the same point on earth at all times. (also see geostationary orbit)

Geostationary operational environmental satellite A NOAA satellite that acquires visible and thermal IR images for meteorological purposes.

Geostationary orbit 'Describes an orbit in which a satellite is always in the same position (appears stationary) with respect to the rotating Earth. The satellite travels around the Earth in the same direction, at an altitude of approximately 35,790 km (22,240 statute miles) because that produces an orbital period equal to the period of rotation of the Earth (actually 23 hours, 56 minutes, 04.09 seconds). A worldwide network of operational geostationary meteorological satellites provides visible and infrared images of Earth's surface and atmosphere'.

Geostatistical layer file A layer file created by the ArcGIS Geostatistical Analyst extension. It can be exported to ESRI GRID for further geoprocessing.

Geostatistics A class of statistics used to analyse and predict the values associated with spatial or spatio-temporal phenomena. Geostatistics provides a means of exploring spatial data and generating continuous surfaces from selected sampled data points.

Geosurvey engine The software module in Survey Analyst - Cadastral Editor that manages the interaction between cadastral fabric jobs and the cadastral fabric.

Geosynchronous Positioned in an orbit moving west to east with an orbital period equal to the earth's rotational period. If a satellite is in a geosynchronous orbit that is circular and lies in the equatorial plane, it is geostationary because it remains over one point on the equator. If not, the satellite appears to make a figure eight once a day between the latitudes that correspond to its angle of inclination over the equator.

Geothermal Refers to heat from sources within the earth.

Geotypical model A symbolic representation for a class of map features, such as government buildings. For example, on a map of the United States, a white building with a dome on top could be used as a geotypical model for all state capitols.

Get report An ArcWeb tool for ArcGIS that enables users to determine characteristics for a specified location and generate a report.

GIF Acronym for Graphic Interchange Format. A low-resolution file format for image files, commonly used on the Internet. It is well-suited for images with sharp edges and reduced numbers of colours.

Gimbal A device with two mutually perpendicular and intersecting axes of rotation, thus giving free angular movement in two directions, on which an object may be mounted.

Giomgr In ArcSDE, a process that listens for client requests to connect to the database. When it receives such a request, it creates a connection by launching a gsrvr process dedicated to that client. The giomgr process is not used if the client makes a direct connection to the ArcSDE geodatabase.

GIS (1) A larger concept than often believed, GIS (geographic information system) is the hardware, software, data, personnel systems, etc. involved with geographical coordinate data. (2) Acronym for geographic information system. An integrated collection of computer software and data used to view and manage information about geographic places, analyse spatial relationships, and model spatial processes. A GIS provides a framework for gathering and organizing spatial data and related information so that it can be displayed and analysed.

GIS coordinate In Survey Analyst for field measurements, the single coordinate for a survey point that is the best overall representation for that survey point's location, defined by one or more projects. Feature geometry is always linked to the GIS coordinate.

GIS data reviewer An application used to manage data quality control, visually check data and run batch checks for attribute and geometry defects. Defects are recorded in an error table that can be used to resolve errors and verify corrections.

GIS modelling The process of using GIS in building models with spatial data.

GIS server The components of ArcGIS Server that host and run services. A GIS server consists of a server object manager and one or more server object containers.

GIScience Abbreviation for geographic information science. The field of research that studies the theory and concepts that underpin GIS. It seeks to establish a theoretical basis for the technology and use of GIS, studies how concepts from cognitive science and information science might apply to GIS, and investigates how GIS interacts with society.

Global analysis The computation of an output raster where the output value at each cell location may be a function of all the cells in the input raster.

Global check method In Survey Analyst for field measurements, one of two ways to apply the Coordinate Out of Tolerance command. The Global Check method searches for coordinates out of tolerance within the whole survey dataset.

Global functions The computation of an output raster where the output value at each cell location may be a function of all the cells in the input raster.

Global ID A field of type UUID (Universal Unique Identifier) in which values are automatically assigned by the geodatabase when a row is created. The GlobalID field is necessary for maintaining object uniqueness across replicas. All feature classes and tables participating in one-way or two-way replication must contain the GlobalID field. This field is not editable and is automatically populated when it is added for existing data.

Global mode A navigation mode in ArcGlobe during which the camera target is always at the centre of the globe.

Global navigation satellite system (GLONASS) This is the Russian counterpart to GPS. GLONASS provides worldwide coverage; however, its performance accuracy is optimized for the northern latitudes and is specified as identical to that of GPS SPS.

Global polynomial interpolation In ArcGIS Geostatistical Analyst, a deterministic interpolation method. The interpolated surface is not required to conform to the sample data points and the method does not have standard errors associated with it.

Global positioning system (GPS) (1) Global Positioning System, the broadcasting satellites, receiving hardware and software used to locate coordinates on or near the surface of the earth. (2) A system of radio- emitting and -receiving satellites used for determining positions on the earth. The orbiting satellites transmit signals that allow a GPS receiver anywhere on earth to calculate its own location through trilateration. Developed and operated by the US Department of Defense, the system is used in navigation, mapping, surveying and other applications in which precise positioning is necessary.

Global spatial data infrastructure A global framework of technologies, policies, standards and human resources necessary to acquire, process, store, distribute and improve the use of geospatial data across multiple countries and organizations.

Globally unique identifier A string used to uniquely identify an interface, class, type library, component category or record.

Globe A sphere on which a map of the earth or a celestial body is represented. Since the earth's natural shape is similar to a sphere, globes distort the earth's features far less than flat maps.

Globe document A disk-based representation of the globe view or views contained in ArcGlobe. Globe documents have a .3dd extension.

Globe properties In ArcGlobe, properties that can be set for a globe document. These include vertical exaggeration, background colour or sun position.

Globe terrain In ArcGlobe, a globe surface with base heights supplied from an elevation layer.

Globe view In ArcGlobe, the display window in which a globe can be viewed.

GLONASS Acronym for Global Navigation Satellite System. The Russian counterpart of the United States Global Positioning System.

Glue A Java-based toolkit for accessing SOAP Web services. Glue is developed by webMethods.

Glyph The geometric shape of a character in a font.

GML Acronym for Geography Markup Language. An OpenGIS Implementation Specification designed to store and transport geographic information. GML is a profile (encoding) of XML.

GMT (1) Greenwich mean time. This international 24-h system is used to designate the time at which Landsat images are acquired. (2) The time at the prime meridian, which runs through the Royal Observatory in Greenwich, England. From 1884 to 1928, Greenwich mean time was the official name (and is still the popular name) for universal time. It sometimes also refers to coordinated universal time.

Gnomonic projection A planar projection, tangent to the earth at one point, projected from the centre of the globe. All great circles appear as straight lines on this projection, so that the shortest distance between two points is a straight line. The gnomonic projection is useful in navigation. The gnomonic projection was used by Thales of Miletus, an ancient Greek astronomer and philosopher, to chart the heavens. It is possibly the oldest map projection.

Goddard space flight centre The NASA facility at Greenbelt, Maryland, that is also a Landsat ground receiving station.

GOES Geostationary Operational Environmental Satellite.

Gon A unit of angular measurement in which the angle of a full circle is 400 gradians and a right angle is 100 gradians. The common abbreviation for gradian is grad.

Goodness of fit In modelling, the degree to which a model predicts observed data; a measure of predictive power.

Google earth Google Earth is a virtual globe program that was originally called Earth Viewer and was created by Keyhole, Inc. It maps the earth by the superimposition of images obtained from satellite imagery, aerial photography and GIS over a 3D globe. It is available under three different licenses: Google Earth, a free version with limited functionality; Google Earth Plus, which includes a few more features; and Google Earth Professional, intended for commercial use.

Gore A map, shaped like the area between a pair of parentheses, of an area that lies between two lines of longitude. A gore can be fitted to the surface of a globe with little distortion.

Gossan Surface occurrence of iron oxide formed by the weathering of metallic sulphide ore minerals.

GPS Acronym for Global Positioning System. A system of radio-emitting and -receiving satellites used for determining positions on the earth. The orbiting satellites transmit signals that allow a GPS receiver anywhere on earth to calculate its own location through trilateration. Developed and operated by the US Department of Defense, the system is used in navigation, mapping, surveying and other applications in which precise positioning is necessary.

Grad A unit of angular measurement in which the angle of a full circle is 400 gradians and a right angle is 100 gradians. The common abbreviation for gradian is grad.

Gradian A unit of angular measurement in which the angle of a full circle is 400 gradians and a right angle is 100 gradians. The common abbreviation for gradian is grad.

Gradient of gravity The direction of the maximum increase in gravity in a horizontal plane.

Gradient The ratio between vertical distance (rise) and horizontal distance (run), often expressed as a percentage. A 10 per cent gradient rises 10 feet for every 100 feet of horizontal distance.

Graduated colour map A map on which a range of colours indicates a progression of numeric values. For example, increases in population density might be represented by the increased saturation of a single colour or temperature differences by a sequence of colours from blue to red.

Graduated symbol map A map with symbols that change in size according to the value of the attribute they represent. For example, denser populations might be represented by larger dots, or larger rivers by thicker lines.

Grain tolerance In ArcInfo Workstation, a parameter controlling the number of vertices and the distance between them on lines that represent curves. The smaller the grain tolerance, the closer the vertices can be. Unlike densify tolerance, grain tolerance can affect the shape of curves.

Granularity (1) Graininess of developed photographic film that is determined by the texture of the silver grains. (2) The coarseness or resolution of data. Granularity describes the clarity and detail of data during its capture and visualization.

Graph A map used to plot a course for air or water navigation.

Graph theory A subfield of mathematics that used diagrams or graphs to study the arrangements of objects and the relationships between objects.

Graphic An image produced by and stored in a computer as data for display.

Graphic component In MOLE, the most elementary part of a graphic. Icon, frame, and fill are examples of components that make up MOLE graphics.

Graphic text Text added in ArcMap layout view that exists in page space and is stored in a map document. Graphic text does not move if the extent or scale is changed.

Graphical device interface (GDI) A standard for displaying and transmitting text and graphical objects output devices, such as monitors and printers. GDI generally refers to the Windows GDI API.

Graphical user interface A software display of program options that allows a user to choose commands by pointing to icons, dialog boxes and lists of menu items on the screen, typically using a mouse. This contrasts with a command line interface in which control is accomplished via the exchange of strings of text.

Graticule A network of longitude and latitude lines on a map or chart that relates points on a map to their true locations on the earth.

Graticule alignment of labels A label positioning method in which labels are oriented along the graticule of the data frame. This is useful for maps of large areas, for cartographic or stylistic reasons.

Gravimeter A device used to measure small variations in the earth's gravitational field between two or more points.

Gravimetric geodesy The science of deducing the size and shape of the earth by measuring its gravitational field.

Gravity model A model that assumes that the influence of phenomena or populations on each other varies inversely with the distance between them.

Great circle Any circle or near circle produced by the intersection of the surface of a sphere and a flat plane that passes through the centre of the sphere. The equator and all lines of longitude are great circles. Great circles are used in navigation, since the shortest path between any two points on the earth's surface lies on a great circle.

Greenwich mean time (GMT) (1) The mean solar time of the meridian of Greenwich, England, used as the prime basis of standard time throughout the world. (2) The time at the prime meridian, which runs through the Royal Observatory in Greenwich, England. From 1884 to 1928, Greenwich Mean Time was the official name (and is still the popular name) for universal time. It sometimes also refers to coordinated universal time.

Greenwich meridian The meridian adopted by international agreement in 1884 as the prime meridian, the 0-degree meridian from which all other longitudes are calculated. The Greenwich prime meridian runs through the Royal Observatory in Greenwich, England.

Grey scale (1) A sequence of grey tones ranging from black to white. (2) All the shades of grey from white to black.

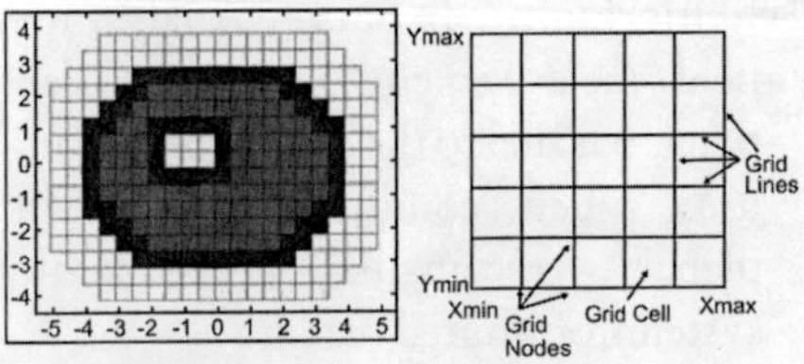

Grid (1) Also 'Coordinate System'. A coordinate system that projects the earth on a flat surface, using square zones for position measurements. Network of uniformly spaced parallel lines intersecting at right angles. When superimposed on a map, it usually carries the name of the projection used for the map – that is, Lambert grid, transverse Mercator grid, universal transverse Mercator grid. (2) In cartography, any network of parallel and perpendicular lines superimposed on a map and used for reference. These grids are usually referred to by the map projection or coordinate system they represent, such as universal transverse Mercator grid.

Grid cell The smallest unit of information in raster data, usually square in shape. In a map or GIS dataset, each cell represents a portion of the earth, such as a square metre or square mile, and usually has an attribute value associated with it, such as soil type or vegetation class.

Grid format The result of interpolation from values of a variable measured at irregularly distributed points, or along survey lines, to values referring to square cells in a rectangular array. It forms a step in the process of contouring data, but can also be used as the basis for a raster format to be displayed and analysed digitally after the values have been rescaled to the 0–255 range.

Grid meridian In Survey Analyst for field measurements, any meridian that is parallel to the central meridian, used when computing points in planar rectangular coordinate systems of limited extent.

Grid north The direction north along the north-south grid lines of a map projection.

Grid stack A mechanism for storing multivariate raster data in ESRI software, consisting of an ordered set of spatially overlapping grids (referred to as layers) referenced by an INFO file or geodatabase. A stack is treated as a single entity for multivariate analysis. Cluster analysis, classification and principal component analysis all work on the layers in a stack.

Ground control A system of points with known positions, elevations, or both, used as fixed references in georeferencing map features, aerial photographs or remotely sensed images.

Ground control point (GCP) (1) An accurately surveyed coordinate location for a physical feature that can be identified on the ground. Control points are used in least-squares adjustments as the basis for improving the spatial accuracy of all other points to which they are connected. (2) A geographic feature of known location that is recognizable on images and can be used to determine geometric corrections. In case of range image (Radar image) the scale in the range direction is constant.

Ground range On radar images, the distance from the ground track to an object.

Ground receiving station Communications equipment for receiving and transmitting signals to and from satellites such as Landsat.

Ground resolution cell Area on the terrain that is covered by the IFOV of a detector.

Ground speed The velocity you are travelling relative to a ground position.

Ground station Data collection, receiving or processing facilities located on the ground.

Ground swath Width of the strip of terrain that is imaged by a scanner system.

Ground truth The accuracy of remotely sensed or mathematically calculated data based on data actually measured in the field.

Group A collection of ArcWeb Services users who share the same activation code. Group members can transfer credits and share data.

Group administrator The person who sets permissions for the members of an ArcWeb Services group and who is usually the chief contact for the organization.

Group layer A group of several layers that appear and act as a single layer. Group layers make it easier to organize a map, assign advanced drawing order options, and share layers for use in other maps.

GRS80 Acronym for Geodetic Reference System of 1980. The standard measurements of the earth's shape and size adopted by the International Union of Geodesy and Geophysics in 1979.

GRS 80 spheroid Geodetic Reference system reference spheroid of 1980, which is a satellite determined spheroid.

GSDI Acronym for global spatial data infrastructure. A global framework of technologies, policies, standards and human resources necessary to acquire, process, store, distribute, and improve the use of geospatial data across multiple countries and organizations.

GSFC Goddard Space Flight Centre is a major NASA space research laboratory established on 1 May 1959 as NASA's first space flight centre.

Gsrvr A process that connects to a relational database management system (RDBMS), using the RDBMS server libraries, and becomes the dedicated link for a GIS client to the database. A gsrvr process is only used in three-tiered geodatabase connections. The GIS client sends instructional commands to the gsrvr process on how to build appropriate SQL statements to perform the requested spatial action (pan, zoom, and so on).

GUI Acronym for graphical user interface. A software display of program options that allows a user to choose commands by pointing to icons, dialog boxes and lists of menu items on the screen, typically using a mouse. This contrasts with a command line interface in which control is accomplished via the exchange of strings of text.

GUID Acronym for globally unique identifier. A string used to uniquely identify an interface, class, type library, component category, or record.

Gyrocompass A nonmagnetic compass that uses a motor-driven gyroscope to indicate true north.

H

Hachure (1) A short line on a map that indicates the direction and steepness of a slope. Hachures that represent steep slopes are short and close together; hachures that represent gentle slopes are longer, lighter, and farther apart. Contours, shading, and hypsometric tints have largely replaced hachuring on modern maps. (2) Any series of lines used on a map to indicate the general direction and steepness of slopes. The lines are short, heavy, and close together for steep slopes; longer, lighter, and more widely spaced for gentle slopes.

Hachured contour On a topographic map, concentric contour lines drawn with hachures to indicate a closed depression or basin. Concentric contour lines drawn without hachure marks indicate a hill.

Halftone A picture in which the gradations of light are obtained by the relative darkness and density of tiny dots produced by photographing the subject through a fine screen.

Halftone image A continuous tone image photographed through a fine screen that converts it into uniformly spaced dots of varying size while maintaining the gradations of highlight and shadow. The size of the dots varies in proportion to the intensity of the light passing through them.

Hamiltonian circuit/path A path through a network that visits each junction in the network only once and then returns to its point of origin. Hamiltonian circuits are named after the Irish mathematician, physicist, and astronomer William Rowan Hamilton.

Hardware key A small hardware device that provides the unique number used in the generation of a license file. The key is plugged into either the parallel or USB port on the License Manager Server. UNIX users of ArcGIS software do not need a hardware key. Hardware keys are also called dongles.

Harmonic Refers to waves in which the component frequencies are whole-number multiples of the fundamental frequency.

HARN Acronym for High Accuracy Reference Network. A regional or statewide resurvey and readjustment of NAD 1983 control points using GPS techniques. The resurvey date is often included as part of the datum name: NAD 1983 (1991) or NAD 91.

Hatch class In linear referencing, a group or category of hatch definitions.

Hatch definition In linear referencing, a specification for where hatch marks are drawn on a line feature. Each hatch definition has its own set of properties that include the multiple of the hatch interval at which the hatches in the hatch definition will be placed, the line or marker symbol of the hatches, and whether the hatches will be labelled. The use of multiple hatch definitions allows for the design of complex hatching schemes.

Hatch style In linear referencing, an organized collection of symbols and settings for the hatch definitions that make up a hatch class. Hatch styles are stored in a style file (.style) and created by the user to maintain standards for displaying hatches on multiple maps with multiple data sources.

Hatches In linear referencing, a series of vertical line or marker symbols displayed on top of features at an interval specified in route measure units.

Hatching In linear referencing, a type of labelling that posts and labels hatches or symbols at a regular interval along measured line features.

HCMM Heat Capacity Mapping Mission, the NASA satellite launched in 1978 to observe thermal properties of rocks and soils. It remained in orbit for only a few months.

HDOP Acronym for horizontal dilution of precision. A measure of the geometric quality of a GPS satellite configuration in the sky. HDOP is a factor in determining the relative accuracy of a horizontal position. The smaller the DOP number, the better the geometry.

Heading The direction of a moving object, expressed as an angle from a known direction, usually north.

Heads-up digitizing Manual digitization by tracing a mouse over features displayed on a computer monitor, used as a method of vectorizing raster data.

Heap In computer programming, a variation on the binary tree data structure in which each node's value is greater than the value of its leaf nodes. Sorting data in a heap allows an element to be located more quickly than it could be found in an ordinary list.

Heat capacity (symbol C) Ratio of heat absorbed or released by a material to the corresponding temperature rise or fall. Expressed in calories per gram per degree centigrade. Also called thermal capacity.

Heat capacity mapping mission (HCMM) NASA satellite orbited in 1978 to record daytime and night-time visible and thermal IR images of large areas.

Hectare A metric areal unit of measure equal to 10,000 square meters. One hectare is equal to 100 ares or 2.47 acres.

Height The vertical distance between two points, or above a specified datum.

Helmert transformation A geometric transformation that scales, rotates, or translates images or coordinates between any two Euclidean spaces. It is commonly used in GIS to transform maps between coordinate systems. In a Helmert transformation, parallel lines remain parallel. The midpoint of a line segment remains a midpoint and all points on a straight line remain on a straight line. The Helmert transformation is named for the German mathematician and geodesist Friedrich Robert Helmert (1843–1917).

Help Node In ArcGIS 8.3 and previous versions, a branch in the

ArcToolbox tree providing overview help for ArcToolbox toolsets.

Hemisphere Half of a celestial body, such as the earth.

Heuristic In computer science, an algorithm that incorporates a shortcut or simplification for solving a programming problem, such as searching. While a heuristic may run faster than a more rigorous algorithm, there is no guarantee that it will find the best solution.

Hexadecimal A number system using base 16 notation, usually comprised of the digits 0–9 and the letters A–F or a–f.

Hierarchical database A database that stores related information in a tree-like structure, where records can be traced to parent records, which in turn can be traced to a root record.

Hierarchy A type of network attribute for a network element in a network dataset. Hierarchy can be used during network analysis to assign priority to a network element. For example, in a transportation network dataset, a 'road class' hierarchy can be assigned to edges to favour major roads instead of local streets.

High accuracy reference network A regional or statewide resurvey and readjustment of NAD 1983 control points using GPS techniques. The resurvey date is often included as part of the datum name: NAD 1983 (1991) or NAD 91.

High gain/low gain antenna A high gain antenna is highly focused, whereas a low gain antenna receives or transmits over a wide angle.

High precision geodetic network A regional or statewide resurvey and readjustment of NAD 1983 control points using GPS techniques. The resurvey date is often included as part of the datum name: NAD 1983 (1991) or NAD 91.

High water Maximum height reached by a rising tide. The height may be due solely to the periodic tidal forces or it may have superimposed upon it the effects of prevailing meteorological conditions. Use of the term 'high tide' is discouraged.

High water line Intersection of the land with the water surface at an elevation of high water.

High water mark Line or mark left upon tidal flats, beach or along shore objects indicating the elevation or the intrusion of high water.

High-altitude sensor A sensor being operated in a modified U2 aircraft, i.e. at altitudes of 15–20 kilometres. Compare with *low-altitude sensor*.

High-level language A programming language that uses keywords and statements that are similar to expressions in human language or mathematics and is, therefore, easier for people to comprehend and use. A high-level language is named for the high level of abstraction it affords developers over low-level processor functions such as memory access and register storage, meaning such operations do not demand the developer's attention.

Highlights Areas of bright tone on an image.

High-pass filter In digital image processing (DIP), a spatial filter that blocks low-frequency (long-wave) radiation, resulting in a sharpened image. It is a spatial filter which selectively enhances contrast variations with high spatial frequencies in an image. It improves the sharpness of images and is a method of edge enhancement.

Hillshading Shadows drawn on a map to simulate the effect of the sun's rays over the varied terrain of the land.

HIRIS High Resolution Imaging Spectrometer, possibly to be carried by the Space Shuttle.

HIRS High Resolution Infrared Spectrometer, carried by NOAA satellites.

Histogram (1) A graph showing the distribution of values in a set of data. Individual values are displayed along a horizontal axis and the frequency of their occurrence is displayed along a vertical axis. (2) A means of expressing the frequency of occurrence of values in a data set within a series of equal ranges or bins, the height of each bin representing the frequency at which values in the data set fall within the chosen range. A cumulative histogram expresses the frequency of all values falling within a bin and lower in the range. A smooth curve derived mathematically from a histogram is termed the probability density function (PDF).

Histogram equalization The redistribution of pixel values in an image so that each range contains approximately the same number of pixels. A histogram showing this distribution of values would be nearly flat.

Historic parcel In Survey Analyst - Cadastral Editor, a parcel that has been replaced with a new parcel by subdivision, merge or other means due to a change in the survey record.

Historical marker In ArcMap, a user-created reference to a time and date stamp. Historical markers can be used to easily connect to a historical version for a specific date and time.

Historical version In ArcMap, a version that a user connects to by using a historical marker or indicating a particular date and time. Once connected, the version provides a read-only view of the geodatabase for the chosen time.

History model A model created, dated and saved when the application is closed to document the tools and parameter values used for each session. The history model is contained within the history toolbox and can be viewed when the application is reopened.

HKCR HKEY_CLASSES_ROOT registry hive. A Windows registry root key that points to the HKEY_LOCAL_MACHINE\Software\Classes registry key. It displays essential information about OLE and association mappings to support drag-and-drop operations, Windows shortcuts and core aspects of the Windows user interface.

Hole A small gap in a raster line feature, usually considered to be an error caused by the poor quality of a source document or by the scanning process.

Honor In network datasets, a type of junction connectivity policy in which junctions connect to other junctions based on the edge connectivity policy set by the user; junctions 'honor' the existing edge connectivity policy.

Horizon The apparent or visible junction of land and sky.

Horizon circle The circle containing all points equidistant from the centre of an azimuthal projection.

Horizontal angle The angle formed by the intersection of two lines in a horizontal plane.

Horizontal control A network of known horizontal geographic positions, referenced to geographic parallels and meridians or to other lines of orientation such as plane coordinate axes.

Horizontal control datum A geodetic datum for any extensive measurement system of positions, usually expressed as latitude-longitude coordinates, on the earth's surface. A horizontal geodetic datum may be local or geocentric. If it is local, it specifies the shape and size of an ellipsoid representing the earth, the location of an origin point on the ellipsoid surface, and the orientation of x- and y-axes relative to the ellipsoid. If it is geocentric, it specifies the shape and size of an ellipsoid, the location of an origin point at the intersection of x-,y- and z-axes at the centre of the ellipsoid, and the orientation of the x-,y- and z-axes relative to the ellipsoid. Examples of local horizontal geodetic datums include the North American Datum of 1927, the European Datum of 1950 and the Indian datum of 1960; examples of geocentric horizontal geodetic datums include the North American Datum of 1983 and the World Geodetic System of 1984.

Horizontal datum A geodetic datum for any extensive measurement system of positions, usually expressed as latitude-longitude coordinates, on the earth's surface. A horizontal geodetic datum may be local or geocentric. If it is local, it specifies the shape and size of an ellipsoid representing the earth, the location of an origin point on the ellipsoid surface and the orientation of x- and y-axes relative to the ellipsoid. If it is geocentric, it specifies the shape and size of an ellipsoid, the location of an origin point at the intersection of x-,y- and z-axes at the centre of the ellipsoid, and the orientation of the x-,y- and z-axes relative to the ellipsoid. Examples of local horizontal geodetic datums include the North American Datum of 1927, the European Datum of 1950 and the Indian datum of 1960; examples of geocentric horizontal geodetic datums include the North American Datum of 1983 and the World Geodetic System of 1984.

Horizontal dilution of precision (HDOP) A measure of the geometric quality of a GPS satellite configuration in the sky. HDOP is a factor in determining the relative accuracy of a horizontal position. The smaller the DOP number, the better the geometry.

Horizontal geodetic datum A geodetic datum for any extensive measurement system of positions, usually expressed as latitude-longitude coordinates, on the earth's surface. A horizontal geodetic datum may be local or geocentric. If it is local, it specifies the shape and size of an ellipsoid representing the earth, the location of an origin point on the ellipsoid surface, and the orientation of x- and y-axes relative to the ellipsoid. If it is geocentric, it specifies the shape and size of an ellipsoid, the location of an origin point at the intersection of x-,y- and z-axes at the centre of the ellipsoid, and the orientation of the x-,y- and z-axes relative to the ellipsoid. Examples of local horizontal geodetic datums include the North American Datum of 1927, the European Datum of 1950 and the Indian datum of 1960; examples of geocentric horizontal geodetic datums include the North American Datum of 1983 and the World Geodetic System of 1984.

Horizontal polarization Linear polarization with the lone electric vector oriented in the horizontal direction in antenna coordinates.

Horizontal transmit-horizontal receive polarization (HH) A mode of radar polarization where the microwaves of the electric field are oriented in the horizontal plane for both signal transmission and reception by means of a radar antenna.

Horizontal transmit-vertical receive polarization (HV) A mode of radar polarization where the microwaves of the electric field are oriented in the horizontal plane for signal transmission, and where the vertically polarized electric field of the backscattered energy is received by the radar antenna.

Host In a computer network, the computer that contains data being accessed by other computers.

Hot Link In ArcView 3.x, a tool for associating external files with a feature in a view. When a user clicks a feature in the view with the Hot Link tool, an image, text file or ArcView document or project displays on screen.

HPGN Acronym for High Accuracy Reference Network. A regional or statewide resurvey and readjustment of NAD 1983 control points using GPS techniques. The resurvey date is often included as part of the datum name: NAD 1983 (1991) or NAD 91.

HRESULT A 32-bit integer returned from any member of a COM interface indicating success or failure, often written in hexadecimal notation. An HRESULT can also give information about the error that occurred when calling a member of a COM interface. Visual Basic translates HRESULTS into errors; Visual C++ developers work directly with HRESULT values.

HTML Acronym for Hypertext Markup Language. A markup language used to create Web pages for publication on the Internet. HTML is a system of tags that define the function of text, graphics, sound and video within a document, and is now an Internet standard maintained by the World Wide Web Consortium.

HTML document A computer file formatted with HTML tags so that it may be viewed in a Web browser and published on the World Wide Web. An HTML document may incorporate text, images, sound, video, and other media components. Characteristically, it also has embedded references, called hypertext links, to other HTML documents. These links enable a person viewing a document in a Web browser to open other documents—which may be stored on other computers anywhere in the world—by clicking on the link using a mouse.

HTML viewer An ArcIMS viewer that uses a single ArcMap image or standard image service. The HTML viewer does not require a Java plug-in.

HTTP Acronym for Hypertext Transfer Protocol. The protocol maintained by the World Wide Web Consortium for communicating between servers and clients to exchange HTML documents across the Internet.

HTTPS Acronym for Hypertext Transfer Protocol (Secure). A variant of HTTP enhanced by a security mechanism. It allows transactions such as e-commerce and data sharing to take place on the World Wide Web in a protected way.

Hub A central node in a network for routing goods to their destinations.

Hue (1) In the IHS system, represents the dominant wavelength of a colour. (2) The dominant wavelength of a colour, by which it can be distinguished as red, green, yellow, blue and so forth.

Human geography The field of geography concerning a range of social, cultural and political aspects of human life as related to their distribution through physical space.

Hydrographic datum A plane of reference for depths, depth contours and elevations of foreshore and offshore features.

Hydrographic survey (1) A survey of a water body, particularly of its currents, depth, submarine relief and adjacent land. (2) Survey of water area, with particular reference to submarine relief, and any adjacent land. See: oceanographic survey.

Hydrography (1) Science that deals with the measurement and description of the physical features of the oceans, seas, lakes, rivers and their adjoining coastal areas, with particular reference to their use for navigation. (2) The measurement and description of water features and their related land areas for the purposes of safe marine navigation.

Hydrologic cycle The circulation of water from the earth through the atmosphere and back again. Its major stages are evaporation, condensation, precipitation, run-off, transpiration, infiltration and percolation.

Hydrology (1) Scientific study of the waters of the Earth, especially with relation to the effects of precipitation and evaporation upon the occurrence and character of ground water. (2) The study of water, its behaviour and its movements across and below the surface of the earth, and through the atmosphere.

Hyperdimensional data cube A data set for which there are more than three different axes of data. For example, observations of changes in temperatures at various oceanic depths over time, for different points in an area, would represent in an (x,y,z,T) hyperdimensional data cube, where (x,y,z) indicate the spatial coordinates for the aquatic location, the time is indicated by T. Compare with *data cube.*

Hyperlink A reference (link) from one point in an electronic document to another document or another location in the same document (the target). Activating the link, usually by clicking it with the mouse, causes the browser to display the target of the link.

Hyperspectral Data product in which the data are binned into approximately fifteen or more spectral bins. Compare with *multispectral*.

Hypsography (1) The study and representation of elevation and the earth's topography. (2) Topography referred to the national geodetic vertical datum of 1929. The science or art of describing heights of land surfaces with reference to this datum.

Hypsometric curve A curve showing the relationship of area to elevation for specified terrain. A hypsometric curve is plotted on a graph on which the x-axis represents surface area and the y-axis represents elevation above or below a datum (normally sea level). The curve shows how much area lies above and below marked elevation intervals.

Hypsometric map A map showing relief, whether by contours, hachures, shading or tinting. See: map, hypsometric.

Hypsometric tinting Relief or depth depicted by a gradation of colours, usually between contour lines. Each colour represents a different range of elevation.

Hypsometry (1) Science or art of determining terrain relief, by any method. (2) The science that determines the spatial distribution of elevations above an established datum, usually sea level.

I

ICAO Acronym for International Civil Aviation Organization. A member organization that represents the worldwide body of nations for standardizing flight rules, regulations and requirements.

Icon In MOLE, the innermost graphic component of a graphic. Icons represent the Function ID position of a Symbol ID code.

ID A unique character string or numeric value associated with a particular object.

IDE Acronym for integrated development environment. A software development tool for creating applications, such as desktop and Web applications. IDEs blend user interface design and layout tools with coding and debugging tools.

Identifier A unique character string or numeric value associated with a particular object.

Identify In ArcGIS, a tool that, when applied to a feature (by clicking it), opens a window showing that feature's attributes.

Identity In geoprocessing, a topological overlay that computes the geometric intersection of two datasets. The output dataset preserves all the features of the first dataset plus those portions of the second (polygon) dataset that overlap the first. For example, a road passing through two counties would be split into two arc features, each with the attributes of the road and the county it passes through.

Identity link An anchor that prevents the movement of features during rubber sheeting.

IDispatch A generic COM interface that has methods allowing clients to ask which members are supported. Classes that implement IDispatch can be used for late binding and DispID binding, which is a form of early binding.

IDL Acronym for Interface Definition Language. A language used to define COM interfaces. The Microsoft implementation of IDL may also be referred to as MIDL or Microsoft IDL.

IDW An interpolation technique that estimates cell values in a raster from a set of sample points that have been weighted so that the farther a sampled point is from the cell being evaluated, the less weight it has in the calculation of the cell's value.

IFOV Instantaneous field of view.

IFSAR Acronym for interferometric synthetic aperture radar. A dual-antenna radar sensor mounted on an airborne or space-borne platform that collects a remotely sensed radar image, called an interferogram. There is a measured energy shift between the signals received by each antenna, and this interference can be colourized to measure elevation or changes in the topography on the earth's surface.

IHS Intensity, hue and saturation system of colours.

IID Acronym for interface identifier. A string that provides the unique name of an interface. An IID is a type of globally unique identifier (GUID).

IKONOS a satellite that produces high resolution, *multispectral* imagery.

Illumination The light incident on a surface or object, either natural or artificial, as determined by the surface's slope and aspect and by the sun's azimuth and altitude.

Image A pictorial representation or description of a scene, typically produced by an optical or electronic device, such as a camera or a scanning radiometer (called as remote sensing). Common examples include remotely sensed data (for example, satellite data), scanned data and photographs.

Image catalog A collection of raster datasets defined in a table of any format, in which the records define the individual raster datasets that are included in the catalog. Raster catalogs can be used to display adjacent or overlapping raster datasets without having to mosaic them together into one large file. In ArcView 3.x, raster catalogs were called image catalogs.

Image contrast The difference in tone between light and dark areas of the image.

Image coordinate An x,y coordinate pair specifying the location of a pixel, or cell, in terms of its row and column position. The x-coordinate gives the column number (commonly starting from 0 at the left edge of the data), and the y-coordinate gives the row number (commonly starting from 0 at the top of the data).

Image data Data produced by scanning a surface with an optical or electronic device. Common examples include scanned documents, remotely sensed data (for example, satellite images) and aerial photographs. An image is stored as a raster dataset of binary or integer values that represent the intensity of reflected light, heat or other range of values on the electromagnetic spectrum.

Image dissection The breaking down of a continuous scene into discrete spatial elements, either by the receptors on the retina or in the process of capturing the image artificially.

Image division A digital image processing technique for increasing the contrast between features in an image by dividing the pixel values in the image by the values of corresponding pixels in a second image. Image division is normally used to identify concentrations of vegetation.

Image plane The (approximately) two-dimensional surface that indicates the location of the image produced by a lens. This is where you should put your detector in order to record a nice, focused image to take home and play with.

Image processing It encompasses all the various operations which can be applied to photographic or image data. These include, but are not limited to image compression, image restoration, image enhancement, prepocessing, quantization, spatial filtering and other image pattern recognition techniques.

Image scale The ratio between a distance in an image and the actual distance on the ground, calculated as focal length divided by the flying height above mean ground elevation. Image scale can vary in a single image from point to point due to surface relief and the tilt of the camera lens.

Image server A public ArcIMS virtual server for image services.

Image service In ArcIMS, a service that uses the spatial server image rendering capabilities. When a request is received, a map is generated on the server and the response is sent back as a JPG, PNG, or GIF image. A new map image is generated each time a client requests new information.

Image service definition In ArcGIS Image Server, a workspace created using the Image Service Definition Editor toolbar that defines the key properties of a service, including name and default spatial reference system. The image service definition workspace has a .ISDef extension.

Image service reference file In Arc GIS Image Server, a file that stores properties for re-establishing a connection to a service and setting up user-defined image service properties. Image service reference files have a .ISRef extension.

Image size the dimensions of an image in either *pixel* dimensions or in bytes of disk space.

Image space The x,y coordinate space defined by the number of columns and rows in a raster dataset. The origin of image space is commonly the centre of the top left pixel of the data and is labelled (0,0). The x-axis corresponds to the number of columns in the raster and the y-axis to the number of rows. For raster data to be used in GIS software, image space must be transformed to a real-world coordinate system through georeferencing.

Image striping A defect produced in line scanner and pushbroom imaging devices produced by the non-uniform response of a single detector, or amongst a bank of detectors. In a line-scan image the stripes are perpendicular to flight direction, but parallel to it in a pushbroom image.

Image swath See ground swath.

Imager Any satellite or aerial instrument that measures and maps the earth and its atmosphere.

Imagery Visible representation of objects and (or) phenomena as sensed or detected by cameras, infrared and multispectral scanners, radar and photometers. Recording may be on photographic emulsion (directly as in a camera or indirectly after being first recorded on magnetic tape as an electrical signal) or on magnetic tape for subsequent conversion and display on a cathode ray tube.

Imaging Radar Most imaging radars produce two-dimensional images. The two dimensions are called range and azimuth.

Impedance A measure of the amount of resistance or cost, required to traverse a path in a network, or to move from one element in the network to another. Resistance may be a measure of travel distance, time, speed of travel multiplied by distance and so on. Higher impedance values indicate more resistance to movement, and a value of zero indicates no resistance. An optimum path in a network is the path of lowest impedance, also called the least-cost path.

Impedance model In ArcInfo, a routing model that determines the best route by finding the path of least resistance.

Impersonation A process by which a Web application assumes the identity of a particular user and thus gains all the privileges to which that user is entitled.

Implement In programming an interface, to provide code for each of the methods of an interface (the interface is defined separately).

Import To bring data from one computer system or application into another. Importing often involves some form of data conversion.

Impulse response Also known as the point spread function, impulse response is the two-dimensional brightness pattern in an image (after processing) corresponding to the signal reflected by an object whose sigma falls within the dynamic range of the system, and for which the width of the imaged pattern is determined by the radar and processor rather than by the size of the object. A trihedral corner reflector is the most commonly used object for generating an impulse response in a test image. A good impulse response has a relatively large value for the pixel that maps the point scatterer location, and very small values for all surrounding pixels. The impulse response is a basic building block in describing a given radar's imaging performance, since an image is built up from the linear combination of impulse responses from all individual scatterers illuminated by the radar. The impulse response width (IRW, or resolution) of the central peak is the most important characteristic of the impulse response, together with the shape of the impulse distribution, both close to and remote from its centre.

Impulse Response to a Point Target

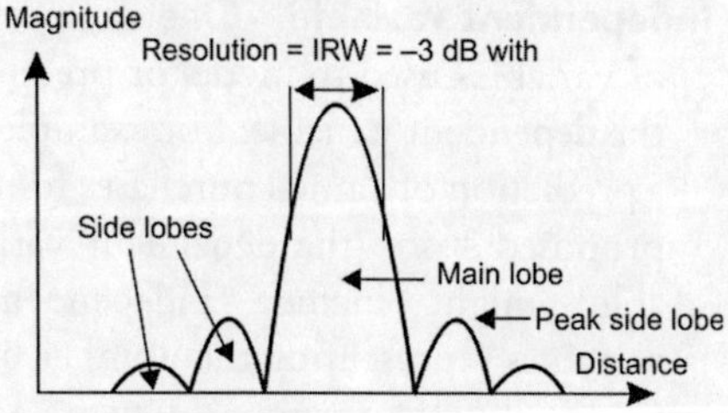

Inbound interface An interface implemented by a class, on which a client can call members.

Inbuilt parcel In Survey Analyst - Cadastral Editor, a cadastral fabric editing command that undoes the parcel generated from the build parcel command.

Incidence angle In radar, the angle formed between an imaginary line normal to the surface and another connecting the antenna and the target.

Incident In (ArcGIS) Network Analyst, a network location used in closest facility analysis. Car accidents, crime scenes and fires are examples of incidents.

Incident energy (1) Electromagnetic radiation impinging on a surface. (2) Electromagnetic radiation that strikes a surface.

INCITS Acronym for International Committee for Information Technology Standards. An ANSI-accredited forum that creates and maintains information and communications technology standards through the participation and consensus of its industry members.

Incoherent Property of a signal or data set in which the phases of the constituents are not statistically correlated, or systematically related in any fashion. Also known as non-coherent.

Independent variable One or a set of variables used to model or predict the dependent variable. For example, a prediction of annual purchases for a proposed store (the dependent variable) might include independent variables representing the number of potential customers, distance to competition, store visibility and local spending patterns. In the regression equation, independent variables appear on the right side of the equal sign and are often referred to as explanatory variable.

Indeterminate flow direction In networks, a flow direction that is unknown or undiscoverable. Indeterminate flow direction occurs when flow direction cannot be determined from the connectivity of the network, the locations of sources and sinks, and the enabled or disabled states of features.

Index A data structure, usually an array, used to speed the search for records in a database or for spatial features in geographic datasets. In general, unique identifiers stored in a key field point to records or files holding more detailed information.

Index contour line On a topographic map, a contour line that is thicker than the rest and usually labelled with the elevation that it represents. Depending on the contour interval, every fourth or fifth contour line may be an index contour.

Index map A schematic map used as a reference for a collection of map sheets, outlining the total area covered along with the coverage extent of, and usually a name or reference for, each map sheet.

Index of refraction (n) Ratio of the wavelength or velocity of electromagnetic radiation in a vacuum to that in a substance.

Industry An organization with specific GIS needs. Examples of industries include government, transportation, health care, homeland security, and public safety.

INFO database A tabular database management system used by ArcInfo Workstation software to store and manipulate attributes of a GIS dataset in ArcInfo Workstation format. INFO databases are stored inside a workspace folder with subdirectories containing files that represent the geometry and topology that make up a coverage.

Information space A geometric representation of relationships between elements in a data domain, in which relative position indicates the degree of similarity between elements. Information spaces are often based on geographic metaphors and are used to provide more intuitive views of a complex, multidimensional data domain.

Informix A commercial relational database management system (RDBMS) supported by ArcSDE.

Infrared band (IR) The band of electromagnetic wavelengths lying between the extreme of the visible (approximately 0.70 micrometre) and the shortest microwaves (approximately 100 micrometres).

Infrared radiation Electromagnetic radiation with wavelengths between about 0.7 to 1,000 micrometres. Infrared waves are not visible to the human eye. Longer infrared waves are thermal infrared waves.

Infrared scanner (thermal mapper) Instrument that detects infrared radiation and converts the detected energy to an electrical signal for recording on photographic film or magnetic tape.

Infrastructure The system of human-made physical structures, such as roads, bridges, canals, cables, wires, communications towers, hospitals, pipes, reservoirs and sewers, that provide communication, transportation, public services, utilities, or all of the above to a populace.

Inheritance In object-oriented programming, the acquisition of methods and properties by child classes or interfaces from their previously existing parent classes or interfaces. Inheritance is one of the defining characteristics of an object-oriented system.

Inner orientation Inner orientation is a process by which the image-forming bundles of rays for each photo is reconstructed, true in its geometry. The IO is the process to find the positions and orientation of the film coordinate system of the image obtained by scanner/camera. During this process fiducial points are measured. Before starting IO process, calibration data (i.e. calibrating focal length, radial lens distortion values) is entered in the system.

In-Phase (I) The real component of the signal that has the same phase as the complex reference frequency. In-phase is represented by the constant I.

In-process Within the process space of a client application, a class contained in a DLL is in-process, as objects are loaded into the process space of the client EXE. A component contained in a separate EXE is out-of-process.

Input data Data that is entered into a computer, device, program or process.

Input event record In geocoding, a piece of information such as a customer address and location of an incident. Input event record types vary by application. They include customer addresses, location of the event or incident, location of equipment and facilities, and the monument offset.

Input feature In geoprocessing, data put into the system for processing, usually specified by a path in a dialog box, script or at the command line.

Input table In geoprocessing, tabular data put into the system for processing, usually specified by a path in a dialog box, script, or at the command line.

Inset map A small map set within a larger map. An inset map might show a detailed part of the map at a larger scale, or the extent of the existing map drawn at a smaller scale within the context of a larger area.

Instance In object oriented programming, a single object created based on the template or definition of the class to which it belongs.

Instantaneous field of view (IFOV) Solid angle through which a detector is sensitive to radiation. In a scanning system, the solid angle subtended by the detector when the scanning motion is stopped.

Instantiation In programming, the process of creating a single object based on the template or definition of the class to which it belongs.

Instrument setup field In the Survey Analyst for field measurements Survey Explorer, a field that allows the user to select the instrument setup or enter the name of a new instrument setup.

Integer data Data that represents phenomena with distinct boundaries. Property lines and streets are examples of discrete data.

Integrated development environment (IDE) A software development tool for creating applications, such as desktop and Web applications. IDEs blend user interface design and layout tools with coding and debugging tools.

Integrated feature dataset In geodatabases, a feature dataset that stores topologically associated feature classes. The topological editing tools in ArcMap can be used to maintain the topological associations of features in an integrated feature dataset. Network feature classes do not participate in the topological associations within an integrated feature dataset.

Integrated side lobe ratio (ISLR) Given a single point scatterer on the ground, the processed SAR image consists of a narrow, strong peak at the location of the scatterer (the mainlobe), surrounded by smaller peaks called sidelobes. The sum of energy in the sidelobes, divided by the sum of energy in the mainlobe, is the Integrated Side Lobe Ratio (ISLR).

Integration A high degree of interconnection between two or more programs or datasets, in which they share a common schema, ontology, semantic approach, or method that allows information to be passed between them without being fully processed.

Intensity (1) In the IHS (intensity, hue, saturation) colour model, the relative brightness of a colour. (2) In the IHS system, brightness ranging from black to white.

Interactive processing Method of image processing in which the operator views preliminary results and can alter the instructions to the computer to achieve desired results.

Interactive vectorization A manual process for converting raster data into vector features that involves tracing raster cells.

Interchange format A file format that allows the easy exchange of data between different software programs.

Interface definition language A language used to define COM interfaces. The Microsoft implementation of IDL may also be referred to as MIDL or Microsoft IDL.

Interferogram A radar image that records interference patterns captured by two antennae a short distance apart.

Interferometer Device such as an imaging radar that uses two different paths for imaging and deduces information from the coherent interference between the two signals. In SAR applications, spatial interferometry has been demonstrated to measure terrain height, and time delay interferometry is used to measure movement in the scene such as oceanic currents.

Interferometric synthetic aperture radar (InSAR) SAR interferometry is a technique involving phase measurements from successive satellite SAR images to infer differential range and range changes for the purpose of detecting very subtle changes on, or of, the earth's surface with unprecedented scale, accuracy and reliability.

SAR interferometry has been demonstrated successfully in a number of applications, including topographic mapping, measurement of terrain displacement as a result of earthquakes and measurement of flow rates of glaciers or large ice sheets. The term InSAR, is most commonly associated with repeat-pass interferometry, as discussed in the section entitled 'Interferometrey' 1.1.5.4. in the User Guide. In contrast, D-InSAR is used to describe differential interferometery.

Interferometry A technique that uses the measured differences in the phase of the return signal between two satellite passes to detect slight changes on the earth's surface. The combination of two radar measurements of the same point on the ground, taken at the same time, but from slightly different angles, to produce stereo images. Using the cosine rule from trigonometry to calculate the distance between the radar and the earth's surface, these measurements can produce very accurate height maps, or maps of height changes. Mapping height changes provides information on earthquake damage, volcanic activity, landslides and glacier movement.

Interior feature weight One of two weights that allow control of how labels are placed relative to polygon features in ArcMap. Higher feature weights prevent labels from being placed over features. A high interior weight prevent labels from occupying the interior of polygon features.

Intermediate data Any data in a process that did not exist before the process existed and that will not be maintained after the process executes.

International date line An imaginary line, generally following the meridian of longitude lying 180 degrees east and west of the Greenwich meridian, where the date changes. The time zone east of the international date line is twelve hours ahead of Greenwich mean time; the time zone west of the international date line is twelve hours behind Greenwich mean time. A traveler going west across the date line adds a day; a traveler going east across it subtracts a day.

International meridian The meridian adopted by international agreement in 1884 as the prime meridian, the 0-degree meridian from which all other longitudes are calculated. The Greenwich prime meridian runs through the Royal Observatory in Greenwich, England.

International organization for standardization A federation of national standards institutes from 145 countries that works with international organizations, governments, industries, businesses and consumer representatives to define and maintain criteria for international standards.

Internationalization The process of creating software that can be adapted to the requirements of different languages and cultures without substantive changes to the source code.

Internet The global network of computers that communicate through common protocols, such as TCP/IP.

Interoperability The capability of components or systems to exchange data with other components or systems, or to perform in multiple environments. In GIS, interoperability is required for a GIS user using software from one vendor to study data compiled with GIS software from a different provider.

Interpolation The estimation of surface values at unsampled points based on known surface values of surrounding points. Interpolation can be used to estimate elevation, rainfall, temperature, chemical dispersion or other spatially-based phenomena. Interpolation is commonly a raster operation, but it can also be done in a vector environment using a TIN surface model. There are several well-known interpolation techniques, including spline and kriging.

Interpretation The process in which a person extracts information from an image.

Interpretation key Characteristic or combination of characteristics that enable an interpreter to identify an object on an image.

Interrupted projection A world projection that reduces distortion by dividing the projected area into gores, each with its own central meridian.

Intersect A geometric integration of spatial datasets that preserves features or portions of features that fall within areas common to all input datasets.

Intersection The point where two lines cross. In geocoding, most often a street crossing.

Intersection connector A character used in address data to indicate that an address is located at an intersection. For example, in the address 'S. Huntington Dr and E. Clark Blvd' the ampersand (&) character is the intersection connector. The intersection connector delimits the address into two parts and assigns intersection searches to the address.

Interval The time duration between the start and end of an imaging operation (land observation) by the ETM+ instrument on board the Landsat 7 spacecraft. The raw wideband data collected during an interval consists of a contiguous set of WRS scenes.

Intervalometer An automatic device used to trigger an operation at regular intervals, specially such a device operating the shutter of a camera.

Intranet A computer network, often using the same software and serving the same functions as those found on the Internet, that is restricted to users within an organization.

Intrinsic stationarity In spatial statistics, the assumption that a set of data comes from a random process with a constant mean and a semivariogram that depends only on the distance and direction separating any two locations.

Inverse distance One divided by distance, often raised to some power

(1/D or 1/D2, for example), where D is a distance value. By inverting the distance among spatial features and using that inverted value as a weight; near things have a larger weight or influence than things that are farther away.

Inverse distance weighted interpolation An interpolation technique that estimates cell values in a raster from a set of sample points that have been weighted so that the farther a sampled point is from the cell being evaluated, the less weight it has in the calculation of the cell's value.

Inverse synthetic aperture radar (ISAR) A SAR system that makes use of the motion of the target to synthesize a large aperture antenna.

Inversor An automatic device part of a photogrammetric rectifier that permits the number of elements installed in the instrument to be reduced.

IP address Acronym for Internet protocol address. A unique number, such as 10.48.6.8, that identifies each computer on the Internet. IP addresses are similar to phone numbers, and allow data to travel between one computer and another via the Internet.

IR colour photograph Colour photograph in which the red-imaging layer is sensitive to photographic IR wavelengths, the green-imaging layer is sensitive to red light, and the blue-imaging layer is sensitive to green light. Also known as camouflage detection photographs and false-colour photographs.

Irregular triangular mesh/surface model A vector data structure that partitions geographic space into contiguous, nonoverlapping triangles. The vertices of each triangle are sample data points with x-, y- and z-values. These sample points are connected by lines to form Delaunay triangles. TINs are used to store and display surface models.

Isanomal A line on a map connecting points of equal difference from a normal value, usually a meteorological value such as average temperature.

Isarithm An isoline drawn according to values that can occur at points; an isometric line.

ISO Abbreviation for International Organization for Standardization. A federation of national standards institutes from 145 countries that works with international organizations, governments, industries, businesses and consumer representatives to define and maintain criteria for international standards.

ISO index Index of the International Standards Organization, designating film speed in photography. Higher values indicate higher sensitivity.

Isobar A line on a weather map connecting places of equal barometric pressure.

Isocentre The point on the aerial photo platform that falls on a line half-way between the Principal point and the Nadir point.

Isochrone A line on a map connecting points of equal elapsed time; especially, travel time to or from a given location.

Isogonic chart Chart showing isogonic lines properly labelled with their magnetic declination.

Isogonic line Line joining points on the earth's surface having equal magnetic declination as of a given date.

Isohyet A line on a map connecting points of equal rainfall.

Isolation level A setting in a database management system (DBMS) that defines how much an application process is isolated from other concurrently executing processes in a DBMS. It specifies the degree to which the rows read and updated by the application are available to other concurrently executing processes. It also specifies the degree to which updates from other concurrently executing application processes are available to the application.

Isoline (1) A line connecting points of equal value on a map. Isolines fall into two classes: those in which the values actually exist at points, such as temperature or elevation values, and those in which the values are ratios that exist over areas, such as population per square kilometre or crop yield per acre. The first type of isoline is specifically called an isometric line or isarithm; the second type is called an isopleth. (2) A line connecting points of equal value, especially on a map. The prefix 'iso' means equal. For example, The 1,000 metre isoline on the map connects points at the 1,000 metre elevation.

Isometric line An isoline drawn according to known values, either sampled or derived, that can occur at points. Examples of sampled quantities that can occur at points are elevation above sea level, an actual temperature or an actual depth of precipitation. Examples of derived values that can occur at points are the average of temperature over time for one point or the ratio of smoggy days to clear days for one point.

Isopleth (1) An isoline drawn according to known values that can only be recorded for areas, not points. Examples include population per square mile or the ratio of residential land to total land for an area. (2) A line on a map connecting points of equal temperature.

Isotherm Contour line connecting points of equal temperature. Isotherm maps are used to portray surface-temperature patterns of water bodies.

Isotropic Having uniform spatial distribution of movement or properties, usually across a surface.

Isotropy A property of a natural process or data where spatial dependence (autocorrelation) changes only with the distance between two locations – direction is unimportant.

Item An element in the Catalog tree. Items include data sources, such as shapefiles and geodatabases, and nonspatial elements, such as folders.

Iterative procedure A repetitive or recurring procedure.

IUnknown All COM interfaces inherit from the IUnknown interface. The default implementation of IUnknown controls object lifetime and provides runtime type support.

J

J2ME (java 2 platform, micro edition) The Java platform edition developed by Sun Microsystems for small, stand-alone or connectable devices. Java ME enables development, deployment and management of applications that can scale from mobile devices to desktop computers. In 2007, Sun replaced J2ME with Java ME.

Jaggies The jagged appearance of curves and diagonal lines in a raster image. Aliasing becomes more apparent as the size of the raster pixels is increased or the resolution of the image is decreased.

Java An object-oriented cross-platform programming language developed by Sun Microsystems.

Java connector An ArcIMS Application Server Connector. It can be used with Java Server Pages (JSP) or as part of a Java applet or application.

Java ME wireless toolkit A toolkit for building applications that run on devices compliant with the Java specification for wireless devices.

Java server faces A framework for building user interfaces for Java Web applications.

Java server pages A Java technology that enables rapid development of platform- independent, Web-based applications. JSP separates the user interface from content generation, enabling designers to change the overall page layout without altering the underlying dynamic content.

Java server pages standard tag library A Java technology that encapsulates core functionality common to many Web-based applications as simple tags. JSTL includes tags for structural tasks such as iteration and conditionals, manipulation of XML documents, internationalization and locale-sensitive formatting, and SQL.

Java viewer An ArcIMS viewer that uses a Java 2 applet. The Java viewer can be used with feature or image services. It allows client-side drawing and editing and requires a Java plug-in.

JavaScript A scripting language that runs within a Web browser and interacts with HTML code to enable Web developers to add functionality to their Web sites.

JDK Acronym for Java Development Kit. A set of Java development tools from Sun Microsystems that provide basic tools needed for writing, testing and debugging Java applications and applets.

Jenks' optimization A method of statistical data classification that partitions data into classes using an algorithm that calculates groupings of data values based on the data distribution. Jenks' optimization seeks to reduce variance within groups and maximize variance between groups.

Job A task scheduled on a computer for immediate or future processing. A job can involve a single task or a batch mode operation.

Job book In Survey Analyst - Cadastral Editor, a tool that facilitates the viewing and management of cadastral fabric jobs.

Job tracking for ArcGIS ESRI software used to track and manage database production workflow. It provides the ability to define tasks, set up production environments, manage geodatabase versioning and track database changes at the feature level. JTX also provides tools for allocating resources and tracking the status and progress of jobs.

JOG Acronym for joint operations graphic. A 1:2,50,000-scale topographic map used by militaries worldwide. Joint operations graphics use a common base graphic to facilitate operations involving air, ground, and naval forces.

Johnson space flight centre A NASA facility in Houston, Texas.

Joined parcel In Survey Analyst - Cadastral Editor, a parcel that is connected to the cadastral fabric and shares common points with neighbouring parcels.

Joining Appending the fields of one table to those of another through an attribute or field common to both tables. A join is usually used to attach more attributes to the attribute table of a geographic layer.

Joint operations graphic A 1:2,50, 000-scale topographic map used by militaries worldwide. Joint operations graphics use a common base graphic to facilitate operations involving air, ground and naval forces.

JPEG Acronym for Joint Photographic Experts Group. A lossy image compression format commonly used on the Internet. JPEG is well-suited for photographs or images that have graduated colours.

JPL Jet Propulsion Laboratory, a NASA facility at Pasadena, California, operated under contract by the California Institute of Technology.

JSF A framework for building user interfaces for Java Web applications.

JSON Acronym for JavaScript Object Notation. A lightweight, human-readable data interchange format. An alternative to XML, JSON is language independent but relies on common programming language structures such as objects and arrays.

JSP Acronym for Java Server Pages. A Java technology that enables rapid development of platform-independent, Web-based applications. JSP separates the user interface from content generation, enabling designers to change the overall page layout without altering the underlying dynamic content.

JSTL A Java technology that encapsulates core functionality common to many Web-based applications as simple tags. JSTL includes tags for structural tasks such as iteration and conditionals, manipulation of XML documents, internationalization and locale-sensitive formatting, and SQL.

JTX Acronym for Job Tracking for ArcGIS. ESRI software used to track and manage database production workflow. It provides the ability to define tasks, set up production environments, manage geodatabase versioning and track database changes at the feature level. JTX also provides tools for allocating resources and tracking the status and progress of jobs.

Junction For network data models in a geodatabase, a point at which two or more edges meet.

Junction connectivity policy In network datasets, a connectivity policy that defines how a junction may connect to an edge. There are two junction connectivity policies: honor and override.

Junction element For network data models in a geodatabase, a point at which two or more edges meet.

K

Ka band Radar wavelength region from 0.8 to 1.1 cm.

Kernel Two-dimensional array of digital numbers used in digital filtering. On a raster, an analysis boundary or processing window within which cell values affect calculations and outside which they do not. Filters are used mainly in cell-based analysis where the value of a centre cell is changed to the mean, the sum, or some other function of all cell values inside the filter. A filter moves systematically across a raster until each cell has been processed. Filters can be of various shapes and sizes, but are most commonly three-cell by three-cell squares.

Key An attribute or set of attributes in a database that uniquely identifies each record.

Key attribute An attribute or set of attributes in a database that uniquely identifies each record. A primary key allows no duplicate values and cannot be null.

Keyboard shortcut A keystroke combination that executes a command that might otherwise require the use of a mouse. For example, Ctrl+C is a well-known keyboard shortcut for copying a selection in Windows.

Keycode A unique number in the feature line of a license or authorization file that controls access to software. Keycodes are based on a unique identifier. In ArcEditor Concurrent Use on Windows, for example, the hardware key number provides this unique identifier. The License Manager compares the keycodes in the license file and the unique identifier for a computer to allow access to the software. If the keycode and the unique identifier agree, then software access is granted.

Keyframe In an animation in ArcMap, ArcScene or ArcGlobe, a snapshot of an object's properties at a certain time.

Keyword A significant word from a document that is used to index or search content.

Kinematic positioning (1) Determining the position of an antenna on a moving object such as a ship or an automobile. (2) Kinematic positioning refers to applications in which the position of a non-stationary object (automobile, ship, bicycle) is determined.

Kinetic energy The ability of a moving body to do work by virtue of its motion. The molecular motion of matter is a form of kinetic energy.

Kinetic temperature Internal temperature of an object determined by random molecular motion. Kinetic temperature is measured with a contact thermometer.

Knockout In offset printing, an area that has been defined to cut through or mask specific layers of coloured inks. Knockouts are used to ensure that certain ink colours are not mixed with inks that are laid down after them.

Knowledge base A database of information about a subject, used in expert systems.

Known point A surveyed point that has an established x,y coordinate value. Known points are used in survey operations to extend survey computations into a project area.

Kodachrome A Kodak colour positive film.

Kohonen map A map that uses a neural network algorithm to classify and illustrate associations in complex datasets and reveal multidimensional patterns. A similar set of methods produces maps referred to as self-organizing maps (SOMs). Kohonen maps are named for the Finnish engineer, Teuvo Kohonen.

Kriging An interpolation technique in which the surrounding measured values are weighted to derive a predicted value for an unmeasured location. Weights are based on the distance between the measured points, the prediction locations and the overall spatial arrangement among the measured points. Kriging is unique among the interpolation methods in that it provides an easy method for characterizing the variance or the precision of predictions. Kriging is based on regionalized variable theory, which assumes that the spatial variation in the data being modelled is homogeneous across the surface. That is, the same pattern of variation can be observed at all locations on the surface. Kriging was named for the South African mining engineer, Danie G. Krige (26 August 1919 – 3 March 2013).

L

L1 frequency 1,575.42 MHz GPS carrier frequency which contains only encrypted P code, used primarily to calculate signal delays caused by the ionosphere.

Label In cartography, text placed on or near a map feature that describes or identifies it.

Label class In ArcMap, a category of labels that represents features with the same labelling properties. For example, in a roads layer, label classes could be created to define information and style for each type of road: interstate, state highway, county road and so on.

Label engine In ArcMap, the software used to place labels.

Label expression A statement that determines the label text. Label expressions typically concatenate or modify the contents of one or more fields and may add additional text strings to create more informative labels. They can contain Visual Basic script or JScript to add logic, text processing, and formatting for the labels.

Label fitting strategy tab In Maplex for ArcGIS, a tab on the Label Placement dialog box that allows control of the ways the label engine can fit more labels into a limited area. Methods that can increase label placement are stacking labels, reducing the font size of labels in congested areas or abbreviating labels.

Label manager In ArcMap, the tool used to display and set labelling properties for the currently active data frame. The Label Manager is accessible through the Labelling toolbar.

Label offset The distance a label should be from the feature it labels. A label offset and a maximum label offset can be set for point features. Maximum label offsets are expressed as a percentage of the label offset. For line features, a label offset can be set from the line (similar to the label offset for point features) and along the line (which controls the position of the label relative to the ends of the line). Label offsets are not available for all label position options.

Label offset constraint The maximum distance away from a point feature that a label may be placed, beyond the specified offset.

Label orientation The angle or direction of alignment for feature labels. Labels for features are usually placed horizontally, but they may also be oriented to an angle stored as an attribute, an angle defined by the orientation of the feature geometry, or along the graticule of the data frame.

Label placement option A parameter used to define a placement property for a label. Label placement properties include such properties as label offset, label placement zone, label fitting strategy, label prioritiy, label stacking and label weight.

Label placement property A parameter used to define a placement property for a label. Label placement properties include such properties as label offset, label placement zone, label fitting strategy, label prioritiy, label stacking, and label weight.

Label placement zone One of eight designated areas on a map, radiating from a point, in which labels may be placed. The user can indicate in which of eight zones labels should be placed, relative to the point. These preferences are taken into account when placing point labels using the Best Position placement option.

Label point In a coverage, a feature class used to represent points or identify polygons. When representing points, the x,y location of the point describes the location of the feature. When identifying polygons, the point can be located anywhere within the polygon.

Label position tab In Maplex for ArcGIS, a tab on the Label Placement dialog box that allows control of how labels are placed relative to features. The position of a label is determined by such parameters as: the orientation, offset and position style for a given feature geometry.

Label priority In ArcGIS, a ranking system that determines the order in which labels will be placed on a map. Labels with higher priority will be placed before labels with lower priority. Labels placed last will have a greater chance of being crowded out or placed in an alternate position.

Label rule A parameter used to define a placement property for a label. Label placement properties include such properties as label offset, label placement zone, label fitting strategy, label prioritiy, label stacking, and label weight.

Label stacking The splitting of long labels to place the text on two or more lines. Maplex for ArcGIS allows specification of which characters trigger a split and whether or not they show up in the label.

Label weight An ESRI Standard Label Engine ranking system that indicates whether labels from a given label class may be covered by another label in cases in which label placement conflicts occur. Labels with higher weight are less likely to be overlapped than labels with lower weight.

LACIE Large Area Crop Inventory Experiment.

Lag In the creation of a semivariogram, the sample distance used to group or bin pairs of points. Using an appropriate lag distance can be helpful in revealing scale-dependent spatial correlation.

Lambert conformal conic projection A common map projection, which is the basis for the State Plane Coordinate (SPC) system for many states.

LAN Acronym for local area network. Communications hardware and software that connect computers in a small area, such as a room or a building. Computers in a LAN can share data and peripheral devices, such as printers and plotters, but do not necessarily have a link to outside computers.

Land cover The classification of land according to the vegetation or material that covers most of its surface; for example, pine forest, grassland, ice, water, or sand.

Land information system A geographic information system for cadastral and land-use mapping, typically used by local governments.

Land use The classification of land according to what activities take place on it or how humans occupy it; for example, agricultural, industrial, residential, urban, rural or commercial.

Land use classification system Coding system of categories and subcategories designed for use on a map to designate land or water use.

Landform Any natural feature of the land having a characteristic shape, including major forms such as plains and mountains and minor forms such as hills and valleys.

Landmark (1) Any prominent natural or artificial object in a landscape used to determine distance, bearing or location. (2) Monument of material mark or fixed object used to designate a land boundary on the ground: any prominent object on land that may be used to determine a location or a direction in navigation or surveying.

Landsat (1) Land Remote Sensing Satellite Program managed by the US Geological Survey. The Landsat Satellite series began in 1972 to gather information about land surface features of the planet. Landsat 5, launched in 1984 and Landsat 7, launched in 1999, are still operational. Data from the satellites have been used for monitoring land cover conditions, geological/mineralogical exploration, urban growth and cartography. Global coverage is available and data sets are provided by the USGS at the cost of reproduction. (2) Multispectral, earth-orbiting satellites developed by NASA (National Aeronautics and Space Administration) that gather imagery for land-use inventory, geological and mineralogical exploration, crop and forestry assessment, and cartography.

Landsat 7 contact period The time duration between the start and end of raw wideband data transmissions from the Landsat 7 spacecraft to a ground station, e.g. the Landsat 7 Ground Station (LGS).

Landscape ecology The study of spatial patterns, processes, and change across biological and cultural structures within areas encompassing multiple ecosystems.

Laplacian filter A form of nondirectional digital filter.

Large scale Generally, a map scale that shows a small area on the ground at a high level of detail.

Large-format camera (LFC) An experiment first carried on the Space Shuttle in October 1984.

Large-format printer A printing device capable of producing an image on large paper or other media sized between 36 and 87 inches (91 and 220 centimetres) wide. Modern large format printers typically use inkjet printing technology to print an image on a roll of paper that is automatically cut to the desired length. Large-format printers may also be called plotters or wide-format printers.

LAS An industry-standard binary file format that maintains information related to lidar data.

Laser Light artificially stimulated electromagnetic radiation: a beam of coherent radiation with a single wavelength.

Laser distance meter A device used for measuring distance that uses Laser or Light Amplification by Stimulated Emission of Radiation

Late binding A COM technique that an application uses for determining an object's properties and methods at run time, rather than when the code is compiled. Late binding is generally used by scripting languages.

Latent image Invisible image produced by the photochemical effect of light on silver halide grains in the emulsion of film. The latent image is not visible until after photographic development.

Latitude (1) The angular distance North or South from the earth's equator measured in degrees with the equator at 0° and the poles 90° N and 90° S. (2) The angular distance, usually measured in degrees north or south of the equator. Lines of latitude are also referred to as parallels.

Latitude of centre The latitude value that defines the centre and sometimes the origin of a projection.

Latitude of origin The latitude value that defines the origin of the y-coordinate values for a projection.

Latitude-longitude A reference system used to locate positions on the earth's surface. Distances east–west are measured with lines of longitude (also called meridians), which run north–south and converge at the north and south poles. Distance measurements begin at the prime meridian and are measured positively 180 degrees to the east and negatively 180 degrees to the west. Distances north–south are measured with lines of latitude (also called parallels), which run east–west. Distance measurements begin at the equator and are measured positively 90 degrees to the north and negatively 90 degrees to the south.

Lattice A representation of a surface using an array of regularly spaced sample points (mesh points) that are referenced to a common origin and have a constant sampling distance in the x and y directions. Each mesh point contains the z-value at that location, which is referenced to a common base z-value, such as sea level. Z-values for locations between lattice mesh points can be approximated by interpolation based on neighbouring mesh points.

Layer The visual representation of a geographic dataset in any digital map environment. Conceptually, a layer is a slice or stratum of the geographic reality in a particular area and is more or less equivalent to a legend item on a paper map. On a road map, for example, roads, national parks, political boundaries and rivers might be considered different layers.

Layer file In ArcGIS, a file with a .lyr extension that stores the path to a source dataset and other layer properties, including symbology.

Layout The arrangement of elements on a map, possibly including a title, legend, north arrow, scale bar and geographic data.

Layout view In ArcMap and ArcReader, a view that shows the virtual page upon which geographic data and map elements, such as titles, legends, and scale bars, are placed and arranged for printing.

Layover In radar images, the geometric displacement of the top of objects toward the near range relative to their base.

Lband Radar wavelength region from 15 to 30 cm. The group of radio frequencies that carry data from GPS satellites to GPS receivers.

LBS Information or a physical service delivered to multiple channels, exclusively based on the determined location of a wireless device. Some location-based applications include emergency services, information services and tracking services.

Lead line Line weighted with lead for making depth soundings in water.

Leader In MOLE, typically two or more force elements grouped together and placed on a line based on user-specified rules. Leaders are often used to clean up the map display in cases where many symbols overlap, to group related units together and to define perimeters or areas of interest for formations.

Least convex hull The smallest convex polygon that encloses a group of objects, such as points. In ArcGIS, TIN boundaries are convex hulls by default.

Least-cost path The path between two locations that costs the least to traverse, where cost is a function of time, distance, or some other criteria defined by the user.

Least-squares adjustment A statistical method for providing a best fit for survey point locations and detecting measurement error by minimizing the sum of the squares of measurement residuals. The method allows many measurements to participate simultaneously in a single computation.

Least-squares corrections The final measurement residuals of a least squares adjustment.

Left-right topology The topological data structure in an ArcInfo coverage that stores, for each arc, the identity of the polygons to the left and right of it. Left-right topology supports analysis functions, such as adjacency.

Legend The description of the types of features included in a map, usually displayed in the map layout. Legends often use graphics of symbols or examples of features from the map with a written description of what each symbol or graphic represents.

Legend patch shape The geometric shape of either a line or a polygon that is used to represent a specific kind of feature in a legend and in the ArcMap table of contents.

Lens One or more pieces of glass or other transparent material shaped to form an image by refraction of light.

Lens distortion Small effects due to the flaws in the optical components (i.e. Lens) of camera systems leading to distortions (which are typically more serious at the edges of photos).

Levee Artificial bank confining a stream channel or limiting adjacent areas subject to flooding; an embankment bordering a submarine canyon or channel, usually occurring along the outer edge of a curve.

Level of confidence In a statistical test, the risk, expressed as a percentage, that the null hypothesis will be incorrectly rejected because of sampling error when the null hypothesis is true. For example, a confidence level of 95 per cent means that if the same test were performed 100 times on 100 different samples, the null hypothesis would be incorrectly rejected five times.

Level of detail An abstraction of a layer in ArcGlobe portraying the layer at some degree of resolution between simplified and unsimplified.

Level of significance In statistical testing, the probability of an incorrect rejection of the null hypotheses.

Level 0R files The generic term used to denote the grouping of band, MSCD, PCD, and calibration data files for a single subinterval.

Level 0R product A US data product in which the data has been spatially reformatted but the data values remain unchanged. No radiometric or geometric corrections have been performed on the data. The reformatting is fully reversible. The data is band sequential. Attached to the image data are radiometric calibration image data, payload correction data, quality data, and metadata.

Level 0R quality and accounting data The data quality and accounting information collected by the Landsat 7 Processing System (LPS), on a subinterval basis, from processing of the ETM+ major frames constructed from the wideband Virtual Channel Data Units (VCDUs) received during a Landsat 7 contact period.

Level surface Surface which at every point is perpendicular to the plumb-line or the direction in which gravity acts.

Leveling In surveying, the measurement of the heights of objects and points according to a specified elevation, usually mean sea level.

LFC Large-format camera.

LIBID Acronym for Library Identifier. A type of GUID consisting of a unique string assigned to a type library.

Library In object-oriented programming, a logical grouping of classes, usually with a header section that lists the classes in the library.

License The grant to a party of the right to use a software package or component. A license differs from a sale in that the user does not necessarily purchase the software but is granted the legal right to use it.

License file A file that contains License Manager license data. Each license file contains information such as the SERVER, ESRI_SENTINEL_KEY number (Windows only), Version, the number of seats, and so on.

Lidar Light intensity detection and ranging, which uses lasers to stimulate fluorescence in various compounds and to measure distances to reflecting surfaces.

Light Electromagnetic radiation ranging from 0.4 to 0.7Fm in wavelength that is detectable by the human eye.

Light meter Device for measuring the intensity of visible radiation and determining the appropriate exposure of photographic film in a camera.

Lighting normal In ArcScene and ArcGlobe, vectors normal to a geometry's surface, stored in that geometry to help define how lighting affects it.

Limits In Survey Analyst for field measurements, restrictions that define an acceptable level of measurement error for each computation.

Line On a map, a shape defined by a connected series of unique x,y coordinate pairs. A line may be straight or curved. Aa spatial feature line is represented by a series of points and has the geometric properties of location and length.

Line connection A procedure that combines groups of individual lines with the same name into a single line for the label engine. This is often necessary because lines such as roads and rivers are usually digitized as many small segments that must be connected together to represent a single real-world feature.

Line copy (line drawing) Map copy suitable for reproduction without the use of a screen; a drawing composed of lines as distinguished from continuous- tone copy.

Line densification the process of adding new vertices to a line.

Line drop out The loss of data from a scan line caused by malfunction of one of the detectors in a line scanner.

Line event In linear referencing, a description of a portion of a route using a from- and to-measure value. Examples of line events include pavement quality, salmon spawning grounds, bus fares, pipe widths, and traffic volumes.

Line feature A map feature that has length but not area at a given scale, such as a river on a world map or a street on a city map.

Line of sight When two objects (such as a satellite and a receiving station) have nothing in between them, then they are in 'line-of-sight' of each other. When a satellite is on the other side of the Earth from a receiving station, the Earth is in between them, so the satellite and the receiving station are not in 'line-of sight' of each other.

Line scanner An imaging device which uses a mirror to sweep the ground surface normal to the flight path of the platform. An image is built up as a strip comprising lines of data.

Line simplification A generalization technique in which vertices are selectively removed from a line feature to eliminate detail while preserving the line's basic shape.

Line smoothing The process of adding extra points to lines to reduce the sharpness of angles between line segments, resulting in a smoother appearance.

Lineage A collection of states representing the changes that have occurred over time in a versioned geodatabase.

Lineament Linear topographic or tonal feature on the terrain and on images, maps, and photographs that may represent a zone of structural weakness.

Linear Adjective that describes the straight line – like nature of features on the terrain or on images and photographs.

Linear dimension A measurement of the horizontal or vertical dimension of a feature. Linear dimensions may not represent the true distance between beginning and ending dimension points because they do not take angle into account as aligned dimensions do.

Linear feature A map feature that has length but not area at a given scale, such as a river on a world map or a street on a city map.

Linear frequency modulation (FM) A linear FM signal has a quadratic phase variation with time, so the instantaneous frequency varies linearly with time.

Linear interpolation The estimation of an unknown value using the linear distance between known values.

Linear polarization A polarization state in which one of the perpendicular components of the electric field has zero magnitude. In this case, the polarization ellipse collapses to a straight line; the tip of the electric field vector traces a straight line on a plane that is transverse to the wave propagation direction.

Linear range cell migration correction (RCMC) A shift and interpolation operation is performed in the range direction to correct for linear Range Cell Migration (RCM).

Linear referencing A method for storing geographic data by using a relative position along an already existing line feature; the ability to uniquely identify positions along lines without explicit x,y coordinates. In linear referencing, location is given in terms of a known line feature and a position or measure, along the feature. Linear referencing is an intuitive way to associate multiple sets of attributes to portions of linear features.

Linear unit The unit of measurement on a plane or a projected coordinate system, often metres or feet.

Linearity Property according to which an operation on a sum of signals is equivalent to the same operation applied to each of the signals individually and the resulting numbers added together. Linearity, over the dynamic range of the system, is an essential attribute of most measurement devices such as imaging radar.

Lineation The one-dimensional alignment of internal components of a rock that cannot be depicted as an individual feature on a map.

Line-on-line overlay In linear referencing, the overlay of two line event tables to produce a single line event table. The new event table can be the logical intersection or union of the input tables.

Line-on-point overlay In linear referencing, the overlay of a line event table and a point event table to produce a single point event table. The new event table can be the logical intersection or union of the input tables.

Line-pair Pair of light and dark bars of equal widths. The number of such line-pairs aligned side by side that can be distinguished per unit distance expresses the resolving power of an imaging system.

Link In georeferencing, connections added between known points in a dataset being georeferenced and corresponding points in the dataset being used as a reference.

Link command In Survey Analyst for field measurements, a command that finds nearby survey points for each feature vertex and automatically creates links. The command allows the user to specify the search tolerance for finding the survey points. With the Link command, users may perform batch links; it is useful to use if there are many unlinked features that need to be associated with nearby survey points.

Link lines In Survey Analyst for field measurements, lines displayed on a map after a survey point and a feature vertex are linked.

Link tool In Survey Analyst for field measurements, a tool that allows the user to make a link between a survey point and a feature vertex by snapping and clicking a feature vertex, then snapping to and clicking the related survey point. With the Link tool, users must make each individual link manually.

LIS A geographic information system for cadastral and land-use mapping, typically used by local governments.

List page In Survey Analyst for field measurements, one of two types of pages in the Survey Explorer. The List page lists multiple survey objects.

Little endian A computer hardware architecture in which, within a multibyte numeric representation, the least significant byte has the lowest address and the remaining bytes are encoded in increasing order of significance.

Load balance The act of distributing application, network, and/or server resources to optimize performance.

Load distribution The act of distributing application, network and/or server resources to optimize performance.

Local analysis The computation of an output raster where the output value at each location is a function of the input value at the same location.

Local area network (LAN) Communications hardware and software that connect computers in a small area, such as a room or a building. Computers in a LAN can share data and peripheral devices, such as printers and plotters, but do not necessarily have a link to outside computers.

Local check method In Survey Analyst for field measurements, one of two ways to apply the Coordinate Out of Tolerance command. The Local check method searches for coordinates out of tolerance within each survey project.

Local datum A horizontal geodetic datum that serves as a basis for measurements over a limited area of the earth; that has its origin at a location on the earth's surface; that uses an ellipsoid whose dimensions conform well to its region of use; and that was originally defined for land-based surveys. A local datum in this sense stands in contrast to a geocentric datum. Examples include the North American Datum of 1927 and the Australian Geodetic Datum of 1966.

Local functions The computation of an output raster where the output value at each location is a function of the input value at the same location.

Local polynomial interpolation In ArcGIS Geostatistical Analyst, a deterministic interpolation method. The interpolated surface is not required to conform to the sample data points and the method does not have standard errors associated with it.

Localization The process of adapting software to the requirements of a different language or culture, including translating user interfaces, documentation and help systems;

customizing features; and accommodating different character sets.

Location An identifier assigned to a region or feature.

Location aware services Services that allow IBM WebSphere Everywhere Access (WEA) application providers to use location-based services from multiple vendors, by providing an application programming interface (API).

Location errors Errors related to the location of map features such as missing lines or missing polygons.

Location monument It is a fixed point in the earth's surface whose position is known through surveying and marked by a permanent structure such as concrete and steel.

Location query A statement or logical expression that selects geographic features based on location or spatial relationship. For example, a spatial query might find which points are contained within a polygon or set of polygons, find features within a specified distance of a feature or find features that are adjacent to each other.

Location utility service An OpenLS ArcWeb service used to find geographical coordinates of an address and to find an address based on the geographical coordinates (geocoding/reverse geocoding).

Location-allocation The process of finding the best locations for one or more facilities that will service a given set of points and then assigning those points to the facilities, taking into account factors such as the number of facilities available, their cost and the maximum impedance from a facility to a point.

Location-based services Information or a physical service delivered to multiple channels, exclusively based on the determined location of a wireless device. Some location-based applications include emergency services, information services and tracking services.

Locked parcel In Survey Analyst - Cadastral Editor, a parcel that has been locked for editing. Locked parcels cannot be edited simultaneously in a multiuser environment.

Locomotion The movements of a person following a route. Locomotion is the physical component of navigation.

Log file A database file that records changes in data, often used as part of a database restoration.

Logarithm The power to which a fixed number (the base) must be raised to equal a given number. The three most frequently used bases for logarithms are base 10, base e, and base 2.

Logical expression A string of numbers, constants, variables, operators and functions that returns a value of true or false.

Logical network An abstract representation of a network, implemented as a collection of hidden tables. A logical network contains edge, junction and turn elements, the connectivity between them and the weights necessary for traversing the network. It does not contain information about the geometry or location of its elements; this information is one of the components of a network system.

Logical operator An operator used to compare logical expressions that returns a result of true or false. Examples of logical operators include less than (<), greater than (>), equal to (=) and not equal to (≠).

Logical query The process of using mathematical expressions to select features from a geographic layer based on their attributes; for example, 'select all polygons with an area greater than 16,000 units' or 'select all street segments named Green Apple Run'.

Logical selection The process of using mathematical expressions to select features from a geographic layer based on their attributes; for example, 'select all polygons with an area greater than 16,000 units' or 'select all street segments named Green Apple Run'.

Long transaction An edit session on a feature dataset that may last from a few minutes to several months. Long transactions are managed by the ArcSDE versioning mechanism.

Longitude (1) The angular distance East or West between the meridian of a particular place on earth and that of Greenwich, England, expressed in degrees or time. (2) The angular distance, usually expressed in degrees, minutes and seconds, of the location of a point on the earth's surface east or west of an arbitrarily defined meridian (usually the Greenwich prime meridian). All lines of longitude are great circles that intersect the equator and pass through the North and South Poles.

Longitude of centre The longitude value that defines the centre and sometimes, the origin of a projection.

Longitude of origin The longitude value that defines the origin of the x-coordinate values for a projection.

Long-range variation In a spatial model, coarse-scale variation that is usually modelled as the trend.

Look angle The angle between the vertical plane containing a radar antenna and the direction of radar propagation. Complementary to the depression angle.

Look direction The radar look direction defines the angle in the horizontal plane in which the radar antenna is pointing when transmitting a pulse and receiving the return signal from the ground or from an object. The look direction is an angular measurement (in degrees) and is usually made with respect to true North. In side-looking imaging radar (SLAR), the look direction is often orthogonal (normal) to the flight trajectory (azimuth) of the platform carrying the radar and is thus synonymous with the range direction. The radar look direction is an important parameter when analysing features with a preferred orientation, for example, fracture patterns in rock formations, regular street patterns or ocean waves, as these may be enhanced through choice of appropriate radar illumination direction.

Looks Radar terminology refers to individual looks as groups of signal samples in a SAR processor that splits the full synthetic aperture into several sub-apertures, each representing an independent look of the identical scene. The resulting image formed by incoherent summing of these looks is characterized by reduced speckle and degraded spatial resolution. The SAR signal processor can use the full synthetic aperture and the complete signal data history in order to produce the highest possible resolution, albeit very speckled, single-look complex (SLC) SAR image product. Multiple looks may be generated by averaging over range and/or azimuth resolution cells. For an improvement in radiometric resolution using multiple looks there is an associated degradation in spatial resolution. Note that there is a difference between the number of looks physically implemented in a processor and the effective number of looks as determined by the statistics of the image data.

Look-up table (LUT) A mathematical formula used to convert one distribution of data to another, most conveniently remembered as a conversion graph.

Loop traverse In surveying, a traverse that starts and ends with the same survey point.

Loose coupling A relatively unstructured relationship between two software components or programs that work together to process data, which requires little overlap between methods, ontologies, class definitions and so on.

Loosely coupled replication A replication model that does not require the parent and child replicas to be directly connected for synchronization to occur. Loosely coupled replication is an asynchronous model, so edits made in one replica have no effect on other related replicas until synchronization. Synchronization can be executed manually or it can be automated.

Loss of signal The inability to receive a satellite signal because the satellite's orbital path has taken it below the antenna's horizon.

Lossless compression Data compression that has the ability to store data without changing any of the values, but is only able to compress the data at a low ratio (typically 2:1 or 3:1). In GIS, lossless compression is often used to compress raster data when the pixel values of the raster will be used for analysis or deriving other data products.

Lossy compression Data compression that provides high compression ratios (for example 10:1 to 100:1) but does not retain all the information in the data. In GIS, lossy compression is used to compress raster datasets that will be used as background images, but is not suitable for raster datasets used for analysis or deriving other data products.

Low water Minimum height reached by a falling tide. The height may be due solely to the periodic tidal forces or it may have superimposed upon it the effects of meteorological conditions.

Low water line Intersection of the land with the water surface at an elevation of low water. Not to be confused with mean low water line.

Low-altitude sensor A sensor being operated in a low-flying aircraft, i.e. altitudes of a few kilometres or less. Compare with *high-altitude sensor*.

Lower scene corners The corners associated with the leading edge (last scan) of a scene. For descending path scenes, the lower left corner corresponds to the south–west corner of a scene and the lower right corner corresponds to the south–east corner of a scene. For ascending path scenes, the lower left corner corresponds to the north–east corner of a scene and the lower right corner corresponds to the north–west corner of a scene. These mappings hold for the band file geolocation fields and the metadata file.

Low-level language A programming language that uses keywords and statements that are little more complex than the ones and zeros of machine language. Low-level language technically includes machine language, but more commonly refers to an assembly language that uses symbols to make machine instructions easier for programmers to read and understand. Each statement in assembly language represents a single command to the processor, affording the developer only a low level of abstraction in regard to mundane functions such as memory access and register storage, meaning such operations demand the developer's close attention.

Low-pass filter A spatial filter that blocks high-frequency (shortwave) radiation, resulting in a smoother image.

Low-sun-angle photograph Aerial photograph acquired in the morning, evening, or winter when the sun is at a low elevation above the horizon.

Loxodrome (rhumb line) A complex curve on the earth's surface that crosses every meridian at the same oblique angle. A rhumb line path follows a single compass bearing; it is a straight line on a Mercator projection, or a logarithmic spiral on a polar projection. A rhumb line is not the shortest distance between two points on a sphere.

Luminance Quantitative measure of the intensity of light from a source.

M

Mach band An optical illusion of dark and light fringes within adjacent areas of contrasted tone. It is a psychophysiological phenomenon which aids human detection of boundaries or edges.

Macro A computer program, usually a text file, containing a sequence of commands that are executed as a single command. Macros are used to perform commonly used sequences of commands or complex operations.

Magnetic bearing A bearing measured relative to magnetic north.

Magnetic declination The angle between magnetic north and true north observed from a point on the earth. Magnetic declination varies from place to place and changes over time, in response to changes in the earth's magnetic field.

Magnetic north The direction from a point on the earth's surface following a great circle toward the magnetic north pole, indicated by the north-seeking end of a compass.

Magnetometer An instrument used to measure variations in the strength and direction of the earth's magnetic field.

Magnifier window A secondary window in ArcMap data view that shows a magnified view of a small area without changing the map extent. Moving the magnifier window around will not affect the current map display.

Maintenance license A license that has a current maintenance contract with ESRI Customer Service. ESRI software users who have maintenance licenses are eligible for upgrade to the latest version of the software.

Maintenance renewal The date a maintenance contract expires for a particular product. Multiple copies of the same product may expire on different dates.

Major axis The longer axis of an ellipse or spheroid.

Majority resampling A technique for resampling raster data in which the value of each cell in an output is calculated, most commonly using a 2x2 neighbourhood of the input raster. Majority resampling does not create any new cell values, so it is useful for resampling categorical or integer data, such as land use, soil or forest type. Majority resampling acts as a type of low-pass filter for discrete data, generalizing the data and filtering out anomalous data values.

Make permanent In ArcGIS Spatial Analyst, an option that creates a permanent raster (one that is saved to disk) from a temporary result.

Managed raster catalog A raster catalog in which the raster datasets are copied to a location assigned by a geodatabase. When a row is deleted from a managed raster catalog, the data is deleted as well.

Many-to-many relationship (1) An association between two linked or joined tables in which one record in the first table may correspond to many records in the second table and vice versa. (2) An association between two linked or joined tables in which many records in the first table may correspond to a single record in the second table.

Map (1) A graphic representation of the spatial relationships of entities within an area. (2) In Photogrammetry, an orthographically projected rendering of existing land features produced with a stereoplotting instrument. Graphic representation of the physical features (natural, artificial or both) of a part or the whole of the earth's surface, by means of signs and symbols or photographic imagery, at an established scale, on a specified projection and with the means of orientation indicated.

Map algebra A language that defines a syntax for combining map themes by applying mathematical operations and analytical functions to create new map themes. In a map algebra expression, the operators are a combination of mathematical, logical or Boolean operators (+, >, AND, tan and so on) and spatial analysis functions (slope, shortest path, spline and so on) and the operands are spatial data and numbers.

Map annotation In ArcGIS, text or graphics stored within the map data frame in an annotation group. Map annotation may be manually entered or generated from labels, and can be individually selected, positioned and modified.

Map cache A setting used in ArcMap that allows temporary storage of geodatabase or ArcIMS feature service features from a given map extent in the desktop computer's RAM, which may result in performance improvements in ArcMap for editing, feature rendering and labelling.

Map collar Any of the supporting objects or elements that help a reader interpret a map. Typical map surround elements include the title, legend, north arrow, scale bar, border, source information, other text and inset maps.

Map configuration file In ArcIMS, the file that contains the core site information. ArcIMS configuration files contain all the basic information about the content to be delivered, such as location of the data and layer symbology. Typically, a configuration file contains data that defines map content and has a file extension of .axl, but it can also be used to deliver metadata or route data (as .axl files) and to serve maps created in ArcMap (.mxd or .pmf files). Regardless of their type, configuration files contain content that the service registers to the ArcIMS spatial server and Web server for processing.

Map digitization Conversion of map data from graphic to digital form.

Map display A graphic representation of a map on a computer screen.

Map document In ArcMap, the file that contains one map, its layout, its associated layers, tables, charts and reports. Map documents can be printed or embedded in other documents. Map document files have a .mxd extension.

Map element In digital cartography, a distinctly identifiable graphic or object in the map or page layout. For example, a map element can be a title, scale bar, legend or other map-surround element. The map area itself can be considered a map element; or an object within the map can be referred to as a map element, such as a roads layer or a school symbol.

Map extent The limit of the geographic area shown on a map, usually defined by a rectangle. In a dynamic map display, the map extent can be changed by zooming and panning.

Map feature A representation of a real-world object on a map.

Map generalization Decreasing the level of detail on a map so that it remains uncluttered when its scale is reduced.

Map hypsographic Map showing relief with elevations referred to the national geodetic vertical datum of 1929.

Map image web service A SOAP ArcWeb service that provides dynamic map content for Internet applications. With this service, users input a geographic extent and several optional items (such as themes, type of map, size of image, map annotation) and receive the location of an output image file. Map Image Web Service provides access to services from a variety of leading data publishers.

Map library In ArcInfo Workstation Map Librarian, a collection of geographic data partitioned spatially as a set of tiles and thematically as a set of layers, indexed by location for rapid access.

Map production system (MPS) The cartographic component of the Production Line Tool Set (PLTS). MPS is designed to facilitate large-volume cartographic production by providing the ability to create dynamic map layouts, batch symbology and enhanced cartographic editing tools.

Map projection A method by which the curved surface of the earth is portrayed on a flat surface. This generally requires a systematic mathematical transformation of the earth's graticule of lines of longitude and latitude onto a plane. Some projections can be visualized as a transparent globe with a light bulb at its centre (though not all projections emanate from the globe's centre) casting lines of latitude and longitude onto a sheet of paper. Generally, the paper is either flat and placed tangent to the globe (a planar or azimuthal projection) or formed into a cone or cylinder and placed over the globe (cylindrical and conical projections). Every map projection distorts distance, area, shape, direction or some combination thereof.

Map query A statement or logical expression that selects geographic features based on location or spatial relationship. For example, a spatial query might find which points are contained within a polygon or set of polygons, find features within a specified distance of a feature or find features that are adjacent to each other.

Map reading The activity of viewing a map in a way that allows the viewer to make sense of or gain information from it. Map reading involves interpreting the meanings of codes and cartographic representations used on the map.

Map scale The ratio or relationship between a distance or area on a map and the corresponding distance or area on the ground, commonly expressed as a fraction or ratio. A map scale of 1/100,000 or 1:100,000 means that one unit of measure on the map equals 100,000 of the same unit on the earth.

Map series (1) A collection of maps usually addressing a particular theme. (2) Family of maps conforming generally to the same specifications and designed to cover an area or a country in systematic pattern.

Map service A type of Web service that generates maps.

Map sheet A single map or chart in a map series, such as any one of the approximately 57,000 USGS 7.5-minute topographic maps of the United States and its territories.

Map style An organized collection of predefined colours, symbols, properties of symbols and map elements. Styles promote standardization and consistency in mapping products.

Map surround Any of the supporting objects or elements that help a reader interpret a map. Typical map surround elements include the title, legend, north arrow, scale bar, border, source information, other text and inset maps.

Map template In ArcMap, a kind of map document that provides a quick way to create a new map. Templates can contain data, a custom interface and a predefined layout that arranges map elements, such as north arrows, scale bars and logos on the virtual page. Map templates have a .mxt file extension.

Map topology A temporary set of topological relationships between coincident parts of simple features on a map, used to edit shared parts of multiple features.

Map unit The ground unit of measurement – for example, feet, miles, metres, or kilometres – in which coordinates of spatial data are stored.

Map, base Map on which information may be placed for purposes of comparison or geographical correlation. The term 'base map' was at one time applied to a class of maps now known as outline maps. It may be applied to topographic maps, also termed 'mother maps' that are used in the construction of other types of maps by the addition of particular data.

Map, bathymetric Maps delineating the form of the bottom of a body of water, or a portion thereof, by the use of depth contours (isobaths).

Map, cadastral Map showing the boundaries of subdivisions of land, often with the bearings and lengths thereof and the areas of individual tracts, for purposes of describing and recording ownership. It may also show culture, drainage and other features relating to land use and value. See: plat.

Map, choropleth Thematic map in which areas are coloured, shaded, dotted or hatched to create darker or lighter areas in proportion to the density of distribution of the theme subject.

Map, engineering Map showing information that is essential for planning an engineering project or development and for estimating its cost. It usually is a large-scale map of a small area or of a route. It may be entirely the product of an engineering survey or reliable information may be collected from various sources for the purpose and assembled on a base map.

Map, flood control Map designed for studying and planning control projects in areas subject to flooding.

Map, forestry Map prepared principally to show the size, density, kind and value of trees in a designated area.

Map, geologic Map showing the structure and composition of geologic features.

Map, hypsometric Map showing relief by any convention, such as contours, hachures, shading, or tinting.

Map, isopleth Map consisting of lines connecting places of equal value of distribution for a given theme such as rainfall or temperature.

Map, land use Map showing by means of a coding system the various purposes for which parcels of land are being used by man.

Map, line Map composed of lines as distinguished from photographic imagery.

Map, orthophotographic See: orthophotographic map.

Map, photographic See: photomap.

Map, planimetric Map that presents only the horizontal positions for features represented; distinguished from a topographic map by the omission of relief in measurable form. The features usually shown on a planimetric map include rivers, lakes and seas; mountains, valleys and plains; forests and prairies; cities, farms transportation routes and public utility facilities; and political and private boundary lines. A planimetric map intended for special use may present only those features essential to the purpose to be served.

Map, slope (clinometric map) Map showing the degree of steepness of the earth's surface by the use of various colours or shading for critical ranges of slope.

Map, soil Map that shows the constitution, structure and texture of the soil and identifies ongoing erosion.

Map, storm evacuation Map designed to identify coastal areas subject to flooding, to indicate recommended areas of refuge and to emphasize available evacuation routes.

Map, thematic Map designed to provide information on a single topic, such as geology, rainfall, population.

Map, topographic Map that present the horizontal and vertical positions of the features represented; distinguished from a planimetric map by the addition of relief in measurable form.

MapServer An ArcGIS Server software component that provides programmatic access to the contents of a map document on disk and creates images of the map contents based on user requests. It is designed for use in building map-based Web services and Web applications using ArcGIS Server.

Maptip In ArcGIS, a user-assistance component that displays an on-screen description of a map feature when the mouse is paused over that feature.

Marginalia Any of the supporting objects or elements that help a reader interpret a map. Typical map surround elements include the title, legend, north arrow, scale bar, border, source information and other text; and inset maps.

Marker symbol A symbol used to represent a point location on a map.

Market area A geographic zone containing the people who are likely to purchase a firm's goods or services.

Market penetration analysis A process that determines the percentage of a market area being reached based on the number of customers within an area divided by the total population in that area.

Marsh, coastal Area of salt-tolerant vegetation in brackish and (or) saline-water habitants subject to tidal inundation.

Marsh, freshwater Tract of low wet ground, usually miry and covered with rank vegetation.

Marshalling The process that enables communication between a client object and server object in different apartments of the same process, between different processes or between different processes on different machines by specifying how function calls and parameters are to be passed over these boundaries.

Mashup In Internet mapping, the combination of content from more than one data source into one dynamic map service.

Mask In digital cartography, a means of covering or hiding features on a map to enhance cartographic representation. For example, masking is often used to cover features behind text to make the text more readable.

Mask angle The minimum acceptable satellite elevation above the horizon to avoid blockage of line-of-sight.

Mass point An irregularly distributed sample point, with an x-, y- and z-value, used to build a triangulated irregular network (TIN). Ideally, mass points are chosen to capture the more important variations in the shape of the surface being modelled.

Master checkout version In ArcGIS 9.1 and previous versions, the data version in the master geodatabase, created when data is checked out, that represents the state of the data at the time it was checked out.

Master geodatabase In ArcGIS 9.1 and previous versions, a geodatabase from which data has been checked out.

Master site In Business Analyst, the site of a known, well-performing business.

Match score In geocoding, a value assigned to all potential candidates for an address match. The match score is based on how well the location found in the reference data matches with the address data being searched.

Matched filtering The matched filter can be thought of as a filter with a different time delay for each frequency component of the signal passing through the filter. That is, the different frequency components of the linear FM signal are each delayed so that they all arrive at the same time at the output of the filter. This way, all the signal energy is gathered into a narrow peak in the compressed pulse.

Matching Filter Real Part of Point Target Signall

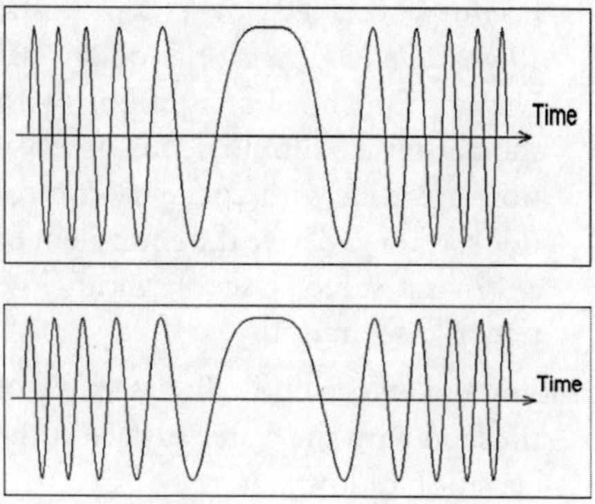

Real Part of Matched Filter

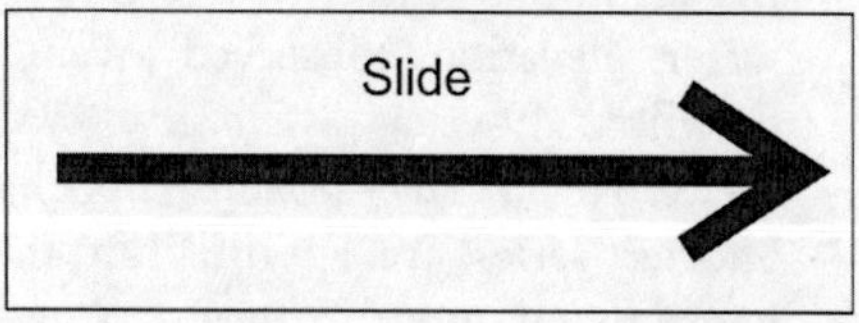

Compressed Target Signal

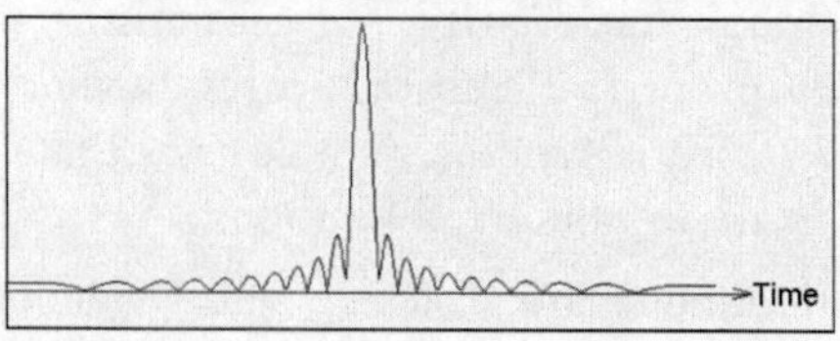

Matching In geocoding, the process of linking a record, such as an address, to a set of reference data. The matched record in the reference data is used to determine the location of the input address.

Mathematical expression A kind of expression that evaluates to a number which is then typically stored in a variable, a field on a table row, or a cell in a raster dataset. Mathematical expressions are generally part of an algebraic equation:Result = Expression (PropertyTax) = ([LandValue] * 0.75) + ([StructureValue] * 0.50).

Mathematical function In ArcGIS Spatial Analyst, a function that applies a mathematical operation to the values of a single input raster. There are four groups of mathematical functions available: logarithmic, arithmetic, trigonometric and powers.

Mathematical model In Survey Analyst for field measurements, a set of relations between measurements and unknown coordinates.

Mathematical operator The symbolic representation of a process or operation performed against one or more operands in an expression, such as '+' (plus or addition) and '>' (greater than). When evaluated, operators return a value as their result. If multiple operators appear in an expression, they are evaluated in order of their operator precedence.

Matrix A rectangular arrangement of data, usually numbers, in rows and columns. In computer science, a two-dimensional array is called a matrix. In GIS, matrices are used to store raster data.

MAUP Acronym for modifiable areal unit problem. A challenge that occurs during the spatial analysis of aggregated data in which the results differ when the same analysis is applied to the same data but different aggregation schemes are used. MAUP takes two forms: the scale effect and the zone effect. The scale effect exhibits different results when the same analysis is applied to the same data but changes the scale of the aggregation units. For example, analysis using data aggregated by county will differ from analysis using data aggregated by census tract. Often this difference in results is valid: each analysis asks a different question because each evaluates the data from a different perspective (different scale). The zone effect is observed when the scale of analysis is fixed, but the shape of the aggregation units is changed. For example, analysis using data aggregated into one-mile grid cells will differ from analysis using one-mile hexagon cells.

Max extent (maximum extent) The maximum bounding rectangle (in x,y coordinates) of an on-screen map. Users cannot zoom out beyond the max extent.

M-coordinate In linear referencing, a measure value that is added to a line feature. M-values are used to measure the distance along a line feature from a vertex (a known location) to an event.

Mean The average for a set of values, computed as the sum of all values divided by the number of values in the set.

Mean centre The location of a single x,y coordinate value that represents the average x-coordinate value and the average y-coordinate value of all features in a study area.

Mean high water Tidal datum that is the arithmetic mean of the high water heights observed over a specific 19-year *Metonic cycle* (National Tidal Datum Epoch). For stations with shorter series, simultaneous observations are made with a primary control tide station to derive the equivalent of a 19-year value. Use of 'mean high tide' is discouraged.

Mean high water line Intersection of the land with the water surface at the elevation of high water. See: shoreline.

Mean low water Tidal datum that is the arithmetic mean of the low water heights observed over a specific 19-year *Metonic cycle* (National Tidal Datum Epoch). For stations with shorter series, simultaneous observations are made with a primary control tide station to derive the equivalent of a 19-year value. Use of 'mean low tide' is discouraged.

Mean low water line Intersection of the land with the water surface at the elevation of low water.

Mean sea level Tidal datum that is the arithmetic mean of the hourly water elevations observed over a specific 19-year *Metonic cycle* (National Tidal Datum Epoch). Shorter series are specified in the name; that is, monthly mean sea level and yearly mean sea level. See: datum.

Mean stationarity In geostatistics, a property of a spatial process in which a spatial random variable has the same mean value at all locations.

Meander line Metes and bounds traverse approximately along the mean high water line of a permanent body of water. By following the sinuosities of the bank or shoreline, the meander line provides data for computing the area of land remaining after the water area has been segregated. A meander line differs from other metes and bounds surveys in that it does not ordinarily determine or fix boundaries.

Meanderable Capable of being depicted by reference to a meander line.

Measure In linear referencing, a value stored along a linear feature that represents a location relative to the beginning of the feature, or some point along it, rather than as an x,y coordinate. Measures are used to map events such as distance, time or addresses along line features.

Measure location fields In linear referencing, either one or two fields in a table that describe the position of an event along a route.

Measure value In linear referencing, a measure value that is added to a line feature. M-values are used to measure the distance along a line feature from a vertex (a known location) to an event.

Measured grid In cartography, any network of parallel and perpendicular lines superimposed on a map and used for reference. These grids are usually referred to by the map projection or coordinate system they represent, such as universal transverse Mercator grid.

Measurement An observed numerical value that is an appraisal of size, extent or amount according to a set criteria.

Measurement error In surveying, the noise that is expected in every measurement. It occurs because the observer makes estimates and uses measuring equipment that is unpredictable in an environment that is also unpredictable.

Measurement residual The difference between a measured quantity and its theoretical true value as determined during each iteration of a least-squares adjustment.

Median The middle value of a set of values when they are ordered by rank. Half the values in a set are higher than the median, and half are lower. When there are two middle values (if the set has an even number of elements) the median is the mean of these two values.

Median centre A location representing the shortest total distance to all other features in a study area.

Median filter A spatial filter, which substitutes the median value of DN from surrounding pixels for that recorded at an individual pixel. It is useful for removing random noise.

Medium-format printer A printing device capable of producing an image on paper or other media sized between 15 and 35 inches (38 and 90 centimetres) wide. Medium-format printers typically use inkjet printing technology to print an image on a roll or sheet-fed media. While most large-format printers are large, free-standing units, most medium-format printers are small enough to fit on a desk.

Members Refers collectively to the properties and methods or functions, of an interface or class.

Memory cache In ArcGlobe, the amount of system memory that the application will use.

Memory leak In computer programming, the loss of computer memory that occurs when an application or component fails to free a section of computer memory when it has finished using it. During a memory leak, the section of memory allocated by one application or component may not be used by any other application.

Mental map A person's perception of a place. A mental map may include the physical characteristics of a place, such as boundaries of a neighbourhood or the attributes of a place, such as a neighbourhood's perceived unsafe areas. A mental map is primarily a psychological construct, although it may also be rendered as an actual map.

Menu A list of available commands or operations displayed on a computer screen from which a user can make a selection.

Menu item An item in a list of commands displayed on a menu.

Mercator Of, relating to, or drawn on the Mercator projection (a conformal map projection with the meridians drawn parallel to each other and the parallels of latitude drawn as straight lines whose distance from each other increases with their distance from the equator).

Mercury US programme of one-man, earth-orbiting spacecraft in 1962 and 1963.

Mereing Establishing a boundary relative to ground features present at the time of a survey.

Merge policy In geodatabases, rules that dictate what happens to the respective attributes of features that are merged together during editing in ArcMap. A merge policy can be set to assign a default value to the new attribute, summarize the values of the merged attributes or create a weighted average from the merged attributes.

Merging Combining features from multiple data sources of the same data type into a single, new dataset.

Meridian (1) A great circle on the earth that passes through the poles, often used synonymously with longitude. Meridians run north–south between the poles. By convention, meridians are labelled with positive numbers that ascend as one moves eastward from the prime meridian, and negative numbers as one moves westward from the prime meridian until the east and west hemispheres meet at the 180-degree line. Meridians can also, however, be labelled with all positive or negative numbers, including positive numbers increasing westward from the prime meridian. (2) Great circle on the surface of the earth passing through the geographical poles and any given point on the earth's surface. All points on a given meridian have the same longitude.

Metabase In Microsoft Internet Information Services (IIS), an XML file where the IIS configuration information is stored.

Metadata (1) A set of descriptive information about the scene data contained in the archive. The information is sufficient for a user, during the process of scene query and selection, to determine at a minimum geographic coverage, date of collection sensor gain mode, time of acquisition, cloud cover, and other quality measurements. (2) Information that describes the content, quality, condition, origin and other characteristics of data or other pieces of information. Metadata for spatial data may describe and document its subject matter; how, when, where and by whom the data was collected; availability and distribution information; its projection, scale, resolution and accuracy; and its reliability with regard to some standard. Metadata consists of properties and documentation. Properties are derived from the data source (for example, the coordinate system and projection of the data), while documentation is entered by a person (for example, keywords used to describe the data).

Metadata element A unit of information within metadata, used to describe a particular characteristic of the data.

Metadata explorer A Web application, developed using the ArcIMS Java Connector, that can be used to view metadata included in an ArcIMS metadata service.

Metadata profile A modification of an existing metadata standard to adapt to data issues, cultural issues or both. A profile is typically a subset of a base standard that tailors the metadata elements in the base standard to better describe the data to the community that uses it. Metadata profiles allow communities to follow a metadata standard, while at the same time enhancing the standard so that it is more appropriate for a particular use or locale.

Metadata server A public ArcIMS virtual server that provides the capability to manage and search a central metadata repository. Data producers can publish their metadata to the repository while data consumers are able to search the repository to locate appropriate data.

Metadata service A service that uses the spatial server metadata capabilities, allowing users to publish and share metadata documents over the Internet or an intranet.

Metes and bounds (1) A surveying method in which the limits of a parcel are identified as relative distances and bearings from landmarks. Metes and bounds surveying often resulted in irregularly shaped areas. (2) Method of describing land by measure of length (metes) of the boundary lines (bounds).

Method In object-oriented programming, an action that an object is capable of performing. Objects that belong to the same class all have the same methods. For example, all Visual Basic form objects can execute a method called 'Show and Hide'.

Metonic cycle Period of 235 lunations or about 19 years devised by Meton, an Athenian astronomer (5th century B.C.) for the purpose of obtaining a period at the end of which the phases of the moon recur in the same order and on the same days as in the preceding cycle.

Metric camera Photogrammetric camera that enables geometrically accurate reconstruction of the optical model of the object scene from its stereo photographs.

Metric system Decimal system of weights and measures based on the metre as a unit length and the kilogram.

Metropolitan statistical area A geographic entity defined by the US Office of Management and Budget for use by federal statistical agencies, including the US Census Bureau. A metropolitan statistical area is based on the concept of a core area with a large population nucleus, plus adjacent communities having a high degree of economic and social integration with that core area. According to the 1990 standards, to qualify as a metropolitan statistical area, the area must include at least one city or urbanized area with 50,000 or more inhabitants and a total metropolitan population of at least 100,000 (75,000 in New England).

Microdensitometer A densitometer that can read densities in minute areas, used particularly for studying spectroscopic and astronomical images.

Micrometer An instrument for measuring minute lengths or angles.

Micron (μ) (1) A distance measure equal to one thousandth of 1 mm. (2) One millionth of a meter, represented by the symbol mu(μ). Microns are used to measure wavelengths in the electromagnetic spectrum.

Micropolitan statistical area A geographic region containing at least one urban area with a population between 10,000 and 50,000, defined by the US Office of Management and Budget for use by federal statistical agencies, including the US Census Bureau. Micropolitan statistical areas include adjacent communities having a high degree of economic and social integration with the core area.

Microwave A very short electromagnetic wave. The portion of the electromagnetic spectrum lying between the far infrared (IR) and the conventional radio frequency portion. While not bounded by definition, it is commonly regarded as extending from 1 mm to 1 m in wavelength (300 GHz to 0.3 GHz frequency). Passive systems operating at these wavelengths sometimes are called microwave systems. Active systems are called radar, although the literal definition of radar requires a distance measuring capability not always included in active systems.

Microwave radiometer A device to measure electromagnetic radiation having wavelengths between 1 m and 1 mm. The most common imaging radars operate at frequencies between 24cm and 0.85 cm.

Mid-infrared (MIR) The range of EM wavelengths from 8 to 14 Fm dominated by emission of thermally generated radiation from materials; also known as thermal infrared.

MIDP Acronym for Mobile Information Device Profile. A set of J2ME APIs for wireless devices.

Mie scattering The scattering of EM energy by particles in the atmosphere with comparable dimensions to the wavelength involved.

Military analyst extension An Arc GIS extension that optimizes the effectiveness of core ArcGIS as a toolset foundation for military planners and intelligence analysts. Military Analyst is COM-compliant and extensible with ArcObjects.

MIL-STD - 2,525B change I T h e military specification for common war fighting symbology; the US military standard that provides guidelines and criteria for the development and display of standard C4I war fighting symbology.

Mimetic symbol A symbol that imitates or closely resembles the thing it represents, such as an icon of a picnic table that represents a picnic area.

Min/max scale The smallest and largest scales at which a layer is visible on a map. Scale ranges are used to prevent detailed layers from displaying when zoomed out and to prevent general layers from displaying when zoomed in.

Minimum bounding rectangle A rectangle, oriented to the x- and y-axes, that bounds a geographic feature or a geographic dataset. It is specified by two coordinate pairs: xmin, ymin and xmax, ymax.

Minimum candidate score In geocoding, the minimum score a potential match record requires to be considered a candidate. This value is adjustable on the Address Locator Properties dialog box.

Minimum ground separation Minimum distance on the ground between two targets at which they can be resolved on an image.

Minimum map unit For a given scale, the size in map units below which a narrow feature can be reasonably represented by a line and an area by a point.

Minimum match score In geocoding, the minimum score a match candidate needs to be considered a match in batch geocoding. This value is adjustable on the Address Locator Properties dialog box.

Minor axis The shorter axis of an ellipse or spheroid.

Minus-blue photographs Black-and-white photographs acquired using a filter that removes blue wavelengths to produce higher spatial resolution.

Minute An angle equal to 1/60 of a degree of latitude or longitude and containing sixty seconds.

Misclosure A discrepancy between existing coordinates and computed coordinates that occurs when the final point of a closed traverse has known coordinates and the final course of a traverse computes different coordinates for the same survey point.

Missions operations centre (MOC) The place where the satellite or spacecraft receives coded instructions and delivers data. The coordination place (usually on earth) for any space mission.

Mixed list In Survey Analyst for field measurements, one of two types of lists in the List page. The mixed list has a mixed set of rows that might have, for instance, survey points, coordinate geometry measurements and a mix of different computations.

Mixed pixel (1) A pixel whose DN represents the average energy reflected or emitted by several types of surface present within the area that it represents on the ground; sometimes called a mixel. (2) In remote sensing, a pixel whose digital number represents the average of several spectral classes within the area that it covers on the ground, each emitted or reflected by a different type of material. Mixed pixels are common along the edges of features.

Mobile toolkit A J2ME-based toolkit for building mobile applications that use ArcWeb Services.

Model An abstraction of reality used to represent objects, processes, or events.

Model generalization The abstraction, reduction and simplification of features and feature classes for deriving a simpler model of reality or decreasing stored data volumes.

Model parameter In ArcGIS, a type of parameter exposed in a geoprocessing model that displays in a model's dialog box and allows for input.

ModelBuilder The interface used to build and edit geoprocessing models in ArcGIS.

Modifiable areal unit problem (MAUP) A challenge that occurs during the spatial analysis of aggregated data in which the results differ when the same analysis is applied to the same data but different aggregation schemes are used. MAUP takes two forms: the scale effect and the zone effect. The scale effect exhibits different results when the same analysis is applied to the same data but changes the scale of the aggregation units. For example, analysis using data aggregated by county will differ from analysis using data aggregated by census tract. Often this difference in results is valid: each analysis asks a different question because each evaluates the data from a different perspective (different scale). The zone effect is observed when the scale of analysis is fixed, but the shape of the aggregation units is changed. For example, analysis using data aggregated into one-mile grid cells will differ from analysis using one-mile hexagon cells.

Modifier In MOLE, text or graphics that display around a symbol or a value that changes the appearance of a symbol. In some military specifications, attributes are referred to as modifiers.

MODIS Acronym for moderate resolution imaging spectroradiometer. A bundle of remote-sensing equipment housed on two NASA (National Aeronautics and Space Administration) satellites, Terra and Aqua, in orbit around the earth. These two MODIS-equipped satellites constantly record multiple images of the globe in various wavelengths and resolutions, imaging the earth's entire surface in less than two days.

Modular op to electric multi spectral scanner (MOMS) An along-track scanner carried on the Space Shuttle that recorded two bands of data.

Modulate To vary the frequency, phase or amplitude of electromagnetic waves.

Modulation The variation of a property of an electromagnetic wave or signal, such as its amplitude, frequency or phase.

Modulation transfer function (MTF) A method of describing spatial resolution.

MOLE Acronym for Military Overlay Editor. An ESRI software application that allows users to create, display and edit military symbology in their maps.

Monitoring Keeping track of how things change over time. For example, with remote sensing, using several images taken over time, you can monitor the result of logging in a forest or how much of an oil slick in the ocean has been cleaned up or how well crops are growing or how much a glacier has melted or how far a plume of sediment travels in a lake, etc.

Monochromatic (1) Related to a single wavelength or a very narrow band of wavelengths. (2) Said of a beam of *photons* when the *bandwidth* of the photons is so narrow that all the photons essentially have the same *wavelength*. This is one of the defining characteristics of a laser beam.

Monoscopic Pertaining to the observation of a single photograph or other view.

Monte carlo method An algorithm for computing solutions to problems that contain a large number of variables by performing iterations with different sets of random numbers until the best solution is found. The Monte Carlo method is usually applied to problems too complex for analysis by anything but a computer.

Monument An object, such as a metal disk, permanently mounted in the landscape to denote a survey station.

Monument (surveying) Permanent physical structure marking the location of a survey point. Common types of monuments are inscribed metal tablets set in concrete posts; and metal rods driven in the ground.

Morphology The structure of a surface.

MOS–1 Marine Observation Satellite, launched by Japan in 1987.

Mosaic (1) A raster dataset composed of two or more merged raster datasets – for example, one image created by merging several individual images or photographs of adjacent areas. (2) A technique whereby multiple satellite images are digitally joined, while correcting for systematic changes in radiometry and geometry thus creating a 'seamless' image product. A large sequential composite of individual photographs showing a continuous overview of a project site from beginning to end.

Mosaic, aerial Assembly of aerial photographs whose edges usually have been torn or cut selectively and matched to the imagery on adjoining photographs to form a continuous representation of a portion of the earth's surface.

Mosaicking The assembling of photographs or other images whose edges are cut and matched to form a continuous photographic representation of a portion of the earth's surface.

Motion compensation Adjustment of a sensing system and/or the recorded data to remove effects of platform motion, including rotation and translation and variations in along-track velocity. Motion compensation is essential for aircraft SARs, but usually is not needed for spacecraft SARs.

Mouse mode A way of using a digitizing tablet in which the digitizer puck behaves like a mouse; the puck is used to point to interface elements rather than to trace shapes on the surface of the digitizing tablet.

MPS Acronym for Map Production System. The cartographic component of the Production Line Tool Set (PLTS). MPS is designed to facilitate large-volume cartographic production by providing the ability to create dynamic map layouts, batch symbology and enhanced cartographic editing tools.

MSA A geographic entity defined by the US Office of Management and Budget for use by federal statistical agencies, including the US Census Bureau. A metropolitan statistical area is based on the concept of a core area with a large population nucleus, plus adjacent communities having a high degree of economic and social integration with that core area. According to the 1990 standards, to qualify as a metropolitan statistical area, the area must include at least one city or urbanized area with 50,000 or more inhabitants and a total metropolitan population of at least 100,000 (75,000 in New England).

MSDN Acronym for Microsoft Developer Network. A set of services designed to help developers write applications using Microsoft products and technologies.

MSS (Multispectral scanner) A device carried on satellites and aircraft that records energy from multiple portions of the electromagnetic spectrum.

MTA Acronym for multiple threaded apartment. An apartment that can have multiple threads running. A process can only have one MTA.

Multiband camera System that simultaneously acquires photographs of the same scene at different wavelengths.

Multichannel receiver A GPS receiver that can simultaneously track more than one satellite signal.

Multidimensional data Data that is comprised of multiple dimensions, such as space and time. For example, a temperature dataset could have dimensions of latitude, longitude, altitude and time.

Multifrequency radar Broadband systems that transmit pulses in a range of frequencies and wavelengths.

Multimodal network A network in which two or more types of transportation modes (such as walking, riding a train or driving a car) are modelled. In a network dataset, multiple connectivity groups are required to create a multimodal network.

Multipart feature In ArcGIS, a digital representation of a place or thing that has more than one part but is defined as one feature because it references one set of attributes. In a layer of states, for example, the state of Hawaii could be considered a multipart feature because its separate geometric parts are classified as a single state. A multipart feature can be a point, line or polygon.

Multipatch In ArcGIS, a type of geometry comprised of planar three-dimensional rings and triangles, used in combination to model objects that occupy discrete area or volume in three-dimensional space. Multipatches may represent geometric objects like spheres and cubes, or real-world objects like buildings and trees.

Multipatch feature In ArcGIS, a real-world geographic feature modelled using multipatch geometry.

Multipath interference caused by reflected GPS signals arriving at the receiver, typically as a result of nearby structures or other reflective surfaces.

Multipath error (1) Errors caused by the interference of a signal that has reached the receiver antenna by two or more different paths. This is usually caused by one path being bounced or reflected. (2) Errors caused when a satellite signal reaches the receiver from two or more paths, one directly from the satellite and the others reflected from nearby buildings or other surfaces. Signals from satellites low on the horizon will produce more error.

Multiple regression Regression in which the dependent variable is measured against two or more independent variables.

Multiple setup page In the Survey Analyst for field measurements

Survey Explorer, one of two types of setup pages. The Multiple Setup page is used for computations that process more than one instrument setup.

Multiple threaded apartment An apartment that can have multiple threads running. A process can only have one MTA.

Multiplex Stereoplotter of the double-projection type characterized by its use of reduced-scale diapositives and stationary lamphouses with condensing lenses.

Multiplexing channel A channel of a GPS receiver that can be sequenced through a number of satellite signals.

Multiplexing channel receiver A receiver that tracks several satellite signals using a single channel.

Multipoint In ArcGIS software, a geometric element defined by an unordered set of x, y coordinate pairs.

Multipoint feature In ArcGIS software, a digital map feature that represents a place or thing that has neither area nor length at a given scale, and that is treated as a single object with multiple locations. For example, the entrances and exits to a prairie dog den might be represented as a multipoint feature. A multipoint feature is associated with a single record in an attribute table.

Multipoint feature class A feature class that can store many points per shape or row, thereby saving storage space and improving read-write performance.

Multipolarization radar A radar capable of simultaneously and coherently acquiring several independent complex polarization measurements for every pixel in the image.

Multispectral (1) Related to two or more frequencies or wavelengths in the electromagnetic spectrum. (2) Sensing in usually 4 distinct wavelength bands (equivalent to colours, not all of which are visible to the human eye). Because the data handling capacity of the sensor is spread over these different wavelengths, this usually translates to lower resolution than panchromatic.

Multispectral classification Identification of terrain categories by digital processing of data acquired by multispectral scanners.

Multispectral image (1) A remote sensing image created using data collected from more than one band. (2) An image created from several narrow spectral bands.

Multispectral scanner (MSS) (1) A line-scanning instrument flown on Landsat satellites that continually scans the earth in a 185 km (100 nautical miles) swath. On Landsats 1, 2, 4, and 5, the MSS had four spectral bands in the visible and near infrared with an IFOV of 80 metres. Landsat 3 had a fifth band in the thermal infrared with an IFOV of 240 metres. The MSS is a non-photographic imaging system that utilizes an oscillating mirror and fiber optic sensor array. The mirror sweeps from side to side, transmitting incoming energy to a detector array that sequentially outputs brightness values (signal strengths) for successive pixels, one swath at a time. The forward motion of the sensor platform carries the instrument to a position along its path where it can image an adjacent swath. (2) A device carried on satellites and aircraft that records energy from multiple portions of the electromagnetic spectrum.

Multiuser geodatabase A geodatabase managed in an RDBMS server by ArcSDE. Multiuser geodatabases can be very large and support multiple concurrent editors. They are supported on a variety of commercial RDBMS, including IBM DB2, IBM Informix, Oracle, Microsoft SQL Server and PostgreSQL.

Multivariate analysis Any statistical method for evaluating the relationship between two or more variables.

Multiversioned view In ArcGIS and ArcSDE, a view that uses stored procedures and triggers to access a specified version of data in a single business table in the geodatabase. A multiversioned view includes all the records in the business table that have been selected and merged with records from the delta tables. The schema of a multiversioned view is identical to that of the business table on which it is based.

M-value In linear referencing, a measure value that is added to a line feature. M-values are used to measure the distance along a line feature from a vertex (a known location) to an event.

MXD In ArcMap, the file that contains one map, its layout and its associated layers, tables, charts and reports. Map documents can be printed or embedded in other documents. Map document files have a .mxd extension.

My ArcWeb locations A feature in the ArcWeb Toolbar for ArcGIS that allows users to save and quickly access selected locations.

My output data folder A computer file that contains all the work done in Business Analyst: study areas; analyses; and customer, store and extracted data.

My services The services that a user creates in Builder to use with ArcWeb Services.

My Toolboxes folder In ArcCatalog, a folder that contains non-system toolboxes created in the ArcToolbox window or directly in the My Toolboxes folder in the ArcCatalog Tree. The My Toolboxes folder points to a location on disk that can be changed in the Geoprocessing tab of the Options dialog box, accessible from the Tools menu.

N

NAD 1927 Acronym for North American Datum of 1927. The primary local horizontal geodetic datum and geographic coordinate system used to map the United States during the middle part of the twentieth century. NAD 1927 is referenced to the Clarke spheroid of 1866 and an origin point at Meades Ranch, Kansas. Features on USGS topographic maps, including the corners of 7.5-minute quadrangle maps, are referenced to NAD27. It is gradually being replaced by the North American Datum of 1983.

North American Datum of 1927, which is based on the Clark 1866 spheroid and has its centre at Meades Ranch, Kansas.

The primary local horizontal geodetic datum and geographic coordinate system used to map the United States during the middle part of the twentieth century. NAD 1927 is referenced to the Clarke spheroid of 1866 and an origin point at Meades Ranch, Kansas. Features on USGS topographic maps, including the corners of 7.5-minute quadrangle maps, are referenced to NAD27. It is gradually being replaced by the North American Datum of 1983.

NAD 1983 Acronym for North American Datum of 1983. A geocentric datum and graphic coordinate system based on the Geodetic Reference System 1980 ellipsoid (GRS80). Mainly used in North America, its measurements are obtained from both terrestrial and satellite data.

North American Datum of 1983, which is based on the WGS84, or the GRS80, spheroid and is measured from the centre of the spheroid.

A geocentric datum and graphic coordinate system based on the Geodetic Reference System 1980 ellipsoid (GRS80). Mainly used in North America, its measurements are obtained from both terrestrial and satellite data.

Nadir (1) In aerial photography, the point on the ground vertically beneath the perspective centre of the camera lens. (2) Point on earth directly beneath a remote sensing system/satellite and the centre of the earth or opposite of zenith.

NAICS Acronym for North American Industry Classification System. A system for classifying individual business locations by their types of economic activity. The statistics agencies of Canada, Mexico and the United States collaborated on NAICS to standardize the industry statistics produced by the three countries. NAICS is used as an identification system by all federal statistical agencies, as well as many state and local agencies, trade associations, private businesses and other organizations. NAICS replaced Standard Industrial Classification (SIC) codes in 1997.

A system for classifying individual business locations by their types of economic activity. The statistics agencies of Canada, Mexico and the United States collaborated on NAICS to standardize the industry statistics produced by the three countries. NAICS is used as an identification system by all federal statistical agencies, as well as many state and local agencies, trade associations, private businesses and other organizations. NAICS replaced Standard Industrial Classification (SIC) codes in 1997.

NaN Acronym for not a number.

Nanometres A distance measure equal to one millionth of 1 mm.

N-ary association A UML term that describes how two classes have a relationship with one or more other classes. In an N-ary association, instances of the classes in question usually exist together but can exist on their own. For example, a three-way valve usually has one size of pipe coming in and two other sized pipes going out. Before the pipes and valve are connected into the water system they are separate parts; therefore, they have an N-ary association.

NASA National Aeronautics and Space Administration (NASA) was established in 1958 through the National Aeronautics and Space Act as an outgrowth of the National Advisory Committee for Aeronautics.

National geodetic vertical datum of 1929 (1) Reference surface established by the US Coast and Geodetic Survey in 1929 as the datum to which relief features and elevation data are referenced in the conterminous United States; formerly called 'mean sea level 1929'. (2) The datum established in 1929 by the US Coast and Geodetic Survey as the surface against which elevation data in the United States is referenced.

National map accuracy standards Specifications promulgated by the US Office of Management and Budget to govern accuracy of topographic and other maps produced by Federal Agencies.

National spatial data infrastructure A federally mandated framework of spatial data that refers to US locations, as well as the means of distributing and using that data effectively. Developed and coordinated by the FGDC, the NSDI encompasses policies, standards, procedures, technology and human resources for organizations to cooperatively produce and share geographic data. The NSDI is developed by the federal governments; state, local and tribal governments; the academic community; and the private sector.

Natural breaks classification A method of manual data classification that seeks to partition data into classes based on natural groups in the data distribution. Natural breaks occur in the histogram at the low points of valleys. Breaks are assigned in the order of the size of the valleys, with the largest valley being assigned the first natural break.

Natural neighbours An interpolation method for multivariate data in a Delaunay triangulation. The value for an interpolation point is estimated using weighted values of the closest surrounding points in the triangulation. These points, the natural neighbours, are the ones the interpolation point would connect to if inserted into the triangulation.

Navigable waters Water usable, with or without improvements, as routes for commerce in the customary means of travel on water.

Navigate To interactively change the observer's or target's position using a tool designed for this purpose, such as the navigate or fly tool. There are three contexts in which a user can navigate: in a scene of ArcScene, in a preview of ArcCatalog, and in a globe of ArcGlobe.

Navigation The combined mental and physical activities involved in travelling to a destination, often a distant or unfamiliar one. Navigation comprises wayfinding and locomotion.

NAVSTAR The name given to GPS satellites. NAVSTAR is an acronym for Navigation Satellite Timing and Ranging.

Near infrared (NIR) In EMR, the shorter wavelength range of the infrared region of the EM spectrum, from 0.7 to 2.5 Fm. It is often divided into very-near infrared (VNIR) covering the range accessible to photographic emulsions (0.7 to 1.0m), and the short-wavelength infrared (SWIR) covering the remainder of the NOR atmospheric window from 1.0 to 2.5m.

Near range Refers to the portion of a radar image closest to the aircraft or satellite flight path.

Nearest neighbour resampling A technique for resampling raster data in which the value of each cell in an output raster is calculated using the value of the nearest cell in an input raster. Nearest neighbour assignment does not change any of the values of cells from the input layer; for this reason it is often used to resample categorical or integer data (for example, land use, soil or forest type), or radiometric values, such as those from remotely sensed images.

Neatline (1) Line separating the body of a map from the map margin. On a standard *quadrangle* map, the neatlines are the meridians and parallels delimiting the quadrangle. (2) The border delineating and defining the extent of geographic data on a map. It demarcates map units so that, depending on the map projection, the neatline does not always have 90-degree corners. In a properly made map, it is the most accurate element of the data; other map features may be moved slightly or exaggerated for generalization or readability, but the neatline is never adjusted.

Negative photograph Photograph on film or paper in which the relationship between bright and dark tones is the reverse of that of the features on the terrain.

Neighbourhood On a raster, an analysis boundary or processing window within which cell values affect calculations and, outside which, they do not. Filters are used mainly in cell-based analysis where the value of a centre cell is changed to the mean, the sum or some other function of all cell values inside the filter. A filter moves systematically across a raster until each cell has been processed. Filters can be of various shapes and sizes, but are most commonly three-cell by three-cell squares.

Neighbourhood functions Methods of defining new values for locations using the values of other locations within a given distance or direction.

Neighbourhood statistics The computation of an output raster where the output value at each cell location is a function of the value at that cell location and the values of the cells within a specified neighbourhood around the cell.

NetBeans IDE An open-source development environment that supports all Java application types.

Network An interconnected set of points and lines that represent possible routes from one location to another. For geometric networks, this consists of edge features, junction features and the connectivity between them. For network datasets, this consists of edge, junction, and turn elements and the connectivity between them. For example, an interconnected set of lines representing a city streets layer is a network.

Network analysis Any method of solving network problems such as traversability, rate of flow or capacity, using network connectivity.

Network analysis class In ArcGIS Network Analyst, a feature class or table containing network analysis objects that is stored within a network analysis layer. Network Analyst solvers read input from, and write output to, network analysis classes.

Network analysis layer A composite layer that contains the properties and network analysis classes used in the analysis of a network problem and the results of the analysis.

Network analysis object In ArcGIS Network Analyst, a feature or row in a network analysis class. Network analysis objects are used as input and written as output during network analysis. A network location is a specific type of network analysis object that has a defined position on a network dataset.

Network ancillary role An auxiliary or additional function performed by a junction feature within a geometric network. Junction features can act as sources or sinks for calculating flow direction. If a junction is a source or a sink, it is said to have an ancillary role in the network.

Network attribute A type of attribute associated with a network element in a network dataset. Network attributes are used to help control flow through a network (similar to a weight in a geometric network). All network elements in a network dataset have the same set of attributes. There are four types of network attributes: cost, descriptor, hierarchy and restriction.

Network dataset A collection of topologically connected network elements (edges, junctions and turns) that are derived from network sources, typically used to represent a linear network, such as a road or subway system. Each network element is associated with a collection of network attributes. Network datasets are typically used to model undirected flow systems.

Network element A component in a network dataset: an edge, junction or turn. All elements in a network dataset share the same set of network attributes. Network elements are used to model topological relationships in undirected flow networks such as traffic flow systems. Network elements are generated from point, line and turn features. When the network dataset is built, point features become junctions, line features become edges and turn features become turn elements.

Network feature A component in a geometric network: an edge or a junction. Features in a geometric network are used to model topological relationships, typically in directed flow networks such as hydrologic or utility systems. Network features are generated from points and lines when the geometric network is built: point features become junctions and line features become edges.

Network layer A layer that references a network dataset. In a geodatabase, a network dataset is a collection of network elements (edges, junctions and turns) that are derived from network sources.

Network location A geographic position in a network system.

Network port A number that is used to specify direct communication over a network to an Internet application.

Network solver A function that performs network analysis based on a set of network data.

Network source Feature classes in a geodatabase that are used to generate and define a network dataset.

Network trace A function that performs network analysis on a geometric network. Specific kinds of network tracing include finding features that are connected, finding common ancestors, finding loops, tracing upstream and tracing downstream.

Neural network A computer architecture modelled after the human brain and designed to solve problems that human brains solve well, such as recognizing patterns and making predictions from past performance. Neural networks are composed of interconnected computer processors that calculate a number of weighted inputs to generate an output. For example, an output might be the approval or rejection of a credit application. This output would be based on several inputs, including the applicant's income, current debt and credit history. Some of these inputs would count more than others; cumulatively, they would be compared to a threshold value that separates approvals from rejections. Neural networks 'learn' to generate better outputs by adjusting the weights and thresholds applied to their inputs.

Neutral format A public format such as SDTS that can be used for data exchange.

New store analysis A process that finds a potential location for a new store by calculating the centroid of a group of customers.

NGVD 1929/29 The datum established in 1929 by the US Coast and Geodetic Survey as the surface against which elevation data in the United States is referenced.

NHAP National High Altitude Photography programme of the US Geological Survey.

NMEA Acronym for National Marine and Electronics Association. A nonprofit association composed of manufacturers, distributors, dealers, educational institutions and others interested in peripheral marine electronics occupations. The NMEA has created a standard that defines an electrical interface and data protocol for communications between marine instrumentation that has been adopted as an industry standard by the GPS industry.

NOAA National Oceanic and Atmospheric Administration.

NoData In raster data, the absence of a recorded value. NoData does not equate to a zero value. While the measure of a particular attribute in a cell may be zero, a NoData value indicates that no measurements have been taken for that cell at all.

Node (1) A connecting point in a geometric network, such as an intersection or interchange of a road network, confluence of streams in a hydrologic network, or switch in a power grid. (2) In a geodatabase, the point representing the beginning or ending point of an edge, topologically linked to all the edges that meet there. (3) The beginning or end point of line.

Nodesnap A tolerance used in ARC/GIS for snapping nodes.

Noise (1) In remote sensing, any disturbance in a frequency band. (2) Random or repetitive events that obscure or interfere with the desired information.

Noise equivalent sigma nought A measure of the sensitivity of a given SAR. It describes the strength of the (additive) system noise in terms of the equivalent (average) power in the image domain that would result from an idealized distributed scatterer of the stated reflectivity. Smaller noise equivalent sigma nought values are better. Within physical limitations, smaller may be achieved by increasing the power of the radar transmitter or by decreasing the noise figure of the electronics.

Noise figure Factor that describes the noise level in a radar receiver relative to the that in a theoretically perfect receiver. The noise figure, which is always larger than one, is typically two or more, and is usually expressed in decibels.

Nominal data Data divided into classes within which all elements are assumed to be equal to each other and in which no class comes before another in sequence or importance; for example, a group of polygons coloured to represent different soil types.

Non-directional filter Mathematical filter that treats all orientations of linear features equally.

Non-selective scattering The scattering of EM energy by particles in the atmosphere which are much larger than the wavelengths of the energy and which causes all wavelengths to be scattered equally.

Nonsimple polygon A polygon that violates topological integrity by crossing its own boundary (usually by making a small loop).

Nonspatial data Data without inherently spatial qualities, such as attributes.

Non-spectral hue A hue which is not present in the spectrum of colours produced by the analysis of white light by a prism of diffraction grating. Examples are brown, magenta and pastel shades.

Non-systematic distortion Geometric irregularities on images that are not constant and cannot be predicted from the characteristics of the imaging system.

Normal colour film Film in which the colours are essentially true representations of the colours of the terrain.

Normal distribution A theoretical frequency distribution of a dataset in which the distribution of values can be graphically represented as a symmetrical bell curve. Normal distributions are typically characterized by a clustering of values near the mean, with few values departing radically from the mean. There are as many values on the left side of the curve as on the right, so the mean and median values for the distribution are the same. Sixty-eight per cent of the values are plus or minus one standard deviation from the mean; 95 per cent of the values are plus or minus two standard deviations; and 99 per cent of the values are plus or minus three standard deviations.

Normal form A set of guidelines for designing table and data structures in a relational database. When followed, normal form guidelines prevent data redundancy, increase database efficiency and reduce consistency errors. A database is said to be in first normal form (1NF), second normal form (2NF), third normal form (3NF) and so on, depending on the level of normal form guidelines followed in its design. In practice, 3NF is commonly used but higher levels are rarely used.

Normal probability distribution A theoretical frequency distribution of a dataset in which the distribution of values can be graphically represented as a symmetrical bell curve. Normal distributions are typically characterized by a clustering of values near the mean, with few values departing radically from the mean. There are as many values on the left side of the curve as on the right, so the mean and median values for the distribution are the same. Sixty-eight per cent of the values are plus or minus one standard deviation from the mean; 95 per cent of the values are plus or minus two standard deviations; and 99 per cent of the values are plus or minus three standard deviations.

Normal template The template that is automatically loaded in ArcMap and contains all the standard toolbar and command default settings. User interface customization that is saved in the Normal template is loaded each time ArcMap is launched.

Normalization The process of organizing, analysing, and cleaning data to increase efficiency for data use and sharing. Normalization usually includes data structuring and refinement, redundancy and error elimination, and standardization.

North arrow A map symbol that shows the direction of north on the map, thereby showing how the map is oriented.

Northing The distance north of the origin that a point in a Cartesian coordinate system lies, measured in that system's units.

NOTAM Acronym for Notice to Airmen. An advisory bulletin containing information about the National Airspace System, typically time-sensitive information between publishing cycles or corrections to published documents and charts.

An advisory bulletin containing information about the National Airspace System, typically time-sensitive information between publishing cycles, or corrections to published documents and charts.

Notice to mariners A periodical update to existing nautical charts, issued by maritime authorities.

Acronym for Notice to Mariners. A periodical update to existing nautical charts, issued by maritime authorities.

NSDI Acronym for National Spatial Data Infrastructure. A federally mandated framework of spatial data that refers to US locations, as well as the means of distributing and using that data effectively. Developed and coordinated by the FGDC, the NSDI encompasses policies, standards, procedures, technology, and human resources for organizations to cooperatively produce and share geographic data. The NSDI is developed by the federal governments; state, local, and tribal governments; the academic community; and the private sector.

NSDI clearinghouse network A community of digital spatial data providers that maintain NSDI Clearinghouse Nodes as part of the US National Spatial Data Infrastructure.

NSDI clearinghouse node A n Internet server that hosts a collection of metadata and data maintained and stored on a computer server by a data provider. An NSDI Clearinghouse Node provides information about geographic data within the data provider's areas of responsibility. Nodes must host FGDC-compliant metadata and data and use a common access protocol.

NSSDC National Space Science Data Centre.

Nugget A parameter of a covariance or semivariogram model that represents independent error, measurement error or micro-scale variation at spatial scales that are too fine to detect. The nugget effect is seen as a discontinuity at the origin of either the covariance or semivariogram model.

Null constraint A DBMS-defined restriction specifying that a column cannot contain a null value.

Null hypothesis A statement that essentially outlines an expected outcome when there is no pattern, no relationship and/or no systematic cause or process at work; any observed differences are the result of random chance alone. The null hypothesis for a spatial pattern is typically that the features are randomly distributed across the study area. Significance tests help determine whether the null hypothesis should be accepted or rejected.

Null value The absence of a recorded value for a field. A null value differs from a value of zero in that zero may represent the measure of an attribute, while a null value indicates that no measurement has been taken.

O

Object (1) An entity such as a land parcel that has a set properties and can perform operations upon requests. (2) In GIS, a digital representation of a spatial or nonspatial entity. Objects usually belong to a class of objects with common attribute values and behaviours.

Object class In a geodatabase, a collection of nonspatial data of the same type or class. While spatial objects (features) are stored in feature classes in a geodatabase, nonspatial objects are stored in object classes.

Object definition language Similar to Interface Definition Language, but used to define the objects contained in an object library.ODL

Acronym for Object Definition Language. Similar to Interface Definition Language, but used to define the objects contained in an object library.

Object library A binary file that stores information about a logical collection of COM objects and their properties and methods in a form that is accessible to other applications at run time. Using a type library, an application or browser can determine which interfaces an object supports and invoke an object's interface methods.

Object model diagram A graphical representation of the types in a library and their relationships.

Object pooling The process of precreating a collection of instances of classes, such that the instances can be shared between multiple application sessions at the request level. Pooling objects allows the separation of potentially costly initialization and acquisition of resources from the actual work the object does. Pooled objects are used in a stateless manner.

Object view A philosophical view of geographic space in which space is seen as empty except where occupied by objects. In this view, every spatial location is either something (an object) or nothing.

ObjectID In ArcGIS, a system-managed value that uniquely identifies a record or feature.

Object-oriented data model A data model that uses objects to organize spatial data. An object is an entity such as a land parcel that has a set of properties and can perform operations upon request.

Object-oriented database A data management structure that stores data as objects (instances of a class) instead of as rows and tables as in a relational database.

Object-oriented programming A programming model in which related tasks, properties and data structures are enclosed inside an object, and work is done when objects make requests and receive results from other objects. For example, a billing program may contain an object that maintains customer records. That object may pass information to another object that handles mailing statements, and another object that handles customer payments may ask it to update a customer record when a payment is received.

Oblate ellipsoid/spheroid An ellipsoid created by rotating an ellipse around its minor axis. The shape of the earth approximates an oblate ellipsoid with a flattening ratio of 1 to 298.257.

Oblique aspect A planar or cylindrical projection whose point of tangency is neither on the equator nor at a pole.

Oblique photograph (1) An aerial photograph taken with the axis of the camera held at an angle between the horizontal plane of the ground and the vertical plane perpendicular to the ground. A low oblique image shows only the surface of the earth; a high oblique image includes the horizon. (2) Photograph acquired with the camera intentionally directed at some angle between horizontal and vertical orientations.

Oblique projection A planar or cylindrical projection whose point of tangency is neither on the equator nor at a pole.

Observer In ArcScene and Arc Globe, the position of the camera in a scene or globe.

Observer offset In ArcScene and ArcGlobe, the height of the observer point above a surface used in analysis when calculating lines of sight and viewsheds.

Oceanic survey Survey or examination of condition in the ocean or any part of it, with reference to animal or plant life, chemical elements present, temperature gradients, etc. See: hydrographic survey

OCX Also known as an ActiveX control, an OLE custom control is contained in a file with the extension .ocx. The ArcGIS controls are ActiveX controls.

OD cost matrix In ArcGIS Network Analyst, a type of network analysis that computes a table containing the total impedance from each origin to each destination. Additionally, it ranks the destinations that each origin connects to in ascending order of the time it takes to travel from that origin to each destination.

OEM Original equipment manufacturer.

Off-nadir Any point not directly beneath a scanner's detectors but rather off at an angle.

Offset In cartography, the displacement or movement of features so that they do not overlap when displayed at a given scale. For example, a road can be offset from a river if the symbols are wide enough that they overlap.

Offset distance The distance a label should be from the feature it labels. A label offset and a maximum label offset can be set for point features. Maximum label offsets are expressed as a percentage of the label offset. For line features, a label offset can be set from the line (similar to the label offset for point features) and along the line (which controls the position of the label relative to the ends of the line). Label offsets are not available for all label position options.

Offshore Comparatively flat zone of variable width that extends from the outer margin of the rather steeply sloping shoreface to the edge of the continental shelf.

OGIS Acronym for Open Geodata Interoperability Specification. A specification, developed by the Open Geospatial Consortium Inc., to support interoperability of GIS in a heterogeneous computing environment.

OID In ArcGIS, a system-managed value that uniquely identifies a record or feature.

OLB Acronym for object library. A binary file that stores information about a logical collection of COM objects and their properties and methods in a form that is accessible to other applications at run time. Using a type library, an application or browser can determine which interfaces an object supports and invoke an object's interface methods.

OLE Acronym for Object Linking and Embedding. A protocol from Microsoft that allows the development of reusable objects that can be exchanged by multiple applications. Applications using OLE can create compound documents that link to data in other applications. The data can be edited from the document without switching between applications.

OLE custom control Also known as an ActiveX control, an OLE custom control is contained in a file with the extension .ocx. The ArcGIS controls are ActiveX controls.

OLE DB provider Object Linking and Embedding database provider. A tool conforming to the OLE standard for sharing data between applications. Each OLE DB provider communicates with and retrieves data from a different database, but a user can work with the data retrieved by any OLE DB provider in a similar way.

OLE View A utility, available as part of Microsoft Visual Studio, that can be used to view type information stored in a type library or object library or inside a DLL.

OMS Orbital manoeuvring system.

On screen digitizing Manual digitizing on the computer monitor by using a data source such as a DOQ as the background.

On the fly Assembled, created, presented or calculated dynamically during a transaction such as a Web page search or data display query.

On-demand cache In ArcGlobe, a temporary layer cache that is placed on disk and built as areas of the layer are viewed.

One-to-many relationship (1) An association between two linked or joined tables in which one record in the first table corresponds to many records in the second table. (2) An association between two linked or joined tables in which one record in the first table corresponds to only one record in the second table.

One-way replication A type of geodatabase replication that allows data changes to be sent multiple times from the parent replica to the child replica. In a one-way replication, the data in the parent replica is editable, and the data in the child replica is read-only. During synchronization, edits are applied directly to the child replica so that no conflicts occur. ArcSDE geodatabases are used to create one-way replicas.

On-screen digitizing Manual digitization by tracing a mouse over features displayed on a computer monitor, used as a method of vectorizing raster data.

Ontology In computer science, a data model that represents a domain by detailing the entities that comprise it and the semantic relationships between them. Ontologies generally include individuals, classes, attributes and relations.

Open geodata interoperability specification A specification, developed by the Open Geospatial Consortium Inc., to support interoperability of GIS in a heterogeneous computing environment.

Open geospatial consortium Acronym for Open Geospatial Consortium. An international consortium of companies, government agencies and universities participating in a consensus process to develop publicly available geospatial and location-based services. Interfaces and protocols defined by OpenGIS specifications support interoperability and seek to integrate geospatial technologies with wireless and location-based services.

Open GIS consortium Acronym for Open Geospatial Consortium. An international consortium of companies, government agencies and universities participating in a consensus process to develop publicly available geospatial and location-based services. Interfaces and protocols defined by Open GIS specifications support interoperability and seek to integrate geospatial technologies with wireless and location-based services.

Open LS Acronym for Open GIS Location Services. A protocol, designed to work across the many different wireless networks and devices, that allows seamless access to multiple content repositories and service frameworks.

Open traverse In surveying, a traverse that does not close back upon itself or on another point of known position. As such, it does not provide a means of checking for errors.

Operand A data value or the symbolic representation of a data value in an expression. Operands may be numbers, character strings, functions, variables, parenthetical expressions in the body of a larger expression and so on. Symbolic representations of operands, such as variables and functions, are evaluated before they are operated upon by the operators in the expression. For example, in the expression '1 + 2', the operands are 1 and 2, and are operated upon by the + (plus) operator, which adds the operands together and returns the value 3.

Operation codes In surveying, an alphanumeric or numeric value included in an instrument vendor's data collector file format. Operation codes are used to describe such elements as new instrument setups and numeric values for horizontal angles, zenith angles, slope distance measurements, height of instrument, and height of target. These basic operation codes are supported in Survey Analyst for field measurements for the following formats: Geoserial Interface (GSI), TDS raw, TDS coordinate, Geodimeter, and Sokkia SDR.

Operator The symbolic representation of a process or operation performed against one or more operands in an expression, such as '+' (plus, or addition) and '>' (greater than). When evaluated, operators return a value as their result. If multiple operators appear in an expression, they are evaluated in order of their operator precedence.

Operator precedence The order in which operators are evaluated in an expression; operators with a higher precedence are evaluated before those with a lower precedence. If all operators in an expression have the same precedence, they are evaluated in the order in which they appear, from left to right. Parentheses may be used to override operator precedence; portions of an expression within parenthesis are evaluated first, and parenthetical expressions may be nested.

Optical centre The point on a rectangular map or image to which the eye is drawn. The visual centre lies slightly (about 5% of the total height) above the geometric centre of the page.

Optical correlator A device that enhances weak signals in noise by performing an optical operation approximating the computation of a correlation function. A radar optical correlator uses the original synthetic aperture radar signal film recording of doppler phase histories to make the radar image by methods that are similar to those used in optical Fourier transformation.

Optimization The process of fine-tuning data, software or processes to increase efficiency, improve performance and produce the best possible results.

Oracle A database company that produces a relational database management system (also called Oracle), which allows data and other objects to be stored in tables. Oracle provides client/server access to data and uses indexes, sequences and other database objects to facilitate rapid data creation, editing and access. ESRI uses Oracle's RDBMS to store vector and raster data for use by ArcSDE.

Orbit Path of a satellite around a body, such as the earth, under the influence of gravity.

Orbital period The time it takes a satellite to complete one revolution (orbit) around the earth. The orbital period of Landsat 7 is about 1.5 hours.

Order In ArcGIS Network Analyst, a network location used to determine routes in vehicle routeing problem (VRP) analysis. Users can specify multiple orders. Orders may represent any combination of service locations or pickup/delivery locations to be visited along a route.

Order pair In ArcGIS Network Analyst, an object used in vehicle routeing problem (VRP) analysis. An order pair defines a relationship between two orders. The first order in the order pair represents a pickup which must then be delivered to the second order in the order pair on the same route.

Ordinal data Data classified by comparative value; for example, a group of polygons coloured lighter to darker to represent less to more densely populated areas.

Ordinary kriging A kriging method in which the weights of the values sum to unity. It uses an average of a subset of neighbouring points to produce a particular interpolation point.

Ordinate In a rectangular coordinate system, the distance of the y-coordinate along a vertical axis from the horizontal or x-axis. For example, a point with the coordinates (7,3) has an ordinate of 3.

Ordnance survey The national mapping and cartographic agency of the United Kingdom. Now a civilian organization, the Ordnance Survey is one of the world's largest producers of maps and was the first national mapping organization in the world to complete a large-scale programme of digital mapping.

Orientation (1) An object's position or relationship in direction with reference to points of the compass. (2) Establishing correct relationship in direction with reference to points of the compass; the state of being in correct relationship in direction with reference to the points of the compass.

Orientation of vertical labels An option that controls whether a label can be placed in an upside-down or sideways position. A small angle prevents labels from switching their reading order in the labelling of nearly vertical lines.

Origin A fixed reference point in a coordinate system from which all other points are calculated, usually represented by the coordinates (0,0) in a planar coordinate system and (0,0,0) in a three-dimensional system. The centre of a projection is not always its origin.

Origin of coordinates Point in a system of coordinates that serves as a zero point in computing the system's elements or in prescribing its use.

Origin-destination cost matrix In ArcGIS Network Analyst, a type of network analysis that computes a table containing the total impedance from each origin to each destination. Additionally, it ranks the destinations that each origin connects to in ascending order of the time it takes to travel from that origin to each destination.

Orthocorrection The process of correcting the geometry of an image so that it appears as though each pixel were acquired from directly overhead. Orthorectification uses elevation data to correct terrain distortion in aerial or satellite imagery.

Orthogonal Intersecting at right angles.

Orthogonal offset A line that is perpendicular to another line at its point of tangency, often used to measure distance from a line to a separate point that does not lie along the original line.

Orthographic projection A planar projection, tangent to the earth at one point, that views the earth's surface from a point approaching infinity, as if from deep space.

Orthographic view In 3D Analyst, a perspective that allows viewing of data in a scene as a two-dimensional plane seen from above. There is no perspective foreshortening in orthographic view, so scale is constant across the entire display.

Orthoimage An image derived from a conventional perspective image by simple or differential rectification so that image displacements caused by sensor tilt and relief of terrain are removed.

Orthomorphic projection A projection that preserves the correct shapes of small areas. In a conformal projection, graticule lines intersect at 90-degree angles and at any point on the map the scale is the same in all directions. A conformal projection maintains all angles at each point, including those between the intersections of arcs; therefore, the size of areas enclosed by many arcs may be greatly distorted. No map projection can preserve the shapes of larger regions.

Orthophotograph A vertical aerial photograph from which the distortions due to varying elevation, tilt and surface topography have been removed, so that it represents every object as if viewed directly from above.

Orthophotographic map Map produced by assembling orthophotographs at a specified uniform scale in a map format.

Orthophotomap Orthophotographic map with contours and cartographic treatment, presented in a standard format and related to standard reference systems.

Orthophotoquad (1) An orthophotograph that has been formatted as a USGS 1:24,000 topographic quadrangle with little or no cartographic enhancement. (2) Monocolour orthophotgraphic map presented in a standard quadrangle format and related to standard reference systems. It has no contours and little or cartographic treatment.

Orthophotoscope (1) An optical-electronic device which converts a normal vertical aerial photograph to an orthophotograph. (2) Photo-mechanical device used in conjunction with a double-projection stereoplotter for producing orthophotograph.

Orthorectification The process of correcting the geometry of an image so that it appears as though each pixel were acquired from directly overhead. Orthorectification uses elevation data to correct terrain distortion in aerial or satellite imagery.

Oscillator A device for producing alternating current.

Outbound interface An interface implemented by a class, on which that object can make calls to its clients; analogous to a callback mechanism.

Outlier An unusual or extreme data value in a dataset. In data analysis, outliers can potentially have a strong effect on results and so must be analysed carefully to determine if they represent valid or erroneous data.

Outline The path that follows the boundary of an object. Outlines are also called strokes.

Outline vectorization A vectorization method that generates lines along the borders of connected cells. It is typically used for vectorizing scanned land-use and vegetation maps.

Out-of-process Within the process space of a client application, a component contained in an EXE is out-of-process; instantiated classes are loaded into the process space of the EXE in which they are defined rather than into that of the client.

Output coordinate system Acronym for output coordinate system. In ArcWeb Services, a user-defined map projection.

Output data Data that is the result of a computer, device, program or process.

Output directory In ArcIMS, the folder designated during installation to hold files being served to users for display in a browser.

Overedge Any portion of a map lying outside the nominal map border (neat line).

Overflow list A list of labels that could not be placed, generated during the process of converting labels to annotation. This list allows the manual placement or deletion of these labels.

Overlap Extent to which adjacent images or photographs cover the same terrain, expressed as a percentage. In aerial photographs, the overlap within the flight line is called the end lap and the overlap in adjacent parallel flight lines is called the side lap.

Overlapping rings A method of defining the rings in an analysis so that the values inside the rings are cumulative. For example, if you had an analysis with three concentric rings and 10 households in each, the total number of households for ring 1 would be 10, the total for ring 2 would be 20 (ring 1 + ring 2) and the total for ring 3 would be 30 (ring 1 + ring 2 + ring 3).

Overlay (1) A spatial operation in which two or more maps or layers registered to a common coordinate system are superimposed, either digitally or on a transparent material, for the purpose of showing the relationships between features that occupy the same geographic space. (2) Printing or drawing on a transparent or translucent medium intended to be placed in register on a map or other graphic and which shows details not appearing or requiring special emphasis on the base material.

Overlay events In linear referencing, an operation that produces a route event table that is the logical intersection or union of two input route event tables. Event overlay is one way to perform line-on-line, line-on-point, and event point-on-point overlays.

Overlay wizard In ArcGIS 8.3 and previous versions, an ArcToolbox wizard that uses overlay operations to create a new coverage by computing the geometric intersection of two coverages: an input coverage and an overlay coverage.

Overprint New material printed on a map or chart to show data of importance or special use, in addition to those data originally printed.

Overprinting In cartography, portraying cartographic updates on a map by printing new or modified information over the original cartography, usually in a distinctive colour.

Override In network datasets, a type of junction connectivity policy in which the way junctions connect to other junctions is not based on the existing edge connectivity policy; junctions 'override' the edge connectivity policy.

Overshoot (1) one type of digitizing error that results in a overextended arc. (2) The portion of an arc digitized past its intersection with another arc.

Overview map A generalized, smaller-scale map that shows the limits of another map's extent along with its surrounding area.

Overview terrain Coarsest representation of the terrain dataset, intended for fast drawing at small scales.

Overview window A secondary window in ArcMap data view that shows the full extent of the data, without changing the map extent. A red box in the window represents the current map extent.

Owner In ArcWeb Services, the person responsible for creating and maintaining a layer or service. The owner determines who has access to the content and the content permissions.

P

Packet In Survey Analyst-Cadastral Editor, an XML stream or file containing the portion of the cadastral fabric that has been extracted by a cadastral fabric job for editing.

Page unit The unit of measure, usually millimetres or inches, used to arrange map elements on a page for printing, as opposed to the coordinate system on the ground that the map represents.

Pan To shift a map image relative to the display window without changing the viewing scale.

Pan sharpening Sharpening a low-resolution multiband image by merging it with a high-resolution panchromatic image.

Panchromatic Sensitive to all or most of the visible spectrum, between 0.4 and 0.7 micrometres. Landsat 7 has a panchromatic band; used for an image that is built from intensity data with a very broad *bandwidth*, such as most or all visible-wavelength photons.

Panchromatic channel A channel of a sensor detector system covering the entire visible part of the electromagnetic spectrum

Panchromatic film Black and white film that is sensitive to all visible wavelengths.

Panchromatic image A single band image generally displayed as shades of grey.

Panchromatic sharpening Sharpening a low-resolution multiband image by merging it with a high-resolution panchromatic image.

Panelled map A map spliced together from smaller maps of neighbouring areas.

Pan-sharpening A computer-enhancement algorithm for improving the resolution of an image. Usually, this involves combining the high spatial resolution of a *panchromatic* image with the colour information from (lower resolution) *multispectral* data. The result is a colour image that has the high resolution of the panchromatic image. Researchers using pan-sharpened images should be aware that they are dealing with data that are probably carrying a certain amount of image-manipulation artefacts – the data may look very good, but they are not as trustworthy as genuinely high-resolution data.

Parallax (1) Displacement of the position of a target in an image caused by a shift in the observation system. (2) The apparent shift in an object's position when it is viewed from two different angles.

Parallax bar A stereoscope containing a micrometer for measuring the effects of parallax in a stereoscopic image.

Parallax difference The difference in the distance on overlapping vertical photographs between two points, which represent two locations on the ground with different elevations.

Parallel An imaginary east–west line encircling the earth, parallel to the equator and connecting all points of equal latitude. Also, the representation of this line on a globe or map.

Parallel of latitude (Parallels) A circle or approximation of a circle, on the surface of the earth, parallel to the equator and connecting points of equal *latitude*; a circle of the celestial sphere parallel to the ecliptic and connecting points of equal celestial latitude.

Parallel processing In computer data communications, a method of storing or sending data side by side, in groups of bits. Parallel data transmission is most often used for printer ports.

Parallel-polarized Describes a radar pulse in which the polarization of the return is the same as that of the transmission. Parallel-polarized images may be HH (horizontal transmit, horizontal return) or VV (vertical transmit, vertical return).

Parameter One of the variables that define a specific instance of a map projection or a coordinate system. Parameters differ for each projection and can include central meridian, standard parallel, scale factor or latitude of origin.

Parametric curve A curve that is defined mathematically rather than by a series of connected vertices. A parametric curve has only two vertices, one at each end.

Parcel A piece or unit of land, defined by a series of measured straight or curved lines that connect to form a polygon.

Parcel construction In Survey Analyst - Cadastral Editor, a set of dimensions entered to create a set construction line work.

Parcel group In Survey Analyst - Cadastral Editor, a set of parcels that exist as a group instead of individually in a cadastral fabric job.

Parcel PIN Acronym for parcel identification number. In Survey Analyst - Cadastral Editor, a unique identifier for a parcel. The format of an identifier is defined by the government's organization and may contain numerical values, alpha characters or both.

Parcel type A classification for parcels, used to provide additional information about them and how they must be treated for least squares adjustment. Standard parcels, blocks and easements are all examples of parcel types.

Parent replica In geodatabase editing, data in a source geodatabase that is replicated.

Parity The even or odd property of an integer. In address matching, parity is used to locate a geocoded address on the correct side of the street (such as odd numbers on the left-hand side, even numbers on the right).

Parse In computing, to divide a sequence of letters and numbers into parts, especially to test their agreement with a set of syntax rules.

Partial address support The ability to return a list of geocoding candidates based on incomplete address information. For example, if a city name but no country is entered in a partial address support search, the result list contains cities whose names match the name entered.

Partial cache In ArcGlobe, an on-demand cache that contains levels of detail for areas that have been visited, or a pre-processed cache that has a specified, incomplete, level of detail range for the entire layer.

Partial scene A partial scene (less than 375 scans) may exist at the beginning or end of a subinterval because imaging events do not always start or end on WRS scene boundaries. If generated, browse and scene metadata for these occurrences accurately reflect their partial scene nature and geographic extent.

Partial sill A parameter of a covariance or semivariogram model that represents the variance of a spatially autocorrelated process without any nugget effect. In the semivariogram model, the partial sill is the difference between the nugget and the sill.

Particle-wave duality the apparent paradox that *photons* can exhibit wave-like behaviour (e.g. make diffraction rings) while they can also exhibit particle-like behaviour (such as the photo-electric effect).

Pass In digital filters, refers to the spatial frequency of data transmitted by the filter. High-pass filters transmit high-frequency data; low-pass filters transmit low-frequency data.

Pass point Points established by aerial photography whose locations were manually selected by reviewing the control.

Passive microwaves Radiation in the 1 mm to 1 m range emitted naturally by all materials above absolute zero.

Passive remote sensing (1) A remote-sensing system, such as an aerial photography imaging system, that only detects energy naturally reflected or emitted by an object. (2) Remote sensing of energy naturally reflected or radiated from the terrain.

Passive sensor (1) A type of remote sensing instrument, a passive sensor picks up radiation reflected or emitted by the earth. ETM+ is a passive remote sensing system. (2) Imaging sensor that can only receive radiation, not transmit it.

Passive system A sensing system that detects or measures radiation emitted by the target.

Password A string of characters that a user must enter to access a computer, program, database or Web site. Passwords are a means of protecting and restricting access to information contained on networks, systems or files.

Patch A single triangular face inside a multipatch geometry. In most cases, many patches (faces) are used together to create a complex 3D model. Examples include geometric shapes, such as spheres, cubes and tubes; geographic features, such as buildings, cars and light poles; and other boundary representations, such as isosurfaces, used to represent geologic structures or environmental plumes. Patches in a multipatch geometry may or may not include an image (texture) displayed on them.

Path The connecting lines, arcs or edges that join an origin to a destination.

Path distance analysis In ArcGIS Spatial Analyst, a description of each cell's least accumulative cost relationship to a source or a set of sources, accounting for surface distance, horizontal cost factors and vertical cost factors.

Path label A label that describes the nature of the association between the objects in a relationship. The forward path label describes the relationship when navigating from the origin to the destination; for example, station points 'have' measurements. The backward path label describes the same relationship navigating from the destination to the origin, which might be 'are taken at' in this example; measurements 'are taken at' stations.

Path-and-row index System for locating Landsat MSS and TM images.

Pathfinding The process of calculating the optimal path between an origin and a destination point or points in a network.

Pattern Regular repetition of tonal variations on an image or photograph.

Pattern recognition In image processing, the computer-based identification, analysis and classification of objects, features or other meaningful regularities within an image.

Payload Sensing equipment carried on a platform.

Payload correction data (PCD) Imaging support data imbedded in the wideband data stream. Includes satellite attitude, ephemeris, time, angular displacement sensor (ADS) data and payload state.

P-Band A frequency range from 0.999 to 0.2998 GHz (30 to 100 cm wavelength) within the microwave (radar) portion of the electromagnetic spectrum. P-band is an experimental SAR frequency that has only been used to-date for research and development purposes. It is part of the NASA JPL AIRSAR multi-frequency (C-, L- & P-band) SAR system designed for earth observation experiments. P-band is not hindered by atmospheric effects and is capable of seeing through heavy rain showers. P-band SAR penetration capabilities are very significant with regard to vegetation canopies, glacier or sea ice, and soil. Its vegetation canopy imaging capability is considered a key element in estimating vegetation biomass by means of remote sensing.

P-code The Precise or Protected code. A very long sequence of pseudo-random binary biphase modulations on the GPS carrier at a chip rate of 10.23 MHz, which repeats about every 267 days. Each 1-week segment of this code is unique to one GPS satellite and is reset each week.

PDF Acronym for Portable Document Format. A proprietary file format from Adobe that creates lightweight text-based, formatted files for distribution to a variety of operating systems.

PDOP Acronym for dilution of precision. An indicator of satellite geometry for a constellation of satellites used to determine a position. Positions with a lower DOP value generally constitute better measurement results than those with higher DOP. Factors determining the total GDOP (geometric DOP) for a set of satellites include PDOP (positional DOP), HDOP (horizontal DOP), VDOP (vertical DOP), and TDOP (time DOP).

Peak The highest point of a mountain or hill.

Pedometer An instrument that gauges the approximate distance travelled on foot by registering the number of steps taken.

Per cent slope A measurement of the rate of change of elevation over a given horizontal distance, in which the rise is divided by the run and then multiplied by one hundred. A 45-degree slope and a 100-per cent slope are the same.

Performance A measure of the speed at which a computer system works. Factors affecting performance include availability, throughput, and response time.

Perigee In an orbit path, the point at which the object in orbit is closest to the centre of the body being orbited.

Periodic line dropout Defect on Landsat MSS or TM images in which no data are recorded for every sixth or sixteenth scan line, causing a black line on the image.

Periodic line striping Defect on Landsat MSS or TM images in which every sixth or sixteenth scan line is brighter or darker than the others. Caused by the sensitivity of one detector being higher or lower than the others.

Permanent dataset A dataset permanently stored on disk.

Permanent license A license that has a time-out date listed as permanent, indicating that the license for a particular product does not expire.

Persistence In computing, the process of saving or storing data; retaining the current state of an object in a memory storage medium such as a database or file on disk.

Personal geodatabase A geodatabase that stores data in Microsoft Access. A personal geodatabase can be read simultaneously by several users, but only one user at a time can edit the same data.

Perspective view A projection mode in 3D applications that allows viewing from a perspective that can be controlled by navigating the scene or globe from a specified location.

Phase preserving When the phase at the peak is correct, the processing algorithm is referred to as phase preserving, regardless of the phase variation across the impulse response.

Phase unwrap In SAR interferometry (InSAR) , the phase delay of the carrier signal at a certain point in the interferogram is a function of the terrain height at that point. However, the phase of the carrier signal can only be measured to within one cycle or 360 degrees. Phase unwrapping refers to converting the measured phase to the absolute phase, by adding the appropriate number of cycles or multiple of 360 degrees, to the measured phase.

Photodetector Device for measuring energy in the visible-light band.

Photogeology (1) Mapping and interpretation of geologic features from aerial photographs. (2) The science of interpreting and mapping geologic features from aerial photographs or remote-sensing data.

Photogrammetry (1) The science and technology of deducting the physical dimensions of objects from measurements on photographs. To obtain reliable information about physical objects and the environment, through processes of recording, measuring and interpreting images. (2) The science of making reliable measurements of physical objects and the environment by measuring and plotting electromagnetic radiation data from aerial photographs and remote-sensing systems against land features identified in ground control surveys, generally in order to produce planimetric, topographic and contour maps.

Photograph Representation of targets on film that results from the action of light on silver halide grains in the film's emulsion.

Photographic IR Short-wavelength portion (0.7 to 0.9 Fm) of the IR band that is detectable by IR colour film or IR black-and-white film.

Photographic UV Long-wavelength portion of the UV band (0.3 to 0.4 Fm) that is transmitted through the atmosphere and is detectable by film.

Photomap (photographic map) (1) Map made by adding marginal information, descriptive data, and a reference system to a photograph or assembly of photographs. (2) An aerial photograph or photographs, referenced to a ground control system and overprinted with map symbology.

Photometer An instrument that records the intensity of light by converting incident radiation into an electrical signal and then measuring it.

Photomosaic Mosaic composed of photographs.

Photon Minimum discrete quantity of radiant energy.

Photons Discrete packets of electromagnetic energy. Ensembles of photons are called light.

Photopic vision Vision under conditions of bright illumination.

Physical geography The field of geography concerning the natural features of the earth's surface.

Physical network One of the two parts of a network system; the actual feature classes that participate in a network system.

Picture element In a digitized image, the area on the ground represented by each digital number. Commonly contracted to pixel.

Picture fill A type of fill pattern created by continuous tiling of either a .bmp (raster image) or a .emf (vector graphic) file.

Pie chart A chart shaped like a circle cut into wedges from a centre point, that represents percentage values as proportionally sized 'slices'. Pie charts are used to represent the relationship between parts and the whole.

Pinch-roller A printing device that draws an image onto large-size paper or transparencies. Although pen and electrostatic plotters have largely been replaced by large-format inkjet printers, the term plotter is still frequently used to refer to all large print devices.

Pinning In ArcGlobe, anchoring a vector graphic element to the underlying globe surface so that it remains fixed.

Pit A depression in the earth's surface.

Pitch Rotation of an aircraft about the horizontal axis normal to its longitudinal axis that causes a nose-up or nose-down attitude.

Pixel (1) (also called as picture element) An abbreviation of picture element. The minimum size area on the ground detectable by a remote sensing device. The size varies depending on the type of sensor. The representation of data detected by a single *detector* during a certain finite collection period. Pixels may be arranged in a two dimensional array to indicate a spatial map of data (e.g. a photograph), in a one-dimensional string to display a *spectrum*, or in some other sort of *data cube* or *hyperdimensional data cube.* (2) The smallest unit of information in an image or raster map, usually square or rectangular. Pixel is often used synonymously with cell.

Pixel size The dimensions on the ground of a single pixel in a raster, measured in map units. Pixel size is often used synonymously with cell size.

Pixel space The x,y coordinate space defined by the number of pixels in a computer's display area, with a pixel being a single unit of colour on the screen. Most computer displays support several pixel configurations (800 x 600, 1024 x 768, 1600 x 1200, and so on). The more pixels there are, the smaller each pixel is for a given display size. Since a pixel is a piece of information, a configuration with more pixels can fit more information into a given display area.

Pixel type The attribute of a variable, field or column in a table that determines the kind of data it can store. Common data types include character, integer, decimal, single, double and string.

Place In census geography, any incorporated or unincorporated city, town or community.

Place finder sample web service A free sample of the Place Finder Web Service. This sample demonstrates the 'find a place' functionality available in the Place Finder Web Service and is provided only to sample ArcWeb Services.

Place finder web service A SOAP ArcWeb service that offers 'find a place' functionality for Internet applications. Users can use the Place Finder Web Service to input a place-name and receive a ranked candidate list of place-names and associated coordinates. The Place Finder Web Service recognizes more than three million place-names, including geographic regions (such as continents, states, cities), bodies of water and major points of interest.

Place-name alias The formal or common name of a location, such as the name of a school, hospital or other landmark. For example, 'Memorial Hospital' is the place name for the address '893 Memorial Drive'. In geocoding, the address locator can be set to accommodate the use of place-name aliases in place of their addresses for matching.

Plain Region of uniform general slope, comparatively level, of considerable extent, and not broken by marked elevations and depressions (it may be an extensive valley floor or a plateau summit); an extent of level or nearly level land; a flat, gently sloping or nearly level region of the sea floor.

Plan In surveying, a high-level organization of parcels; a survey document containing data from a recorded subdivision survey plan or from a legal description. Often, many parcels are defined in one plan. In the

Survey Analyst - Cadastral Editor cadastral fabric, plans hold such information about the subdivision plan record as the date, surveyor, entry units, scale factor and so on. Each parcel contains a reference to a plan.

Planar coordinate system A two-dimensional measurement system that locates features on a plane, based on their distance from an origin (0,0) along two perpendicular axes.

Planar enforcement A set of rules used to define a consistent method of building point, line and polygon features from spaghetti-digitized data. For example, planar enforcement includes rules that polygons of differing soil types cannot overlap, and that lines must be split at intersections.

Planar projection A projection that transforms points from a spheroid or sphere onto a tangent or secant plane. Because its directions are often true, the planar projection is also known as an azimuthal or zenithal projection.

Planarize The process of creating multiple line features by splitting longer features at the places where they intersect other line features. This process is often applied to nontopological line work that has been spaghetti digitized or imported from a CAD drawing.

Planck's Law An expression for the variation of emittance of a blackbody at a particular temperature as a function of wavelength.

Plane survey A survey of a small area that does not take the curvature of the earth's surface into account.

Planetable Instrument consisting essentially of a drawing board on a tripod and some type of sighting device (alidade) with attached straightedge, used for plotting the lines of survey directly from observation in the field.

Planimeter The configuration of a surface and the horizontal positions of its natural and cultural features, depicted by means of lines, symbols and notations on a scaled map or plan sheet without regard to relief.

Planimetric Two-dimensional; showing no relief.

Planimetric base A two-dimensional map that serves as a guide for contour mapping, usually prepared from aerial photographs.

Planimetric map A map that displays only the x,y locations of features and represents only horizontal distances.

Planimetric shift Deviations in the horizontal positions of features in an aerial photograph caused by differences in elevation. Planimetric shift causes changes in scale throughout a photograph.

Planimetry Plan details of a map - those having no indication of relief or contour.

Plat (1) A survey diagram, drawn to scale, of the legal boundaries and divisions of a tract of land. (2) Diagram drawn to scale showing all essential data pertaining to the boundaries and subdivisions of a tract of land, as determined by survey or protraction. As used by the Bureau of Land Management, the drawing which represents the particular area included in a survey, such as township, private land claim or mineral claim, and the lines surveyed, established, or retraced, showing the direction and length of each such line; The relation to the adjoining official surveys; the boundaries, descriptions and area of each parcel of land subdivided; and, as nearly as may be practicable, a representation of the relief and improvements within the limits of the survey.

Platform (1) In computing, the operating system of a machine, such as the UNIX, Linux or Windows operating systems. Platform may also refer to a programming language or development environment, such as COM, Java or ArcGIS 9. (2) The observing structure, i.e. aircraft or satellite, that carries remote sensing apparatus.

Playback mode The time mode in which data is displayed using the ArcGIS Tracking Analyst Playback Manager, replaying either real-time or fixed-time data.

Playback window The span of time defined by the Start and End text boxes in the ArcGIS Tracking Analyst Playback Manager. This window can be set to include the temporal extent of one or more layers.

Plotter A printing device that draws an image onto large-size paper or transparencies. Although pen and electrostatic plotters have largely been replaced by large-format inkjet printers, the term plotter is still frequently used to refer to all large print devices.

PLTS Acronym for Production Line Tool Set. A software package that allows users to prepare and maintain data for maps, perform quality assurance/quality control tasks and create map sheets. Its base product, PLTS Foundation, is composed of three parts: Foundation Tools, GIS Data ReViewer, and Map Production System and MPS-Atlas. The tools can be used with different product specifications and are available as solutions that have been created for nautical, defense, aeronautical and mapping agencies.

PLTS data loader A tool used to batch load a personal or enterprise geodatabase from a variety of sources (coverages, shapefiles or geodatabases). The PLTS data loader uses a cross-reference database to map the input schema to the output schema.

PLTS map gallery In PLTS, a tool for managing layer representation, such as symbology and label expressions.

Plug-in A small software application that extends the functionality of a Web browser.

Plug-in data source An additional read-only data source provided by either ESRI or a third-party developer. It may be a data source forming part of the core ArcObjects or an extension.

Plumb line A line that corresponds to the direction of gravity at a point on the earth's surface; the line along which an object will fall when dropped.

PMC A geodatabase designed to manage database and map production for the Production Line Tool Set (PLTS). The PMC stores metadata and geometries of projects, products and source material.

PNG Acronym for Portable Network Graphics. A bitmapped graphics format similar to GIF.

POI manager web service A SOAP ArcWeb service for uploading a custom set of points. POI Manager is available as a Web interface and a SOAP interface.

Point (1) A geometric element defined by a pair of x,y coordinates. (2) A spatial feature that is represented by a single coordinate and has only the geometric property of location. Also called as node or vertex.

Point and coordinate analysis In Survey Analyst, part of the validation of survey data. This type of analysis is used to authenticate the relationships between survey points, coordinates and the physical locations they represent.

Point digitizing A method of digitizing in which the digitizer selects particular points or vertices, to encode.

Point event In linear referencing, a feature that occurs at a precise point location along a route and uses a single measure value. Examples include accident locations along highways, signals along rail lines, bus stops along bus routes and pumping stations along pipelines.

Point feature A map feature that has neither length nor area at a given scale, such as a city on a world map or a building on a city map.

Point identifier field A field in the Survey Analyst Survey Explorer dialog box that allows a user to specify the name of a particular survey point.

Point mode digitizing A method of digitizing in which the digitizer selects particular points or vertices, to encode.

Point name flag In Survey Analyst, a visual indicator, after each keystroke, of whether or not a point with the typed name already exists in the survey dataset.

Point name prefix In Survey Analyst, a unique part of every survey project. Points have a common name space across all projects of a survey dataset. However, different points in different projects can have the same name. When making use of these points, users can use the prefix of the survey project before the point's name to uniquely identify it. This ensures that the correct point is used.

Point size A unit of measure for fonts, nearly equal to 1/72 of an inch.

Point spread function (PSF) The image of a point source of radiation, such as a star, collected by an imaging device. A measure of the spatial fidelity of the device.

Point thinning Act of reducing point data in a dataset. Point thinning reduces the number of point measurements needed to represent a surface for a given area.

Point-in-polygon overlay A spatial operation in which points from one feature dataset are overlaid on the polygons of another to determine which points are contained within the polygons.

Point-on-line overlay In linear referencing, the overlay of a line event table and a point event table to produce a single point event table. The new event table can be the logical intersection or union of the input tables.

Polar aspect A planar projection with its central point located at either the north or south pole.

Polar flattening A measure of how much an oblate spheroid differs from a sphere. The flattening equals the ratio of the semimajor axis minus the semiminor axis to the semimajor axis.

Polar orbit (1) A satellite orbit with an inclination near 90 degrees that passes over each polar region. (2) An orbit with an orbital inclination of near 90 degrees where the satellite ground track will cross both Polar Regions once during each orbit. The term describes the near-polar orbits of spacecraft.

Polar radius The distance from the earth's geometric dentre to either pole.

Polarimetric active radar calibrator (PARC) Device used to receive and retransmit radar pulses. These devices usually consist of a polarisation sensitive receive and transmit antenna and a stable amplifier which boosts the signal level so that the device being calibrated receives a high signal of a given polarisation.

Polarimetric radar radar which permits measurement of the full polarisation signature of every resolution element.

Polarization The direction of orientation in which the electrical field vector of electromagnetic radiation vibrates or process of confining the vibrations of the magnetic, or electric field, vector of light or other radiation to one plane. Orientation of the plane of the electric field relative to the earth's surface.

Polarization beamwidth For a pair of antennas in a transmit-receive configuration, the polarization beamwidth is defined as the angle between the two directions from the transmitter at which the polarization efficiency is one-half of its maximum value.

Polarization efficiency For an electromagnetic wave that is incident upon a receiving antenna, the polarization efficiency is defined as the ratio of power actually received in an impedance-matched load, to the maximum power that could be received if the antenna had optimum polarization for the received wave.

Polarization ellipse For an elliptically polarized wave, the tip of the electric field vector traces an ellipse on a plane that is transverse to the wave propagation direction. This polarization ellipse describes the polarization properties of the electromagnetic waves, including the ratio of the perpendicular electric field components and their relative phases.

Polarization loss The inverse of polarization efficiency.

Polarized radiation Electromagnetic radiation in which the electrical field vector is contained in a single plane, instead of having random orientation relative to the propagation vector. Most commonly refers to radar images.

Polygon On a map, a closed shape defined by a connected sequence of x,y coordinate pairs, where the first and last coordinate pair are the same and all other pairs are unique.

Polygon feature A map feature that bounds an area at a given scale, such as a country on a world map or a district on a city map.

Polygon overlay The process of superimposing two or more geographic polygon layers and their attributes to produce a new polygon layer.

Polygon-arc topology In a polygon coverage, the list of topologically connected arcs that define the boundary of a polygon feature and the label point that links it to an attribute record in the coverage point attribute table.

Polyhedron A three-dimensional object or volume defined by a number of plane faces or polygons.

Polyline In ArcGIS software, a shape defined by one or more paths, in which a path is a series of connected segments. If a polyline has more than one path (a multipart polyline), the paths may either branch or be discontinuous.

Polyline feature In ArcGIS software, a digital map feature that represents a place or thing that has length but not area at a given scale. A polyline feature may have one or more parts. For example, a stream is typically a polyline feature with one part. If the stream goes underground and later reemerges, it might be represented as a multipart polyline feature with discontinuous parts. If the stream diverges around an island and then rejoins itself, it might be represented as a multipart polyline feature with branching parts. A multipart polyline feature is associated with a single record in an attribute table.

Port number A number that is used to specify direct communication over a network to an Internet application.

Portal A Web resource that provides access to a broad array of related resources and services.

Portlet A Web component that processes requests and generates dynamic content. Portlets are used in portals as pluggable user interfaces to add specialized content, such as weather information, news, or maps, to Web pages. Users can customize the content, appearance, and position of a portlet.

Position (1) The latitude, longitude, and altitude (x,y,z coordinates) of a point, often accompanied by an estimate of error. Position may refer to an object's orientation (facing east, for example) without referring to its location. (2) The latitude, longitude, and altitude of a point. An estimate of error is often associated with a position.

Positional dilution of precision (PDOP) Measure of the geometrical strength of the GPS satellite configuration.

Positive photograph Photographic image in which the tomes are directly proportional to the terrain brightness.

Postal code A series of letters or numbers, or both, in a specific format, used by the postal service of a country to divide geographic areas into zones in order to simplify delivery of mail.

Posting During versioned geodatabase editing, the process of applying the current edit session to the reconciled target version.

Post-processed differential GPS In post-processed differential GPS the base and roving receivers have no active data link between them. Instead, each records the satellite observations that will allow differential correction at a later time. Differential correction software is used to combine and process the data collected from these receivers.

Power The number of times a value is to be multiplied by itself, indicated by an exponent. For example, 2 3 = 2*2*2.

Precise code The PRN code used by United States and allied military GPS receivers.

Precise positioning service (PPS) -The most accurate dynamic positioning possible with GPS, based on the dual frequency P-code.

Precision The closeness of a repeated set of observations of the same quantity to one another. Precision is a measure of the control over random error. For example, assessment of the quality of a surveyor's work is based in part on the precision of their measured values.

Precision code The PRN code used by United States and allied military GPS receivers.

Prediction In spatial modeling, the process of forming a statistic from observed data to assign values to random variables at locations where data has not been collected.

Prediction standard error A value quantifying the uncertainty of a prediction; mathematically, the square-root of the prediction variance. (The prediction variance is the variation associated with the difference between the true and predicted value.) As a rule, 95 per cent of the time the true value will lie within the predicted value plus or minus two times the prediction standard error if data is normally distributed.

Preferences User-specified settings that determine how an application will act or appear when it is used.

Prefix pin Used within the Survey Analyst Survey Explorer's Point Identifier field to separate the prefix and point name strings.

Preliminary topology In coverages, refers to incomplete region topology. Region topology defines region-arc and region-polygon relationships. A topological region has both the region-arc relationship and the region-polygon relationship. A preliminary region has the region-arc relationship but not the region-polygon relationship. In other words, preliminary regions have no polygon topology. Coverages with preliminary topology have red in their icons in ArcCatalog.

Preprocessed cache In ArcGlobe, a cache that is built prior to viewing, using an independent process accessed from the shortcut menu of a layer.

Presentation service An OpenLS ArcWeb service used for generating map images.

Preview A live view of GIS data in ArcCatalog. Users can pan and zoom the preview, query features, and create thumbnail images to store in metadata.

Previsual symptom A vegetation anomaly that is recognizable on IR film before it is visible to the naked eye or on normal colour photographs. It results when stressed vegetation loses its ability to reflect photographic IR energy and its recognizable on IR colour film by a decrease in brightness of the red hues.

Primary colours Colours from which all other colours are derived in a particular colour system. On a display monitor, these colours are red, green, and blue. In printing, they are cyan, magenta, and yellow. In traditional pigments, they are red, blue, and yellow.

Primary key An attribute or set of attributes in a database that uniquely identifies each record. A primary key allows no duplicate values and cannot be null.

Primary reference data In geocoding, the most basic reference material used in an address locator, usually consisting of the geometry of features in a region and an associated address attribute table.

Primary table In geocoding, the attribute table associated with the primary reference data. Based on the address locator style selected, certain address elements must be present in the primary table.

Prime meridian Meridian of longitude 0 degrees, used as the origin for measurements of longitude. The meridian of Greenwich, England, is the internationally accepted prime meridian on most charts. However, local or national prime meridians are occasionally used.

Prime vertical In astronomy and geodesy, the vertical circle that passes through an observer's zenith and through the east and west points of the horizon.

Primitive In MOLE, the most elementary part of a graphic. Icon, frame, and fill are examples of components that make up MOLE graphics.

Principal component analysis The analysis of covariance in a multiple data set so that the data can be projected as additive combinations on to new axes, which express different kinds of correlation among the data.

Principal point The optical centre of an aerial photograph. The principal point is the point where the perpendicular projected through the centre of the lens intersects the aerial photo platform image.

Principal-component (PC) image Digitally processed image produced by a transformation that recognizes maximum variance in multispectral images.

Principle scale same as the scale of the reference globe.

Printout Display of computer data in alphanumeric format.

Private virtual server An ArcIMS virtual server that offers functionality not directly accessible when creating a Web site. The functionality associated with each of the three private ArcIMS virtual servers (geocode server, query server, and extract server) is called from within a service created using a public virtual server. For example, if you incorporate an image service with a geocodable street network into a Web site with the Geocode tool, a client can access the geocode server's functionality. The geocode virtual server works with the image virtual server to perform address matching and other geocoding functions.

PRJ Usually a text file named prj.adf that is associated with a coverage, GRID, or TIN. The PRJ file contains the coordinate system information for the data. In a more general sense, PRJ can refer to the coordinate system of data even if the information is not stored in a prj.adf file. For example, 'The PRJ of the shapefile is WGS 1984 UTM zone 15 north.'

PRN code Acronym for pseudo-random noise code. A repeating radio signal broadcast by each GPS satellite and generated by each GPS receiver. In a given cycle, the satellite and the receiver start generating their codes at the same moment, and the receiver measures how much later the satellite's broadcast reaches it. By multiplying that time by the speed of radio waves, the receiver can compute the distance between the satellite's antenna and its own.

Probability A measure of the likelihood that a particular outcome, such as a spatial pattern or event, will occur given a set of possible outcomes. Probability values range from 0 for impossible outcomes to 1 for completely certain outcomes. The probability that a tossed coin will land heads-up, for example, is 0.5, since landing heads-up is one of two possible outcomes.

Probability density function (PDF) A function indicating the relative frequency with which any measurement may be expected to occur. In remote sensing it is represented by the histogram of DN in one band for a scene.

Probability map A surface that gives the probability that the variable of interest is above or below a specified threshold value.

Procedure In software, a block of code that performs some task. Procedures are commonly used to organize code into reusable units. In object-oriented programming, a procedure that is specific to an object or class is called a method.

Process In geoprocessing in ArcGIS, a tool and its parameter values. One process, or multiple processes connected together, creates a model.

Process chain In ArcGIS Image Server, a list of raster processes that are performed on the fly by the server on raster data. The full process chain contains the raster dataset process applied to the raster data before mosaicking, and the service process applied after mosaicking.

Processing initial stages of data processing.

Production line tool set A software package that allows users to prepare and maintain data for maps, perform quality assurance/quality control tasks, and create map sheets. Its base product, PLTS Foundation, is composed of three parts: Foundation Tools, GIS Data ReViewer, and Map Production System and MPS-Atlas. The tools can be used with different product specifications, and are available as solutions that have been created for nautical, defense, aeronautical, and mapping agencies.

Profile graph A graph of the elevation of a surface along a specified line.

ProgID A string value, stored in the system registry, identifying a class by library and class name, for example esriCarto.FeatureLayer. The ProgID registry key also contains the human-readable name of a class, the current version number of the class, and a unique class identifier. ProgIDs are used in VB object instantiation.

Project In Survey Analyst, a specific task for capturing survey data. This can include anything from a field control survey to data entry from a subdivision plan.

Project and map sheet catalog A geodatabase designed to manage database and map production for the Production Line Tool Set (PLTS). The PMC stores metadata and geometries of projects, products and source material.

Project data Any data in a process that existed before the process existed.

Project file In ArcView 3, a file for creating and storing documents for GIS work. All activity in ArcView 3 takes place within project files, which use five types of documents to organize information: views, tables, charts, layouts, and Avenue scripts. A project file organizes its documents and stores their unique settings in an ASCII format file with the extension .apr.

Project repair In an ArcView 3.x project, the process of updating document paths when referenced document data is moved from one disk location to another. Project repair can be avoided if a system variable is entered as part of the document path and set within the necessary system file(s).

Projected coordinate system A reference system used to locate x, y, and z positions of point, line, and area features in two or three dimensions. A projected coordinate system is defined by a geographic coordinate system, a map projection, any parameters needed by the map projection, and a linear unit of measure.

Projected coordinates A measurement of locations on the earth's surface expressed in a two-dimensional system that locates features based on their distance from an origin (0,0) along two axes, a horizontal x-axis representing east–west and a vertical y-axis representing north–south. Projected coordinates are transformed from latitude and longitude to x,y coordinates using a map projection.

Projection A method by which the curved surface of the earth is portrayed on a flat surface. This generally requires a systematic mathematical transformation of the earth's graticule of lines of longitude and latitude onto a plane. Some projections can be visualized as a transparent globe with a light bulb at its dentre (though not all projections emanate from the globe's dentre) casting lines of latitude and longitude onto a sheet of paper. Generally, the paper is either flat and placed tangent to the globe (a planar or azimuthal projection) or formed into a cone or cylinder and placed over the globe (cylindrical and conical projections). Every map projection distorts distance, area, shape, direction, or some combination thereof.

Projection transformation The mathematical conversion of a map from one projected coordinate system to another, generally used to integrate maps from two or more projected coordinate systems into a GIS.

Projective transformation A transformation used only to transform coordinates digitized directly from high-altitude aerial photographs of relatively flat terrain, assuming there is no systematic distortion in the photographs.

Prolate ellipsoid An ellipsoid created by rotating an ellipse around its major axis.

Property An attribute of an object defining one of its characteristics or an aspect of its behaviour.

Property file A file that stores attributes for a particular ArcIMS item or function. Property files have a .properties file extension.

Property page A user interface component that provides access to change the properties of an object or objects.

Proportional symbol A symbol whose size differs in relation to the phenomenon being mapped.

Protected code The PRN code used by United States and allied military GPS receivers.

Proximity analysis A type of analysis in which geographic features (points, lines, polygons, or raster cells) are selected based on their distance from other features or cells.

Proximity query A form of spatial query in which geographic features within a specified distance of a particular feature are selected.

Proximity web service A SOAP ArcWeb service for finding all point-of-interest (POI) locations within a distance of a specified point or line (for example, find all POIs within five miles of x,y) or for determining the nearest specified number of POI locations to a specified point or line (such as find nearest three POIs to x,y).

Proxy object A local representation of a remote object, supporting the same interfaces as the remote object. All interaction with the remote object from the local process is mediated via the proxy object. A local object makes calls on the members of a proxy object as if it were working directly with the remote object.

Pseudo node (1) A node appearing along a continuous arc. (2) In a geodatabase topology, a temporary feature marking the location where an edge has been split during an edit session. This type of pseudo node becomes a vertex when the edit is saved.

Pseudo-lite A ground-based differential GPS receiver which transmits a signal like that of an actual GPS satellite, and can be used for ranging.

Pseudorandom Being or involving entities (as numbers) selected by a definite computational process that satisfies one or more standard tests for statistical randomness.

Pseudo-random noise (PRN) A signal with random noise-like properties. It is very complicated but repeated pattern of 1's and 0's. Acronym for pseudo-random noise. A signal carrying a code that can be reproduced exactly, but appears to be randomly distributed like noise. Each NAVSTAR satellite has a unique PRN code.

Pseudo-random noise code A repeating radio signal broadcast by each GPS satellite and generated by each GPS receiver. In a given cycle, the satellite and the receiver start generating their codes at the same moment, and the receiver measures how much later the satellite's broadcast reaches it. By multiplying that time by the speed of radio waves, the receiver can compute the distance between the satellite's antenna and its own.

Pseudo-range A distance measurement based on the correlation of a satellite transmitted code and the local receiver's reference code, that has not been corrected for errors in synchronization between the transmitter's clock and the receiver's clock.

Pseudoscopic effect it results from the accidental reversing of two photos when the right eye views the left photo, and the left eye views the right photo. It is also called reversal of relief or pseudoscopic view.

Public land survey system (PLSS) Acronym for Public Land Survey System. The description of the location of land in the United States using a survey system established by the federal government in 1785. The system is based on the concept of a township, a square parcel of land measuring 6 miles on each side. The township's position is described as a number of 6-mile units east of a north–south line (called the meridian) and north or south of an east–west line (called the baseline). Each township is divided into 36 sections, each of which is 1 square mile. A section is divided into quarters equal to 160 acres. The quarter section may be further divided into four 40-acre parcels. The PLSS is also called the rectangular survey.

A land partitioning system used in the United States.

Public land system Public lands are subdivided by a rectangular system of surveys established and regulated by the Bureau of Land Management. The standard format for subdivision is by townships measuring 6 miles (480 chains) on a side. Townships are further subdivided into 36 numbered sections of 1 square mile (640 acres) each.

Public participation The active involvement of stakeholders outside an organization in the decision-making or planning processes of that organization. Public participation in GIS processes may include making GIS tools and data accessible, at an appropriate technical level, to stakeholders, or it may result in knowledge gained from stakeholders being incorporated into GIS analyses.

Public virtual server An ArcIMS virtual server that must be specified when creating a new service. Three public virtual servers install with ArcIMS: image server, feature server, and metadata server. Two additional public virtual servers, ArcMap server and route server, are optional extensions to ArcIMS.

Publish To produce data and information in a distributable format, such as digital, Internet, or print media.

Published map file Acronym for Published Map File. A file exported by the Publisher extension that can be read by ArcReader. Publisher Map Files end with a .pmf extension.

A file exported by the Publisher extension that can be read by ArcReader. Publisher Map Files end with a .pmf extension.

Publisher Software that converts raw information into a more usable format.

Puck The handheld device used with a digitizer to record positions from the tablet surface.

Pull check-in In ArcGIS 9.1 and previous versions, a check-in operation initiated from a master geodatabase.

Pulse A short burst of electromagnetic radiation transmitted by the radar. Also described as a group of waves with a distribution confined to a short interval of time. Such a distribution is described in the time domain or in spatial dimensions, by its width and its amplitude or magnitude, from which its energy may be found. In radar, use is made of modulated or coded pulses which must be processed to decode or compress the original pulse to achieve the impulse response observed in the image. Coded pulses have a time-bandwidth product that is much larger than unity. The resolution that may be achieved after processing is determined by the bandwidth of the original pulse.

Pulse compression In collecting the SAR data, a long-duration linear FM pulse is transmitted. This allows the pulse energy to be transmitted with a lower peak power. The linear FM pulse has the property that, when filtered with a matched filter, the result is a narrow pulse in which all the pulse energy has been collected to the peak value. Thus, when a matched filter is applied to the received echo, it is as if a narrow pulse were transmitted, with is corresponding range resolution and signal-to-noise ratio (SNR).

Pulse length Duration of a burst of energy transmitted by a radar antenna, measured in microseconds.

Pulse repetition frequency (PRF) Rate of recurrence of the pulses transmitted by a radar.

Push broom scanner (1) A remote-sensing tool with a line of many fixed sensors that record reflected radiation from the terrain along a satellite's direction of movement, creating scan-line strips that are contiguous or that overlap slightly, thereby producing an image. (2) An alternate term for an along-track scanner

Push broom system An imaging device consisting of a fixed linear array of many sensors which is swept across an area by the motion of the platform, thereby building up an image. It relies on sensors whose response and reading is nearly instantaneous, so that the image swathe can be segmented into pixels representing small dimensions on the ground.

Push check-in In ArcGIS 9.1 and previous versions, a check-in operation initiated from a checkout geodatabase.

P-value A probability resulting from a statistical test of the coefficient associated with each independent variable in a regression model. The null hypothesis for this statistical test states that the coefficient is not significantly different from zero. Small p-values reflect small probabilities. They suggest that the coefficient is significantly different from zero, and consequently, that the associated explanatory variable is helping to model or predict the dependent variable. Variables with coefficients near zero do not help predict or model the dependent variable; they are almost always removed from the regression equation (unless there are strong theoretical reasons to keep them).

Pyramid In raster datasets, a reduced resolution layer that copies the original data in decreasing levels of resolution to enhance performance. The coarsest level of resolution is used to quickly draw the entire dataset. As the display zooms in, layers with finer resolutions are drawn; drawing speed is maintained because fewer pixels are needed to represent the successively smaller areas.

Q

QQ plot A scatter chart in which the quantiles of two distributions are plotted against each other.

Quad sheet A rectangular map bounded by lines of latitude and longitude, often a map sheet in either the 7.5-minute or 15-minute series published by the US Geological Survey. Quadrangles are also called toposheets.

Quad-centred photograph Middle exposure of a phototriplet (three consecutive aerial photographs) take so that the middle photograph is exposed directly above the centre of the *quadrangle* and the preceding and following photographs are exposed directly above the boundaries of the quadrangle. The flying height is set such that the quad-centred photograph covers the entire quadrangle.

Quadrangle (1) A rectangular map bounded by lines of latitude and longitude, often a map sheet in either the 7.5-minute or 15-minute series published by the US Geological Survey. Quadrangles are also called topo sheets. (2) Four-sided area, bounded by *parallels of latitude* and *meridians* of longitude used as an area unit in mapping (dimensions are not necessarily the same in both directions). Also, a geometric figure of significance in geodetic surveying.

Quadrant In a rectangular coordinate system, any of the quarters formed by the central intersection of x and y axes that divide a plane into four equal parts.

Quadrat In spatial sampling, one of a set of identically-sized areas, often rectangular, within which the members of a population are counted. The size, number and location of quadrats within a study area are chosen by the sampler. Population counts in each quadrat are compared to determine distribution patterns.

Quadrat analysis Comparison of the statistically expected and actual counts of objects within spatial sampling areas (quadrats) to test for distribution patterns such as randomness and clustering.

Quadrature (Q) channel The signal component that is 90° out of phase with respect to the reference frequency. It is represented by the letter Q. It is the imaginary part, which indicates the magnitude of the signal, of the complex number. See Complex Number.

Quadrature polarization radar (quad pol radar) A Radar system designed to simultaneously collect imaging data of a scene in two orthogonal polarization states on transmit and the same two polarization states on receive. From such a data set a complete scattering matrix of the reflectivity of the scene may be synthesised, leading to the concept of polarization signature.

Quadtree A method for encoding raster data that reduces storage requirements and improves access speeds by storing values only for homogeneous regions rather than for every pixel. The raster is recursively subdivided into quadrants until all regions are homogeneous or until some specified level has been reached.

Qualitative data Data classified or shown by category, rather than by amount or rank, such as soil by type or animals by species.

Quality assurance A process used to verify the quality of a product after its production.

Quality beta In ArcGIS Survey Analyst, when used together, the F test and W test are called the B method of testing, for which a power can be defined. This power is the Quality Beta and can be defined as follows: the probability that the null hypothesis is accepted while, in fact, it is false, is equal to 1-b.

Quality control A process used during production of a product to ensure its quality.

Quantile In a data distribution, a value representing a class break, where classes contain approximately equal numbers of observations. The p-th quantile, where p is between 0 and 1, is that value that has a proportion p of the data below the value. For theoretical distributions, the p-th quantile is the value that has p probability below the value.

Quantile classification A data classification method that distributes a set of values into groups that contain an equal number of values.

Quantitative data Data grouped or shown by measurements of number or amount, such as population per unit area.

Quantitative geography The application of mathematical and statistical concepts and methods to the study of geography.

Quantization (1) To subdivide into small but measurable increments. (2) To calculate or express in terms of quantum mechanics.

Quantum The elementary quantity of EM energy that is transmitted by a particular wavelength. According to the quantum theory, EM radiation is emitted, transmitted and absorbed as numbers of quanta, the energy of each quantum being a simple function of the frequency of the radiation.

Query A request to select features or records from a database. A query is often written as a statement or logical expression.

Query expression A type of expression that evaluates to a Boolean (true or false) value, that is typically used to select those rows in a table in which the expression evaluates to true. Query expressions are generally part of a SQL statement.

Query interface A client may request a reference to a different interface on an object by calling the Query Interface method of the IUnknown interface.

Query language A language for storing, retrieving and editing data in a database.

Query server A private ArcIMS virtual server that works with an image server to support query functions. When a viewer requests an identify or find operation, the query server returns the related geographic or tabular data. It can be built against shapefiles, ArcSDE layers and joined external tables.

Query table A table containing results from a query.

Query web service A SOAP Arc Web service for determining the characteristics and features for a specified location. With Query Web Service, users can specify which attributes are available to query, set the level of geography for the query, and generate a list of attributes or field names and their descriptions for a point, area, envelope or dataset.

R

Radar Acronym for radio detection and ranging. A device or system that detects surface features on the earth by bouncing radio waves off them and measuring the energy reflected back.

Radar altimeter (1) A non-imaging device that records the time of radar returns from vertically beneath a platform to estimate the distance to and hence the elevation of the surface; carried by Seasat and the EAS-ERS-1 platforms. (2) An instrument that determines elevation, usually from mean sea level, by measuring the amount of time an electromagnetic pulse takes to travel from an aircraft to the ground and back again.

Radar angles A radar echo from a region where there are no visible targets; may be caused by insects, birds or refractive index variations in the atmosphere.

Radar antenna The radar antenna is a structure for transmitting and receiving radiated energy; it is an important subsystem that defines, to a great extent, a radar's operational capabilities and cost. In radar remote sensing the main function of the antenna is to concentrate a radiated microwave energy into a beam of required shape, referred to as the antenna pattern, to transmit it into the desired direction (look direction), and to receive the returned energy from surfaces or objects. Radar remote sensing antennas provide scene illumination (See Active Remote Sensing System). The main parameters of radar antennas are operating frequency band, antenna pattern shape (directivity), power (or antenna-) gain, beam-width, side-lobe level, polarization, and power handling capability. Moving antennas can be used to form a synthetic aperture, where the physical antenna is small compared to the synthesised antenna, and has a sufficiently wide radiation pattern to illuminate the observed surface over a significant period of platform motion.

Radar beam The vertical fan-shaped beam of electromagnetic energy produced by the radar transmitter.

Radar cross section (RCS) A measure of the intensity of backscattered radar energy from a point target. Expressed as the area of a hypothetica surface which scatters radar equally in all directions and which would return the same energy to the antenna.

Radar interferometry The analysis of interferograms that have been created by IFSAR or artificially. Radar interferometry involves the comparison of two or more images of the same area taken from different positions and calibrated with surveyed ground points to generate three-dimensional digital elevation models (DEMs), or models demonstrating slight movements of surface features.

Radar parallax Apparent change in the position of an object due to an actual change in the point of view of observation. For a SAR, true parallax occurs only with viewpoint changes that are away from the nominal flight path of the radar. In contrast to aerial photography, parallax cannot be created by forward and aft looking exposures. Parallax may be used to create stereo viewing of radar images.

Radar phase Phase is a property of a periodic phenomenon, for example a wave, referring to its starting point or advancement (fraction) relative to an arbitrary origin. In radar remote sensing, the concept of phase is usually applied to the oscillation of electromagnetic waves. When viewed as a cyclical phenomenon, like wave motion or the crankshaft motion of a bicycle pedal, phase can be expressed in degrees. One-quarter cycle represents a phase rotation of 90 degrees; completion of one complete cycle corresponds to a phase rotation of 360 degrees. Waves are considered in-phase, if their origins of phase 0 degrees are perfectly aligned; out-of-phase conditions are met when phase 0 and 180 degrees are aligned. Precise knowledge of phase properties in radar signal data is a key element in interferometric as well as in polarimetric SAR.

Radar radio detection and ranging (RADAR) 'Short for "radio detection and ranging," radar sends out short pulses of microwave energy and records the returned signal's strength and time of arrival.' A method, system or technique, including equipment components, for using beamed, reflected and timed electromagnetic radiation to detect, locate and (or) track objects, to measure altitude and to acquire a terrain image. The radio detection instrument consists of a transmitter that sends out high-frequency radio waves and a receiver that picks them up after they have been reflected by an object. Basic building blocks of a radar are the transmitter, the antenna (normally used for both transmission and for reception), the receiver and the data handling equipment. A synthetic aperture radar system, by implication, includes an image processor, even though it may be remotely located in time or space from the radar electronics. The advantage radar sensors have over other types of sensors, is that microwaves can penetrate clouds, most rain storms and even dry snow. Therefore, for those parts of the world where cloud and rain present a problem in acquiring images (tropics, coastal/maritime regions), radar is highly beneficial.

Radar scattering coefficient A measure of the back-scattered energy from a target with a large area. Expressed as the average radar cross section per unit area in decibels (dB). It is the fundamental measure of the radar properties of a surface.

Radar scatterometer A non-imaging device that records radar energy back-scattered from terrain as a function of depression angle.

Radar shadow Just as with a flash-light, a radar sensor 'illuminates' a scene, and if an object blocks the beam, a shadow area develops behind it. Such shadows can be seen in a radar image. Radar shadows are pure black – they contain absolutely no information.

Radar transmission Energy sent by the radar, normally in the form of a sequence of pulses, to illuminate a scene of interest.

Radar wave energy For a waveform of time-limited duration such as a radar pulse reflected by an object, the pulse energy is given by the power of the signal integrated over the duration of the signal It is expressed in units of watt-seconds = joules = j.

RADARSAT–1 RADARSAT-1 is an advanced earth observation satellite project developed by Canada to monitor environmental change and to support resource sustainability. It is Canada's first earth observation satellite and the world's first operationally-oriented radar sensor. With a planned lifetime of five years, RADARSAT-1 is equipped with a Synthetic Aperture Radar (SAR). Launched in November 1995, this C-band SAR satellite includes a steerable beam, which offers a wide selection of image scales and resolutions. It operates at 5.3 GHz. As of March 2013, the satellite was declared non-operational and is no longer collecting data. RADARSAT-2 was launched in December 2007.

Radial basis functions In ArcGIS Geostatistical Analyst tool box a deterministic interpolation method. The interpolated surface is forced to conform to the sample data points and the method does not have standard errors associated with it. Spline interpolation is a special case of radial basis functions.

Radial relief displacement The tendency of vertical objects to appear to lean radially away from the centre of a vertical aerial photograph. Caused by the conical field of view of the camera lens.

Radial triangulation It is a triangulation based upon lines radiating from the centre of each of two overlapping photographs to certain objects appearing on each photograph.

Radial-line plotting Determination of the location of points by the successive intersection and resection of direction lines radiating from the radial centres of overlapping aerial photographs.

Radian (1) Angle subtended by an arc of a circle equal in length to the radius of the circle. (2) The angle subtended by an arc of a circle that is the same length as the radius of the circle, approximately 57 degrees, 17 minutes and 44.6 seconds. There are 2π radians in one complete rotation.

Radiant energy peak Wavelength at which the maximum electromagnetic energy is radiated at a particular temperature.

Radiant flux Rate of flow of electromagnetic radiation measured in watts per square centimetre.

Radiant temperature Concentration of the radiant flux from a material. Radiant temperature is the kinetic temperature multiplied by the emissivity to the one-fourth power.

Radiation (1) Propagation of energy in the form of electromagnetic waves. (2) The emission and propagation of energy through space in the form of waves. Electromagnetic energy and sound are examples of radiation. (3) The emission and propagation of waves transmitting energy through space or through some medium.

Radio altimetry The science and techniques involved in using a radar altimeter for precise measurements of distance between the sensor and the surface or feature. The basic concept of radar altimetry involves a short radar pulse transmitted toward the surface or feature and accurate measurement of the round trip time to the reflecting surface, which yields precise distance. The accuracy, which is generally on the order of several centimetres or tens of centimetres, depends mainly on the sharpness of the pulse, or signal bandwidth, and the footprint of the radar beam. Measurement techniques involve beam-limited and pulse-limited radar altimeters that provide two dimensional, high-resolution height profiles along the flight line. The synthetic aperture approach is also used by SAR altimeters to reduce the size of the footprint along the flight line. . Prominent geoscientific applications of radar altimetry include measurement of ocean surface dynamics as well as solid surface topographic mapping.

Radio band The range of wavelengths or frequencies of electromagnetic radiation designated as radio waves; approximately 10(4) to 10(9) Hz in frequency.

Radio detection and ranging (RADAR) Radar is an active form of remote sensing that operates in the microwave and radio wavelength regions.

Radio Echo The signal reflected by a radar target, or the trace produced by this signal on the screen of the cathode-ray tube in a radar receiver.

Radio waves radiation with wavelengths longer than about 1 meter.

Radiometer (1) An instrument for quantitatively measuring the intensity of electromagnetic radiation in some band of wavelengths in any part of the electromagnetic spectrum. Usually used with a modifier, such as an infrared radiometer or a microwave radiometer. (2) An instrument that measures the intensity of radiation in a particular band of wavelengths in the electromagnetic spectrum, such as infrared or microwave.

Radiometeric Relating to, using or measured by a radiometer. The measurement of radiation.

Radiometric compensation The sensitivity of the antenna is sensitive to angle and the radiometric compensation adjusts for variation in the received signal due to this effect. See Side Lobes.

Radiometric correction Procedures that correct or calibrate aberrations in data values due to specific distortions from such things as atmosphere effects (such as haze) or instrumentation errors (such as striping) in remotely sensed data.

Radiometric resolution The sensitivity of a sensor to incoming reflectance. Radiometric resolution refers to the number of divisions of bit depth (for example, 255 for 8-bit, 65,536 for 16-bit and so on) in data collected by a sensor.

Radius The distance from the centre to a point on the outer edge of a circle, circular curve or sphere.

Raleigh effect It is the elastic scattering of light or other electromagnetic radiation by particles much smaller than the wavelength of the light, which may be individual atoms or molecules.

Random line dropout In scanner images, the loss of data from individual scan lines in a nonsystematic fashion.

Random noise In a spatial model, variation in the value of a variable that cannot be described by a mathematical function and is not spatially correlated: it includes measurement error and microscale variation (variation at a finer scale than that at which the data has been sampled). Random noise is one of the three main components – along with drift and spatially correlated variation – that contribute to the change in value of a variable over a surface. In a semivariogram, random noise is represented by the nugget. Random noise is sometimes called white noise.

Range (1) A parameter of a variogram or semivariogram model that represents a distance beyond which there is little or no autocorrelation among variables. (2) In radar usage this is the distance in the direction of radar propagation, usually to the side of the platform in an imaging radar system. The slant range is the direct distance from the antenna to the object, whereas the distance from the ground track of the platform to the object is termed the ground range. In GPS, a fixed distance between two points, such as between a starting and an ending waypoint or a satellite and a GPS receiver.

Range direction See look direction.

Range domain A type of attribute domain that defines the range of permissible values for a numeric attribute. For example, the permissible range of values for a pipe diameter could be between 1 and 32 inches.

Range resolution In radar images, the spatial resolution in the range direction, which is determined by the pulse length of the transmitted microwave energy.

Range time The fast time within a received pulse, relative to the pulse transmission time

Rangefinders Small handheld telescopes equipped with a laser that lets you measure the distance from the telescope to the object. Effective at ranges less than about 1 km. Compare with LiDAR.

Rank A method of assigning an accuracy value to feature classes to avoid having vertices from a feature class collected with a high level of accuracy being snapped to vertices from a less accurate feature class. Vertices from higher ranking feature classes will not be moved when snapping with vertices with lower ranked feature classes. The highest rank is one; up to 50 different ranks can be assigned.

Raster (1) A spatial data model that defines space as an array of equally sized cells arranged in rows and columns, and composed of single or multiple bands. Each cell contains an attribute value and location coordinates. Unlike a vector structure, which stores coordinates explicitly, raster coordinates are contained in the ordering of the matrix. Groups of cells that share the same value represent the same type of geographic feature. (2) The scanned and illuminated area of a video display, produced by a modulated beam of electrons sweeping the phosphorescent screen line by line from top to bottom at a regular rate of repetition.

Raster band One layer in a raster dataset that represents data values for a specific range in the electromagnetic spectrum (such as ultraviolet, blue, green, red, and infrared), or radar, or other values derived by manipulating the original image bands. A raster dataset can contain more than one band. For example, satellite imagery commonly has multiple bands representing different wavelengths of energy from along the electromagnetic spectrum.

Raster calculator An ArcGIS Spatial Analyst tool for performing mathematical calculations with operators and functions, setting up selection queries or typing Map Algebra syntax. Inputs to the Raster Calculator can be raster datasets, raster layers, coverages, shapefiles, tables, constants and numbers.

Raster catalog A collection of raster datasets defined in a table of any format, in which the records define the individual raster datasets that are included in the catalog. Raster catalogs can be used to display adjacent or overlapping raster datasets without having to mosaic them together into one large file. In ArcView 3.x, raster catalogs were called image catalogs.

Raster cell The smallest unit of information in raster data, usually square in shape. In a map or GIS dataset, each cell represents a portion of the earth, such as a square metre or square mile, and usually has an attribute value associated with it, such as soil type or vegetation class.

Raster cleanup The process of drawing, filling and erasing raster cells using ArcScan Raster Cleanup and Raster Painting tools.

Raster data An abstraction of the real world where spatial data is expressed as a matrix of cells or pixels, with spatial position implicit in the ordering of the pixels. With the raster data model, spatial data is not continuous but divided into discrete units. This makes raster data particularly suitable for certain types of spatial operations (e.g. overlays or area calculations). Unlike vector data, there are no implicit topological relationships.

Raster data model (1) A representation of the world as a surface divided into a regular grid of cells. Raster models are useful for storing data that varies continuously, as in an aerial photograph, a satellite image, a surface of chemical concentrations or an elevation surface. (2) A spatial data model that uses a grid to represent the spatial variation of a feature.

Raster dataset In ArcGIS, a raster spatial data model that is stored on disk or in a geodatabase. Raster datasets can be stored in many formats, including TIFF, Imagine, ESRI Grid and MrSid.

Raster dataset band One layer in a raster dataset that represents data values for a specific range in the electromagnetic spectrum (such as ultraviolet, blue, green, red and infrared), or radar, or other values derived by manipulating the original image bands. A raster dataset can contain more than one band. For example, satellite imagery commonly has multiple bands representing different wavelengths of energy from along the electromagnetic spectrum.

Raster format A means of representing spatial data in the form of a grid of DN, each line of which can be used to modulate the lines of a video raster.

Raster graphics Graphics in which an image is generated by scanning an entire screen or page and marking every point as black, white or another colour (as opposed to vector graphics).

Raster intersection Three or more lines in a raster that meet at a common point.

Raster layer In ArcGIS, a layer that references a raster as its data source and a raster renderer that defines how the raster data should be rendered and any additional display properties.

Raster model A representation of the world as a surface divided into a regular grid of cells. Raster models are useful for storing data that varies continuously, as in an aerial photograph, a satellite image, a surface of chemical concentrations or an elevation surface.

Raster pattern Pattern of horizontal lines swept by an electron beam across the face of a CRT that constitute the image display.

Raster postprocessing In ArcScan, the automatic correction of vector feature results immediately after batch vectorization is completed. Postprocessing involves generalizing lines, straightening angles and smoothing lines.

Raster preprocessing Simple raster editing that prepares images for viewing and analysis. Preprocessing includes georeferencing, clipping, positioning, resizing, enhancing and mosaicking.

Raster process definition file In Arc GIS Image Server, an XML file that contains properties of the raster dataset, including metadata, as well as the definition of processes to be applied to one or more individual rasters. A raster process definition file has a .RPDef extension.

Raster product format A data format composed of rectangular pixel arrays (compressed or uncompressed), produced by the National Geospatial-Intelligence Agency and US allies for military applications.

Raster snapping An automatic editing operation in which points or features within a specified distance (tolerance) of other points or features are moved to match or coincide exactly with each others' coordinates.

Raster statistics Statistics that are calculated from the cell values of each band in a raster. The statistics that are calculated include the minimum, maximum, mean and standard deviation cell values, and if the dataset is thematic, the number of classes. Statistics are required for some rendering and geoprocessing operations.

Raster tile An internal subsetting of a spatial dataset (commonly raster) into a manageable rectangular set or rows and columns of pixels, typically used to process or analyse a large raster dataset without consuming vast quantities of computer memory.

Raster tracing An interactive vectorization process that involves drawing along the boundary of contiguous raster cells to create vector features.

Rasterization The conversion of points, lines and polygons into cell data.

Rasterize The process of converting vector points, lines and areas into raster image format.

Rasterized feature layer A feature layer in ArcGlobe that exists as points, lines and polygons but is rendered as cell data. When layers are added to ArcGlobe, they may automatically be rendered in raster format to retain their cartographic symbology.

Ratio image An image prepared by processing digital multi-spectral data as follows: for each pixel, the value for one band is divided by that of another. The resulting digital values are displayed as an image.

Ratioing In digital image processing, enhancing the contrast between features in an image by dividing the values of pixels in one image by the values of corresponding pixels in a second image.

Raw data Numerical values representing the direct observations output by a measuring instrument transmitted as a bit stream in the order they were obtained.

Ray tracing A technique that traces imaginary rays of light from a viewer's eye to the objects in a three-dimensional scene to determine which parts of the scene should be displayed from that perspective.

Rayleigh criterion In radar, the relationship between surface roughness, depression angle and wavelength that determines whether a surface will respond in a rough or smooth fashion to the radar pulse.

Rayleigh scattering Selective scattering of light in the atmosphere by particle that are small compared with the wavelength of light.

RDBMS Acronym for relational database management system. A type of database in which data is organized across one or more tables. Tables are associated with each other through common fields called keys. In contrast to other database structures, an RDBMS requires few assumptions about how data is related or how it will be extracted from the database.

Real aperture radar (RAR) (1) A radar system where the antenna beamwidth is controlled by the physical length of the antenna. Also known as brute force or noncoherent radar. A SLAR system in which azimuth resolution is determined by the physical length of the antenna and by the wavelength. The radar returns are recorded directly to produce images. The advantages of RAR is their simple design and data processing. However, its resolution is poor so RAR are limited to short range, low altitude missions, scanning short wavelengths. The use of the data is limited as shorter wavelengths experience a large amount of atmospheric effects, scattering and dispersion, for example. Because the missions are flown at low altitudes, the coverage is small. The resolution is limited by the length of the antenna. The antenna needs to be many times longer than the wavelength to produce narrow bandwidths. However, it is impractical to design an antenna long enough to produce high-resolution data. (2) Radar system in which azimuth resolution is determined by the transmitted beam width, which is in turn determined by the physical length of the antenna and by the wavelength.

Real time kinematic (RTK) The DGPS procedure whereby carrier phase corrections are transmitted in real time from a reference station to the user's roving receiver.

Real-time Time in which reporting on events or recording of events is simultaneous with the events. For example, the real time of a satellite is the time in which it simultaneously reports its environment as it encounters it.

Real-time data Data that is displayed immediately, as it is collected. Real-time data is often used for navigation or tracking.

Real-time differential GPS A base station which computes, formats, and transmits corrections usually through some sort of data link (e.g. VHF radio or cellular telephone) with each new GPS observation. The roving unit requires some sort of data link receiving equipment to receive the transmitted GPS corrections and get them into the GPS receiver so they can be applied to its current observations.

Real-time mode The time mode in ArcGIS Tracking Analyst in which data is automatically displayed on the map after being added.

Receiving station At a receiving station, antennas collect the signals sent by an orbiting satellite. Electronic devices process the signals and the data are stored. Usually the station also converts the data into usable digital and printed images.

Reclassification The process of taking input cell values and replacing them with new output cell values. Reclassification is often used to simplify or change the interpretation of raster data by changing a single value to a new value or grouping ranges of values into single values – for example, assigning a value of 1 to cells that have values of 1 to 50, 2 to cells that range from 51 to 100 and so on.

Recognizability Ability to identify an object on an image.

Reconcile In concurrency management, to merge all modified data in the current database edit session with a second version of the data.

Record A set of related data fields, often a row in a database, containing all the attribute values for a single feature. For example, in an address database, the fields that together provide the address for a specific individual comprise one record. In the SQL query language, a record is analogous to a tuple.

Rectangular survey The description of the location of land in the United States using a survey system established by the federal government in 1785. The system is based on the concept of a township, a square parcel of land measuring 6 miles on each side. The township's position is described as a number of 6-mile units east of a north–south line (called the meridian) and north or south of an east–west line (called the baseline). Each township is divided into 36 sections, each of which is 1 square mile. A section is divided into quarters equal to 160 acres. The quarter section may be further divided into four 40-acre parcels. Also referred to as the Public Land Survey System (PLSS).

Rectification The process of preparing a photographic image from perspective photography by removing displacement points caused by tilt, relief and central projection perspective. A process of making image data conform to a map projection.

Rectification, differential The process of scanning and reprojecting a photograph onto a horizontal plane in differential elements to remove displacements caused by tilt and relief. The process may be accomplished by any one of a number of instruments developed specifically for the purpose.

Rectification, simple Projection of an aerial photograph (mathematically, graphically, or photographically) from its plane onto a horizontal plane by translation, rotation and (or) scale change to remove displacement due to tilt of the camera.

Rectilinear (1) Characterized by straight lines, usually parallel to orthogonal axes. (2) Refers to images with no geometric distortion in which the scales in the horizontal and vertical directions are identical.

Recycling The process by which server instances in a pool are replaced by new instances of services. Recycling allows services that have become unusable to be destroyed and replaced with fresh services, and it reclaims resources taken up by stale services.

Redistricting The process of revising the boundaries of administrative, legislative or election districts.

Redundancy (1) In Survey Analyst, occurs when the number of observed measurements is greater than the number of computed coordinates in a measurement network. (2) Information on an image which is either not required for interpretation or cannot be seen. Redundancy may be spatial or spectral. The term also refers to multispectral data where the degree of correlation between bands is so high that one band contains virtually the same information as all the bands.

Reference A pointer to an object, interface or other item allocated in memory. COM objects keep a running total of the references to themselves via the IUnknown interface methods AddRef and Release.

Reference data In geocoding, material containing the location and address information of specific features. Reference data consists of the spatial representation of the data and the related attribute table.

Reference data source In ArcGIS, a spatial data layer that a geocoding service uses to perform address geocoding. A reference data source can be any point, line or polygon feature class that contains the necessary address attributes. Each address component is stored as a separate attribute. The address attributes can be contained in the reference data source itself or in a table joined to it.

Reference datum Any datum, plane or surface from which other quantities are measured.

Reference ellipsoid An ellipsoid associated with a geodetic reference system or geodetic datum.

Reference globe The reduced model of the earth from which map projections are constructed. Also called a nominal or generating globe.

Reference grid A grid of numbered rows and lettered columns (or vice versa) superimposed on a map, used to find and identify features. Alphanumeric grids are commonly used as a reference system on local street maps.

Reference level A surface to which heights, elevations or depths are referenced.

Reference map A map designed to show where geographic features are in relation to each other.

Reference point In Survey Analyst, a point with known coordinates, used as input to a computation.

Reference scale The scale at which symbols appear on a digital page at their true size, specified in page units. As the extent is changed, text and symbols will change scale along with the display. Without a reference scale, symbols will look the same at all map scales.

Reference spheroid An ellipsoid associated with a geodetic reference system or geodetic datum.

Reference system A method for identifying positions on the globe. This is often constructed with a grid that either refers to the earth's latitude and longitude (graticule), or a planar equivalent that divides grid lines by a fixed length from a predefined point of origin.

Reference theme In ArcView 3.x, a theme used to perform geocoding. A reference theme can be any point, line or polygon theme that contains the necessary address fields. Each address component (such as street name, address range, street prefix and street type) is stored as a separate field. The address fields can be contained in the reference theme itself or in a table joined to it. During geocoding, each address is matched to a feature in the reference theme that has matching address components (either the single address or address range), then coordinates for the new point features are determined.

Referencing In Survey Analyst, occurs when a copy of a coordinate from a different project is added to the survey point for exclusive use in your project. When this referencing happens, you must choose one of the following coordinates: the GIS coordinate or the current coordinate of the owning survey project.

Referential constraint A DBMS-defined restriction that ensures that foreign key values in the rows in the child table always have matching primary key values in the parent table.

Referential integrity A mechanism for ensuring that data remains accurate and consistent as a database changes. When changes are made to a table related to another table by a common key, the changes are automatically reflected in both tables.

Reflectance (1) Ratio of the radiant energy reflected by a body to the energy incident on it. Spectral reflectance is the reflectance measured within a specific wavelength interval. (2) The proportion of incident radiant energy that is reflected by a surface. Reflectance varies according to the wavelengths of the incident radiant energy and the colour and composition of the surface.

Reflected energy The energy reflected from the subject to the sensor.

Reflected energy peak Wavelength (0.5 Fm) at which maximum amount of energy is reflected from the earth's surface.

Reflected IR Electromagnetic energy of wavelengths from 0.7 Fm to about 3 Fm that consists primarily of reflected solar radiation. This is called reflected radiation because it is produced by the sun and is observed after it is reflected off the surface of the ground. Compare with thermal infrared.

Reflection Reflection occurs when radiation (light, radar signals, etc.) bounces off a target. It is very important in remote sensing how that reflection happens, how much is reflected and how the radiation is changed in the process of reflection, because it tells us much about the target that caused the reflection.

Reflectivity Ability of a surface to reflect incident energy.

Refraction Bending of electromagnetic rays as they pass from one medium into another when each medium has a different index of refraction.

Regedit A utility, part of the Windows operating system, that allows you to view and edit the system registry.

Region In geography, an area usually distinguished by common cultural or physical characteristics, such as Southern California, Western Europe or Southeast Asia.

Region data model A data model that is built upon lines and polygons and allows disjoint components and overlapped areas.

Register To align two or more maps or images so that equivalent geographic coordinates coincide.

Registration (1) Process of superposing two or more images or photographs so that equivalent geographic points coincide. (2) The step in the install process during which users are required to contact ESRI Customer Service and notify them of their single use, server or developer product registration number.

Registration number A three-letter, nine-digit number (ABC123456789, for example) that authenticates software with ESRI. Every single use and server product, including their extensions or options, has a unique registration number.

Registry Stores information about system configuration for a Windows machine. COM uses the registry extensively, storing details of COM components including ProgIDs and ClassIDs, file location of the binary code, marshalling information and categories in which they participate.

Registry file A file containing information in Windows Registry format. Double-clicking a .reg file in Windows will enter the information in the file into the system registry. Often used to register components to component categories.

Regression A statistical method for evaluating the relationship between a single dependent variable and one or more independent variables thought to influence the dependent variable. Regression is used to predict the value of the dependent variable or to determine whether an independent variable in fact influences the dependent variable, and to what extent.

Regression coefficient A value associated with each independent variable in a regression equation, representing the strength and type of relationship the independent variable has to the dependent variable. For example, fire frequency might be modelled as a function of solar radiation, vegetation, precipitation and aspect. A positive relationship between fire frequency and solar radiation is likely (the more sun, the more frequent the fire incidents). When the relationship is positive, the sign for the associated coefficient is also positive. A negative relationship between fire frequency and precipitation is also likely (places with more rain have fewer fires). Coefficients for negative relationships have negative signs. If the relationship is strong, the absolute value of the coefficient is large. Weak relationships are associated with coefficients near zero.

Regression equation The mathematical formula applied to independent variables to best predict the dependent variable being modelled. The notation in regression equations is always Y for the dependent variable and X for the independent variables. Each independent variable is associated with a regression coefficient describing the strength and sign of that variable's relationship to the dependent variable. A regression equation might look like this (where b represents a regression coefficient): Y = b0+b1 X1 + b2 X2 + ... bn Xn

RegSvr32 A Windows utility that registers a DLL or similar component library to a system registry. A DLL and its components must be registered before it can be used.

Rehydrate In programming, to reinstantiate an object and its state from persisted storage.

Reject processing Handling unmatched addresses through fine-tuning the geocoding process. After a table of addresses are matched the first time, unmatched addresses can be reviewed or edited. Reject processing attempts to find possible matches by correcting errors or adjusting search criteria for the addresses that fail the first time.

Relate An operation that establishes a temporary connection between records in two tables using a key common to both.

Relate manager An ArcToolbox utility used to build, modify, save and delete connections (relates) between tables. Saved connections can be reactivated in future ArcToolbox sessions.

Relational database A collection of tables, which can be connected to each other by attributes whose values can uniquely identify a record in a table.

Relational database management system A type of database in which data is organized across one or more tables. Tables are associated with each other through common fields called keys. In contrast to other database structures, an RDBMS requires few assumptions about how data is related or how it will be extracted from the database.

Relational join Appending the fields of one table to those of another through an attribute or field common to both tables. A join is usually used to attach more attributes to the attribute table of a geographic layer.

Relational operator In ArcGIS Spatial Analyst, an operator that evaluates specific relational conditions. If a condition is TRUE, the output is assigned a value of 1. If the condition is FALSE, the output is assigned a value of 0.

Relationship An association or link between two objects in a database. Relationships can exist between spatial objects (features), between nonspatial objects (rows in a table), or between spatial and nonspatial objects.

Relationship class An item in the geodatabase that stores information about a relationship. A relationship class is visible as an item in the ArcCatalog tree or contents view.

Relative accuracy A measure of positional consistency between a data point and other near data points. Relative accuracy compares the scaled distance of objects on a map with the same measured distance on the ground.

Relative coordinates Coordinates identifying the position of a point with respect to another point.

Relative mode A way of using a digitizing tablet in which the digitizer puck behaves like a mouse; the puck is used to point to interface elements rather than to trace shapes on the surface of the digitizing tablet.

Relative path In computing, the location of a computer file given in relation to the current working directory.

Relative positioning The determination of relative positions between two or more receivers which are simultaneously tracking the same GPS signals.

Relative replica In geodatabase editing, the other replica in a replica pair of parent and child. The relative replica to a child replica is the parent replica and vice versa.

Reliability diagram A map element that contains a simplified view of the sources used to compile a map.

Relief Elevations and depressions of the earth's surface, including those of the ocean floor. Relief can be represented on maps by contours, shading, hypsometric tints, digital terrain modelling or spot elevations.

Relief displacement Geometric distortion on vertical aerial photographs. The tops of objects appear in the photograph to be radially displaced from their bases outward from the photograph's centrepoint.

Relief shading (1) Shadows drawn on a map to simulate the effect of the sun's rays over the varied terrain of the land. (2) Technique for making hypsography on a map appear three dimensional by the use of graded shadow effects. Generally, the features are shaded as though illuminated from the northwest.

Rematching The process of re-geocoding a feature or features in a geocoded feature class.

Remote sensing (1) In the broadest sense, the measurement or acquisition of information about some property of an object or phenomenon, by a recording device that is not in physical or intimate contact with the object or phenomenon under study. (2) Instruments that record characteristics of objects at a distance, sometimes forming an image by gathering, focusing and recording reflected light from the sun, or reflected radio waves emitted by the spacecraft. (3) Collecting and interpreting information about the environment and the surface of the earth from a distance, primarily by sensing radiation that is naturally emitted or reflected by the earth's surface or from the atmosphere, or by sensing signals transmitted from a device and reflected back to it. Examples of remote-sensing methods include aerial photography, radar and satellite imaging.

Remote-sensing imagery Imagery acquired from satellites and aircraft, including panchromatic, radar, microwave and multispectral satellite imagery.

Renderer A mechanism that defines how data appears when displayed. For example, the hillshade renderer for raster data in ArcMap calculates and applies shading based on existing data values for slope and aspect.

Rendering tab In 3D Analyst, a tab on the Layer Properties dialog box that allows users to control whether or not a layer is displayed during scene navigation, how it is shaded and its drawing priority. It also allows users to adjust how the computer's memory is allocated when rendering large images.

Repeat pass interferometry Method based on two image acquisitions of the same scene from slightly displaced orbits of a satellite. Phase information of the two image data files are superimposed. The two phase values at each pixel are then subtracted, leading to an interferogram that records only the differences in phase between the two original images. Phase differences can be related to the altitude variation at each position in the swath and enable the production of a Digital Elevation Model (DEM). For optimum results, there should be no change in the backscatter to maintain coherence; vegetated sites are, therefore, a problem. For detection of feature movement (e.g. tracking glaciers) orbits should be as close as possible. Ground control points (GCPs) are required to accurately superimpose the two images. And knowledge of the sensor location is critical. With a good baseline and coherence, this technique can be better than stereo (~ 10 m vertical accuracy).

Repeatable read The isolation level in a database management system (DBMS) which ensures that when the same rows are read multiple times during the course of a transaction, the returned values are the same between subsequent reads.

Replica In geodatabase editing, the portion of the data in a geodatabase that is copied from a source geodatabase to a destination geodatabase during the replication process. The replica includes information needed to synchronize changes to the data. There are three types of replicas: one-way replicas, two-way replicas, and check-in/checkout replicas.

Replica geodatabase In geodatabase editing, the portion of the data in a geodatabase that is copied from a source geodatabase to a destination geodatabase during the replication process. The replica includes information needed to synchronize changes to the data. There are three types of replicas: one-way replicas, two-way replicas, and check-in/checkout replicas.

Replica pair In geodatabase editing, each combination child and parent replica.

Replica schema changes file An XML file that describes schema differences between the data in a replica and its relative replica.

Replica schema file A workspace document file that describes the schema of the data in a replica.

Replica version In geodatabase editing, the version being replicated during the replication process. For all replica types, the replica version for a parent replica is the version the user was connected to when the replica was created. For one-way and two-way replicas, the replica version for a child replica is always the default version. Changes made to replica versions may be synchronized between the replicas.

Replication A means of copying and distributing data from one database to local, remote or mobile users and then synchronizing between these databases for consistency.

Report web service A SOAP ArcWeb service for creating a report or a map for a specified area in the United States. For a given study area or trade area, the Report Web Service applies a spatial query (if needed), obtains the necessary raw data for the report, aggregates it and returns the aggregated results in PDF, Excel or XML format.

Repository In ArcGIS, a database that contains the schema information needed to create a geodatabase from a UML model created using a CASE tool.

Representation A method of illustrating data so it can be viewed and understood. In cartography, representation is used to depict likenesses of real-world features in such a way that the depictions symbolize or correspond to the real features. Representation is used to present information in a format that is viewable, storable and transferable.

Representation control point In ArcGIS, a specialized geometry vertex that controls the phases of certain geometric effects in a representation rule.

Representation rule In ArcGIS, the combination of symbology and any geometric effects that together define the appearance of features sharing a common representation rule ID.

Representational fraction (1) Scale of a map or chart expressed as a fraction or ratio that relates unit distance on the map to distance measured in the same unit on the ground. (2) The ratio of a distance on a map to the equivalent distance measured in the same units on the ground. A scale of 1:50,000 means that one inch on the map equals 50,000 inches on the ground.

Reproduction Summation of all processes involved in printing copies from an original drawing. A printed copy of an original drawing made by the processes of reproduction.

Re-projection Projection of spatial data from one coordinate system to another.

Resampling (1) Modifying the geometry of an image (which may be from either a remotely sensed or map data source). This process usually involves rectification and/or registration. (2) The process of interpolating new cell values when transforming rasters to a new coordinate space or cell size.

Reseau marks Pattern of small crosses added to photographs.

Residual In a regression model, the difference between the observed Y value and the predicted Y value; the unexplained portion of the dependent variable. Predicted values rarely match observed values exactly. The residual is one measure of model fit. Large residuals indicate poor model fit.

Resolution (1) A measure of the amount of detail that can be seen in an image; the size of the smallest object recognizable using the detector. (2) Intensity or rate of data sampling. In remotely sensed imagery, resolution is significant in four measurement dimensions: spectral, spatial, radiometric and temporal. (3) The detail with which a map depicts the location and shape of geographic features. The larger the map scale, the higher the possible resolution. As scale decreases, resolution diminishes and feature boundaries must be smoothed, simplified, or not shown at all; for example, small areas may have to be represented as points.

Resolution merging Sharpening a low-resolution multiband image by merging it with a high-resolution panchromatic image.

Resolution target Series of regularly spaced alternating light and dark bars used to evaluate the resolution of images or photographs.

Resolving power A measure of the ability of individual components and of remote sensing systems, to separate closely spaced targets.

Resource centre ESRI Web site providing various online resources – such as online help, user forums, blogs, samples, user communities, developer content and so on – to help users learn about and use a particular ESRI product. The Resource Centre gateway is located at <A HREF=‘http://resources.esri.com’ TARGET= ‘_blank’>http://resources.esri.com</A>.

REST Acronym for Representational State Transfer. An architecture for exchanging information between peers in a decentralized, distributed environment. REST allows programs on different computers to communicate independently of an operating system or platform by sending a Hypertext Transfer Protocol (HTTP) request to a uniform resource locator (URL) and getting back data in some format – for example, XML, or inside a URL. REST is used in Web services.

Restriction A Boolean network element attribute used for limiting traversal through a network dataset. ‘One way street’, ‘no trucks allowed’, and ‘buses only’ are examples of restrictions.

Reststrahlen band In the IR region, refers to absorption of energy as a function of silica content.

Retrograde (1) Of or relating to the orbital revolution or axial rotation of a planetary or other celestial body that moves clockwise from east to west, in the direction opposite to most celestial bodies. (2) Of or relating to the brief, regularly occurring, apparently backward movement of a planetary body in its orbit as viewed against the fixed stars, caused by the differing orbital velocities of earth and the body observed.

Return In radar, a pulse of microwave energy reflected by the terrain and received at the radar antenna. The strength of a return is referred to as return intensity.

Return beam vidicon (RBV) A system in which images are formed on the photosensitive surface o a vacuum tube; the image is scanned with an electron beam and transmitted or recorded. Landsat 3 used a pair of RBVs to acquire images.

Reverse geocoding The process of finding a street address from a point on a map.

Revolution Orbital motion about a point located outside the orbiting body.

RF The ratio of a distance on a map to the equivalent distance measured in the same units on the ground. A scale of 1:50,000 means that one inch on the map equals 50,000 inches on the ground.

RGB A colour model that uses red, green, and blue, the primary additive colours used to display images on a monitor. RGB colours are produced by emitting light, rather than by absorbing it as is the case with ink on paper. Adding 100 per cent of all three colours results in white.

Rhumb line A complex curve on the earth's surface that crosses every meridian at the same oblique angle. A rhumb line path follows a single compass bearing; it is a straight line on a Mercator projection, or a logarithmic spiral on a polar projection. A rhumb line is not the shortest distance between two points on a sphere.

Rich client application An application that stores and retrieves data locally rather than remotely, enabling easy interaction with other internal resources. Client-side applications maintain a consistent look and feel and support a complex user interface but do not offer the security of server-side applications.

RINEX Receiver INdependent EXchange format. A set of standard definitions and formats to promote the free exchange of GPS data and facilities the use of data from any GPS receiver with any software package. The format includes definitions for three fundamental GOS observables: time, phase and range.

Ring (1) A bearing measured relative to a vessel or aircraft's heading. (2) In ArcGIS, a geometric element from which polygons are constructed. A ring is a closed path (one that begins and ends at the same point). (3) The process of drawing to a display; the conversion of the geometry, colouring, texturing, lighting, and other characteristics of an object into a display image.

Ring study The simplest and most widely used type of market-area analysis, in which a circle is generated around an area on a map; then the underlying demographics are extracted from the area delineated by the circle. Generally, a ring study is used to generate a rough visualization of the market area around a point.

Ringing Fringe-like artefacts produced at edges by some forms of spatial-frequency filtering.

River addressing In hydrology applications, another name for linear referencing. River addressing allows objects such as gauging stations to be located by their relative positions along a line feature.

RMSE Acronym for root mean square error. A measure of the difference between locations that are known and locations that have been interpolated or digitized. RMS error is derived by squaring the differences between known and unknown points, adding those together, dividing that by the number of test points and then taking the square root of that result.

Roamer A transparent gauge that represents easting and northing distances at a given map scale, used to locate positions on a map.

Rods The receptors in the retina which are sensitive to brightness variations.

Roll Rotation of an aircraft that causes a wing-up or wing-down attitude.

Roll compensation system Component of an airborne scanner system that measures and records the roll of the aircraft. This information is used to correct the imagery for distortion due to roll.

Roller-feed scanner A type of scanner that moves a document through a roller assembly over camera sensors that capture a digital image.

Root mean square (RMS) error (1) In geometric transformation, the RMS measures that deviation between the actual location and the estimated location and the established location of the tics. (2) A measure of the difference between locations that are known and locations that have been interpolated or digitized. RMS error is derived by squaring the differences between known and unknown points, adding those together, dividing that by the number of test points, and then taking the square root of that result.

Rotation Motion around an axis passing through the rotating body.

Rotor-SAR (ROSAR) A synthetic aperture radar concept whereby antennas are mounted at the tips of the rotor blades (i.e. helicopter) illuminating a circular ring-shaped swath.

Rough criterion In radar, the relationship between surface roughness, depression angle and wavelength that determines whether a surface will scatter the incident radar pulse in a rough or intermediate fashion.

Roughness In radar, the average vertical relief of a small-scale irregularities of the terrain surface. Also called surface roughness.

Route (1) A basic element of the dynamic segmentation model that uses a collection of sections to represent a linear phenomenon such as a highway, bike or stream. (2) Any line feature, such as a street, highway, river, or pipe, that has a unique identifier.

Route analysis In ArcGIS Network Analyst, a type of network analysis that determines the best route from one network location to one or more other locations. It can also calculate the quickest or shortest route depending on the impedance chosen. The order of the stops may be determined by the user. For example, if the impedance is time, then the best route is the quickest route.

Route editing toolbar A set of tools that allows users to create and modify routes in ArcMap.

Route event In linear referencing, linear, continuous or point features occurring along a base route system.

Route event source In linear referencing, the result of the dynamic segmentation process. A route event source serves an event table as a dynamic feature class. Every row in the table is served as a feature whose shape is calculated when needed. For example, a route event source can act as the basis of a feature layer in ArcMap.

Route event table In linear referencing, a table that stores route locations and their attributes. A route event table, at a minimum, consists of a route identifier field and a measure location field (point events) or fields (line events).

Route feature class In linear referencing, a collection of routes with a common system of measurement stored in a single feature class (for example, a set of all highways in a county).

Route finder web service A SOAP ArcWeb service for generating multipoint driving directions between locations. This Web service accepts x,y coordinate inputs for two or more locations, along with routing preferences, and returns a route map and textual driving directions for the suggested route.

Route identifier In linear referencing, a numeric or character value used to identify a route.

Route location In linear referencing, a discrete location along a route (point) or a portion of a route (line). A point route location uses only a single measure value to describe a discrete location along a route. A line route location uses both a from- and to-measure value to describe a portion of a route.

Route measure In linear referencing, a value stored along a linear feature that represents a location relative to the beginning of the feature, or some point along it, rather than as an x,y coordinate. Measures are used to map events such as distance, time or addresses along line features.

Route measure anomalies In linear referencing, route measure values that do not adhere to the expected behaviour. Route measure anomalies can often be fixed with ArcMap route editing tools.

Route reference In linear referencing, a collection of routes with a common system of measurement stored in a single feature class (for example, a set of all highways in a county).

Route renewal In ArcGIS Network Analyst, an object used in vehicle routing problem (VRP) analysis. A route renewal object specifies a depot that can be used by a particular route to load/unload the vehicle along the route as necessary so that the capacity is reset and the route can service more orders.

Route seed point In ArcGIS Network Analyst, a feature used in vehicle routeing problem (VRP) analysis. A route seed point can be used to cluster the orders for a specified route.

Route server A public ArcIMS virtual server that supports routeing, reverse geocoding, and geocoding of Spatial Data Compressed (SDC) data. This server is available as an extension to ArcIMS.

Route service A type of Web service that determines driving directions between a set of route stops.

Route zone In ArcGIS Network Analyst, a feature used in vehicle routeing problem (VRP) analysis. A route zone has a polygon geometry and can be used to define the area of coverage or available service for a specified route.

Rover (1) A portable GPS receiver used to collect data in the field. The rover's position can be computed relative to a second, stationary GPS receiver. (2) Any mobile GPS receiver collecting data during a field session. The receiver's position can be computed relative to another, stationary GPS receiver.

Roving window On a raster, an analysis boundary or processing window within which cell values affect calculations and outside which they do not. Filters are used mainly in cell-based analysis where the value of a centre cell is changed to the mean, the sum or some other function of all cell values inside the filter. A filter moves systematically across a raster until each cell has been processed. Filters can be of various shapes and sizes, but are most commonly three-cell by three-cell squares.

Row (1) The latitudinal (nominal) centre line of a Landsat scene. (2) The scan lines that constitute an image. (3) A record in a table.

Row standardization A technique for adjusting the weights in a spatial weights matrix. When weights are row standardized, each weight is divided by its row sum. The row sum is the sum of weights for a feature's neighbours.

RPF Acronym for Raster Product Format. A data format composed of rectangular pixel arrays (compressed or uncompressed), produced by the National Geospatial-Intelligence Agency and US allies for military applications.

R-squared A statistic computed by the regression equation to quantify model performance. The value of R-squared ranges from 0 to 100 per cent. If a model fits the observed dependent variable values perfectly, the R-squared value is 1.0, although this is highly unlikely. An R-squared value like 0.49, for example, is far more likely and means that the model explains 49 per cent of the variation in the dependent variable.

RSS Acronym for Really Simple Syndication, Resource Description Framework (RDF) Site Summary or Rich Site Summary, depending on the source. A simple, structured XML format for sharing content among different Web sites. RSS documents include key metadata elements such as author, date, title, a brief description and a hypertext link. This information helps a user (or an RSS publisher service) decide what materials are worth further investigation. Examples include news feeds, events lists, news stories, headlines and excerpts from blogs and discussion forums.

RSS feed A text, audio or media clip delivered over the Internet using RSS technology. RSS feeds can be delivered on demand to a browser with RSS-enabled software.

R-tree A tree data structure, similar to a B-tree, used for indexing spatial data within a database. In an R-tree structure, data is sorted into a set of hierarchical nodes that may overlap. Each node has a variable number of entries, each of which includes an identifier for child nodes or actual data elements and a bounding box for all entries within the child node or the data elements. Searching algorithms check the bounding boxes before searching within a child node, thus avoiding extensive searches.

Rubber banding A procedure for adjusting the coordinates of all the data points in a dataset to allow a more accurate match between known locations and a few data points within the dataset. Rubber sheeting preserves the interconnectivity between points and objects through stretching, shrinking or reorienting their interconnecting lines.

Rubber sheeting A procedure for adjusting the coordinates of all the data points in a dataset to allow a more accurate match between known locations and a few data points within the dataset. Rubber sheeting preserves the interconnectivity between points and objects through stretching, shrinking or reorienting their interconnecting lines.

Run time The time during which a program is running or the time it takes to run a program.

Run-length encoding A data compression technique for storing raster data. Run-length encoding stores data by row. If two or more adjacent cells in a row have the same value, the database stores that value once instead of recording a separate value for each cell. The more adjacent cells there are with the same value, the greater the compression.

Runtime environment The host that provides the services required for compiled code to execute. The Service Control Manager (SCM) is effectively the runtime environment for COM. The Visual Basic Virtual Machine (VBVM) is the runtime environment that runs Visual Basic code.

S

SAMII Stratospheric Aerosol Measurement experiment, carried by Nimbus–7.

SAMS Stratospheric and Mesospheric Sounder, carried by Nimbus–7.

SAR focusing In a long synthetic aperture (array), SAR focusing involves the removal and compensation of path length differences from the antenna to the target on the ground. The main advantage of a focused synthetic aperture is that it increases its array length over those radar signals that can be processed and thus increases potential SAR resolution at any range. SAR focusing is a necessary process when the length of a synthetic array is a significant fraction of the range to ground being imaged, as the lines-of-sight (range) from a particular point on the ground to each individual element of the array differ in distance. These range differences or path length differences of the radar signals can affect image quality. In a focused SAR image these phase errors can be compensated for by applying a phase correction to the return signal at each synthetic aperture element. Focusing errors may be introduced by unknown or uncorrected platform motion. In an unfocused SAR image, the usable synthetic aperture length is quite limited.

Satellite Anybody, natural or artificial, in orbit around a planet. The term is used most often to describe moons and spacecraft. A man-made satellite is a spacecraft that orbits another body, such as a planet or the sun. A natural satellite is another term for a moon.

Satellite constellation The arrangement of a set of satellites in space.

Satellite imagery Imagery acquired from satellites and aircraft, including panchromatic, radar, microwave and multispectral satellite imagery.

Saturation (1) The intensity of a colour. A highly saturated colour is a vivid, brilliant colour; to dull a colour (decrease its saturation), you add small amounts of its complement, making it closer to grey. (2) The intensity or purity of a colour; the perceived amount of white in a hue relative to its brightness or how free it is of grey of the same value.

S-Band A nominal frequency range from 4 to 2 GHz (7.5 to 15 cm wavelength) within the microwave (radar) portion of the electromagnetic spectrum. S-band radars are used for medium-range meteorological applications, for example, rainfall measurements, as well as airport surveillance and specialized tracking tasks.

SBUV Solar Back-scatter Ultraviolet Instrument, carried by NOAA satellites.

Scalable The ability to grow in size or complexity without showing negative effects.

Scalable vector graphics An XML-based graphics file format that describes two-dimensional vector images, including animation. SVG images scale to fit the display window without compromising quality.

Scalar reference In Survey Analyst, used to define measurement units based on a common standard.

Scale (1) The ratio or relationship between a distance or area on a map and the corresponding distance or area on the ground, commonly expressed as a fraction or ratio. A map scale of 1/100,000 or 1:100,000 means that one unit of measure on the map equals 100,000 of the same unit on the earth. (2) Ratio of distance on an image to the equivalent distance on the ground.

Scale bar A map element used to graphically represent the scale of a map. A scale bar is typically a line marked like a ruler in units proportional to the map's scale.

Scale factor (1) Ratio of the local scale to the scale of the reference globe. The scale factor is 1.0 along a standard line. (2) The reciprocal of the ratio used to specify scale on a map. For example, if the scale of a map is given as 1:50,000, the scale factor is 50,000.

Scale palette An ArcWeb Services tool that allows a user to view, add, delete and modify scale ranges for feature classes.

Scale range The scales at which a layer is visible on a map. Scale ranges are commonly used to prevent detailed layers from displaying at small scales (zoomed out) and to prevent general layers from displaying at large scales (zoomed in).

Scale viewer A tool in the ArcWeb Services Web site that allows a user to see the current scale and move to different scale views.

Scan An image data line produced from a single detector of a band during a scan.

Scan Line (1) A series of spacecraft pointing in one dimension. (2) Narrow strip on the ground that is swept by IFOV of a detector in a scanning system.

Scan skew Distortion of scanner images caused by forward motion of the aircraft or satellite during the time required to complete a scan.

Scanner A device that captures a print or hard-copy image, such as a text document or map, and records the information in digital format. (2) An imaging system in which the IFOV of one or more detectors is swept across the terrain.

Scanner distortion Geometric distortion that is characteristic of cross-track scanner images.

Scanning The process of capturing data from hard-copy maps or images in digital format using a device called a scanner.

Scanning mirror Landsat TM's scanning mirror collects data on both the forward and reverse scans.

Scanning sensor A sensor with scanning functions

Scanning sonar Active system for acquiring images of the seafloor using pulsed sound waves.

Scanning synthetic aperture radar (ScanSAR) A devise having the capability to illuminate several subswaths by scanning its antenna off-nadir into different positions.

Scatter chart A chart in which each data point is marked against perpendicular x- and y-axes. Scatter charts are frequently used in analysis to find data trends.

Scattering coefficient curves Display of scatterometer data in which relative backscatter is shown as a function of incidence angle.

Scatterometer Non-imaging radar device that quantitatively records backscatter of terrain as a function of incidence angle.

Scene (1) Each Landsat image collected is a scene. Each Landsat scene is 115 x 106 miles long. The globe is divided into 57,784 scenes and each Landsat scene has about 3 billion bytes of data. (2) In 3D Analyst, a document containing 3D data that can be viewed in perspective.

Schema The structure or design of a database or database object, such as a table, view, index, stored procedure or trigger. In a relational database, the schema defines the tables, the fields in each table, the relationships between fields and tables and the grouping of objects within the database. Schemas are generally documented in a data dictionary. A database schema provides a logical classification of database objects.

Schema synchronization During geodatabase editing, the process of applying schema changes from a replica to the relative replica in a replica pair. Examples of schema changes include adding or dropping a field.

Schema-only checkout In ArcGIS 9.1 and previous versions, a type of checkout that creates the schema of the data being checked out in the

ASAR ScanSAR Geometry

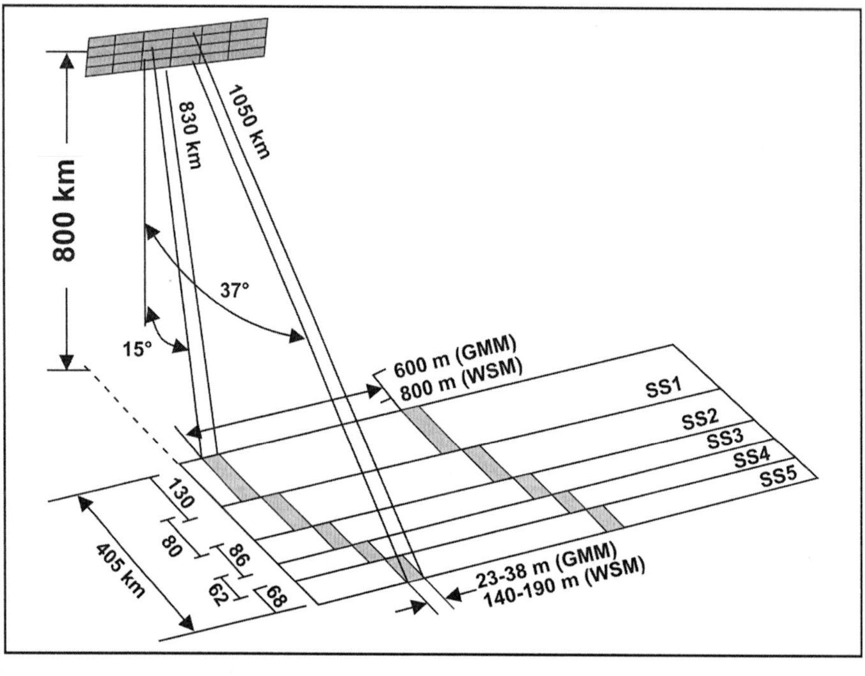

checkout geodatabase but does not copy any data.

SCM Acronym for Service Control Manager. An administrative tool that enables the creation and modification of system services. It effectively serves as the runtime environment for COM.

S-code The standard PRN code used by most civilian GPS receivers.

Scotopic vision Vision under conditions of low illumination, when only the rods are sensitive to light. Visual acuity under these conditions is highest in the blue part of the spectrum.

Scratch file A file, created by either a software user or an operating system that holds temporary data or results during an operation. When the operation is complete, the file is deleted.

Scratch workspace A path to a container for file-based geographic data that can be set in the Environment Settings dialog box or at the command line, into which all automatically generated outputs will be placed.

Script A set of computing instructions, usually stored in a file and interpreted at run time.

Scrubbing Checking the accuracy of data before it is converted into a different format.

SDC dataset A collection of Smart Data Compression (SDC) feature classes sharing attribute information with different geometries. An SDC dataset is stored in a set of related files and contains multiple feature classes. SDC is the core data structure used in ArcGIS StreetMap, ArcIMS RouteServer, RouteMAP IMS, Business Analyst, and BusinessMAP.

SDC feature class A highly compressed, read-only data structure that can store spatial geometry (points, lines and polygons) and attribute data. The SDC structure supports geocoding, routing and most spatial operations. SDC is the core data structure used in ArcGIS Street Map, ArcIMS RouteServer, RouteMAP IMS, ArcGIS Business Analyst and Business MAP.

SDE user The account used to administer ArcSDE. The SDE user requires certain advanced database privileges to configure and manage ArcSDE geodatabases and ArcSDE services.

SDEHOME A system environment variable that defines the location where ArcSDE software is installed.

Sdesetup A program run by a database administrator during ArcSDE installation or upgrade. The sdesetup program creates or upgrades the ArcSDE system tables. It is named differently for different databases; for example, the program used with Oracle 9i is called sdesetupora9i.

SDI Acronym for spatial data infrastructure. A framework of technologies, policies, standards, and human resources necessary to acquire, process, store, distribute and improve the use of geospatial data across multiple public and private organizations.

SDK Acronym for software development kit. A set of code libraries and related tools used to develop platform-specific applications.

SDTS Acronym for Spatial Data Transfer Standard. A data exchange format for transferring different databases between dissimilar computing systems, preserving meaning and minimizing the amount of external information needed to describe the data. All federal agencies are required to make their digital map data available in SDTS format upon request, and the standard is widely used in other sectors.

Sea level (water level) Height of the surface of the sea at any given time.

Sea Satellite (Seasat) NASA ocean research satellite that was in operation July-September of 1978. Seasat was the first (civilian) satellite to carry a SAR. It operated at L-band, using horizontal polarization at 22E incidence angle. Data from Seasat is still important for applications and processing technique development.

Seamline In ArcGIS Image Server, a polygon that defines the mosaic boundary of the raster dataset used in an image service. Overlapping raster datasets can be blended along the seamline by a specified width. Seamlines are created with the Seamline extension tools and are stored as shapefiles within the image service definition.

Search radius The maximum distance in coverage units a feature can be from the current point for consideration as the closest feature. The default is the width or height of the near coverage BND (boundary) divided by 100, whichever is larger.

Search tolerance In network analysis, the threshold distance used to find the closest network element to a network location.

Searching neighbourhood In spatial interpolation, a polygon that forms a subset of data around the prediction location. Only data within the searching neighbourhood is used for interpolation.

Seat In software licensing, the number of simultaneous instances of software that can be used at one time. Most often, seats represent software users at individual computers. Seats may, however, also represent the simultaneous number of servers or connections in use.

SEC The attribute table for the section feature class in a coverage, containing the route number and arc number to which the section belongs, the starting and ending positions expressed as percentages of the arc length, the starting and ending positions expressed as measures along the route, an internal sequence number, a section feature identifier and user-defined attributes.

Secant A straight line that intersects a curve or surface at two or more points.

Secant projection A projection whose surface intersects the surface of a globe. A secant conic or cylindrical projection, for example, is recessed into a globe, intersecting it at two circles. At the lines of intersection, the projection is free from distortion.

Second An angle equal to one sixtieth of a minute of latitude or longitude.

Second normal form The second level of guidelines for designing table and data structures in a relational database. The second-normal-form guideline incorporates the guidelines of first normal form; in addition, it recommends removing data that applies to multiple rows in a table into its own table and using a foreign key to create a relationship to the original table. A database that follows these guidelines is said to be in second normal form.

Secondary reference data All material used as reference data in an address locator beyond the primary reference data. Secondary reference data can consist of an Alternate Name table or a Place-Name Alias table.

Second-order stationarity In geostatistics, the assumption that a set of data comes from a random process with a constant mean, and spatial covariance that depends only on the distance and direction separating any two locations.

Section (1) The arcs or portions of arcs used to define a route. (2) Unit of subdivision of a *township*; normally a quadrangle 1 mile square with boundaries conforming to *meridians* and parallels within established limits, and containing 640 acres as nearly as practicable. In GIS, it means the standard, non-topological vector data format used in ARCVIEW.

Section table The attribute table for the section feature class in a coverage, containing the route number and arc number to which the section belongs, the starting and ending positions expressed as percentages of the arc length, the starting and ending positions expressed as measures along the route, an internal sequence number, a section feature identifier and user-defined attributes.

Secure sockets layer An encryption protocol for the secure transfer of private information over the Internet.

Segment In ArcGIS, a geometric element from which paths are constructed. A segment consists of a start point, an endpoint, and a function that describes a straight line or curve between these two points. Curves may be circular arcs, elliptical arcs or Bézier curves.

Select To choose from a number or group of features or records; to create a separate set or subset.

Selectable layers Layers from which features can be selected in ArcMap with the interactive selection tools. Selectable layers can be chosen using the Set Selectable Layers command in the Selection menu or on the optional Selection tab in the table of contents.

Selected set A subset of features in a layer or records in a table that is chosen by the software user.

Selected values A subset of the features in a layer or records in a table. ArcMap provides several ways to select features and records graphically or according to their attribute values.

Selection anchor In an ArcMap editing session, a small 'x' located in the centre of selected features. The selection anchor is used in the snapping environment, or when rotating, moving and scaling features.

Selection file An ArcInfo Workstation file that stores the specifications of the selection criteria to be applied against specific coverages and INFO tables. It does not store the actual selected features or records, only the selection methods. Selection files are created with the WRITESELECT command in ArcPlot.

Selective availability (S/A) The intentional degradation by the US Department of Defense of the GPS signal for civilian receivers, which could cause errors in position of up to 100 metres. Selective availability (S/A) was removed from the civilian signal in May 2000. Since the lifting of S/A restrictions, position accuracy levels have improved to 20 metres or less. Intentional degradation of the performance capabilities of the NAVSTAR satellite system for civilian use by the US military, accomplished by artificially creating a significant clock error in the satellites. A random error that was imposed on information from satellites relaying positional data; this error was rescinded in 2000 but may be reinstated by the USA military.

Self-organizing map A map that uses a neural network algorithm to classify and illustrate associations in complex datasets, and reveal multidimensional patterns. A similar set of methods produces maps sometimes referred to as Kohonen maps. Kohonen maps are named for the Finnish engineer Teuvo Kohonen.

Semantics The definition of the meaning of concepts within a data model by their relationships to other concepts.

Semicross variogram A function of the distance and direction separating two locations, used to quantify cross correlation. The cross variogram is defined as the variance of the difference between two variables of different types or attributes at two locations. The cross variogram generally increases with distance and is described by nugget, sill and range parameters.

Semimajor axis The equatorial radius of a spheroid, often referred to as 'a'.

Semiminor axis The polar radius of a spheroid, often referred to as 'b'.

Semivariogram The variogram divided by two.

Sense of place One's perception of the essential character of a place in which one resides or has resided, stemming from a personal response to the environment. Sense of place usually refers to perceptions of a neighbourhood or city, but can also describe feelings about a larger region, state or country.

Sensitivity Degree to which a detector responds to electromagnetic energy incident on it.

Sensitivity analysis Analysis designed to test the robustness of model and analytical results to ensure that small changes in model parameters or data structure do not exhibit large changes in the results.

Sensitivity time control (STC) Pre-programmed change in radar amplitude due to weaker backscatter from greater ranges and varying incidence angles across the imaged swath.

Sensor (1) An electronic device for detecting energy, whether emitted or radiated, and converting it into a signal that can be recorded and displayed as numbers or as an image. (2) Device that produces an output (usually electrical) in response to stimulus such as incident radiation. Sensors aboard satellites obtain information about features and objects on the earth by detecting radiation reflected or emitted in different bands of the electromagnetic spectrum.

Analysing the transmitted data provides valuable scientific information about the earth. Weather satellites commonly carry radiometers which measure radiation from snow, ice, clouds and bodies of water. Space-borne radars are used for earth observations, bouncing the radar waves off land and ocean surfaces to study sea-surface conditions, ice thickness and land surface features. A wind scatterometer is a special type of radar designed to measure ocean surface winds indirectly by bouncing signals off the water and measuring them from various angles. Infrared (IR) detectors measure heat generated by earth features in the IR band of the spectrum. Photographic reconnaissance sensors in their simplest form are large telescope-camera systems used to view objects on earth's surface. The bigger the lens, the smaller the object that can be detected. Camera-telescope systems now incorporate all sorts of sophisticated electronics to produce better images, but even these systems need cloudless skies, excellent lighting and good colour contrast between objects and their surroundings to detect objects the size of a basketball. Some of the satellites produce film images that must be returned to earth, but a more convenient method is to record the image as a series of digital code numbers, then reconstruct the image from the electronic code using a computer at a ground station.

Sequential analysis Analysis based on a sample of an unfixed size in which testing continues only until a trend is observed with a predefined level of certainty.

Serializable read The isolation level in a database management system (DBMS) that offers the highest degree of isolation from concurrent transactions. All reads in the transaction only see data committed before the transaction began and never see concurrent transaction changes committed during transaction execution.

Serialization A method of data conversion in which data is encoded as a sequence, stored in a file, memory buffer, or other medium and transmitted across a network connection. Data is often serialized for transmission over phone lines or networks that require data to be sent one bit at a time.

Serialization file A file that stores serialized data.

Server A computer that manages shared resources, such as disks, printers and databases on a network.

Server account The operating system account that the server object manager service runs as. The server account is specified by the GIS server postinstallation utility.

Server context In ArcGIS, the environment on the GIS server within which a service or server object and its associated objects are running. A server context runs within a server container process. A developer gets a reference to a server object through the server object's server context and can create other objects within a server object's context.

Server directory A location on a file system used by a GIS server to retrieve map caches and write different types of output.

Server object In ArcGIS, an object that manages and serves a GIS resource, such as a map or a locator. A server object is a high-level object that simplifies the programming model for doing certain operations and hides the fine-grained ArcObjects that do the work. Server objects also have SOAP interfaces, which makes it possible to expose server objects as Web services that can be consumed by clients across the Internet. At ArcGIS 9.3, there are seven types of server objects: MapServer, GeocodeServer, GeodataServer, GeometryServer, GlobeServer, GPServer, and ImageServer.

Server object isolation In ArcGIS, a condition that describes whether multiple instances of a service can run within the same process. Services with high isolation run dedicated processes, whereas services with low isolation share processes with other services of the same type.

Server product Products that can be used on one enterprise server machine. Each server product requires a unique registration number used to generate the authorization file. The Software Authorization Wizard is then used to enter the authorization file and unlock the software for use.

Server-side address locator A n address locator in which processing is done on one computer with the results accessible to other computers. Server-side address locators consist of services available over the Internet, via local area networks or via an ArcSDE server.

Service A persistent software process that provides data or computing resources for client applications.

Service area analysis In ArcGIS Network Analyst, a type of network analysis for determining the region that encompasses all accessible streets (streets that lie within a specified impedance). For example, the 20-minute service area for a network location (such as a fire station) includes all the streets that can be reached within 20 minutes from that location.

Service configuration file In ArcIMS, the file that contains the core site information. ArcIMS configuration files contain all the basic information about the content to be delivered, such as location of the data and layer symbology. Typically, a configuration file contains data that defines map content and has a file extension of .axl, but it can also be used to deliver metadata or route data (as .axl files) and to serve maps created in ArcMap (.mxd or .pmf files). Regardless of their type, configuration files contain contents that the service registers to the ArcIMS spatial server and Web server for processing.

Service group A bundle of data that covers a geographic area. Service groups are used for OpenLS and Mobile Toolkit requests.

Service provider A software program that receives service requests from clients, processes data and serves it back to the client.

Service table In ArcGIS Image Server, a table defining links to multiple raster datasets and some service-specific attributes, including display range. A service table is stored as part of the footprint shapefile within an image service definition.

Servlet A Java platform technology for extending Web servers that provides a component-based, platform-independent method for building Web-based applications.

Servlet connector In ArcIMS, the default connector used to connect the ArcIMS Application Server to a Web server.

Servlet engine A Java platform technology that interfaces with a Web server and hosts servlets. Servlet engines provide a common application programming interface (API) that servlets use to build Web applications.

Sextant A handheld navigational instrument that measures, from its point of observation, the angle between a celestial body and the horizon or between two objects. The angle is measured on a graduated arc that covers one-sixth of a circle (60 degrees).

SGML Acronym for Standard Generalized Markup Language. A markup language with a predefined vocabulary and structure. SGML is used to structure information in a DTD, or Document Type Definition format, for exchanging information between different programs and machines. SGML uses a tag structure and was standardized by ISO in 1986.

SHA1 Acronym for Secure Hash Algorithm v1. A Federal Information Processing Standards (FIPS) method for encoding sensitive information before sending the information over the Internet. The method jumbles data, such as passwords, so it is virtually impossible to recover the original information.

Shade symbol A colour or pattern used to fill polygons on a map.

Shaded relief image A raster image that shows changes in elevation using light and shadows on terrain from a given angle and altitude of the sun.

Shading Graphic patterns such as cross hatching, lines, or colour or grayscale tones that distinguish one area from another on a map.

Shallowly stateful application An application that uses the session state management capabilities of a Web server to maintain application state and makes stateless use of server objects. Shallowly stateful applications can use pooled server objects.

Shape The characteristic appearance or visible form of a geographic object as represented on a map. A GIS uses points, lines and polygons to represent the shapes of geographic objects.

Shapefile (1) A vector data storage format for storing the location, shape, and attributes of geographic features. A shapefile is stored in a set of related files and contains one feature class. (2) The term used by ESRI for non topological vector data.

Shared boundary A boundary common to two features. For example, in a parcel database, adjacent parcels share a boundary. Another example is a parcel that shares a boundary on one side with a river. The segment of the river that coincides with the parcel boundary shares the same coordinates as the parcel boundary.

Shared content In ArcWeb Services, data files, layers and services that are available to more than one user.

Shared vertex A vertex common to multiple features. For example, in a parcel database, adjacent parcels share a vertex at the common corner.

Sheet-fed scanner A type of scanner that moves a document through a roller assembly over camera sensors that capture a digital image.

Shield A map symbol that serves as a route marker. Shields come in many varieties, but the most common shields in the United States are for interstate highways, US routes, state routes and county routes. A uniform standard exists for interstate highways, US routes and most county routes across the United States, while shields for state routes vary by state.

Shift Adding a constant value to a coordinate.

Shoreline Intersection of the land with the water surface.

Short pulse radar Radar system capable of generating nanosecond pulses of energy at microwave frequencies.

Shortcut key A keystroke combination that executes a command that might otherwise require the use of a mouse. For example, Ctrl+C is a well-known keyboard shortcut for copying a selection in Windows.

Shortcut menu A list menu that pops up when the right mouse button is clicked in Windows applications. Some keyboards also have an application key that opens shortcut menus.

Shortest path The best route or the route of least impedance between two or more points, taking into account connectivity and travel restrictions such as one-way streets and rush-hour traffic.

Short-range variation In a spatial model, fine-scale variation that is usually modelled as spatially-dependent random variation.

Showcase application A complete application that demonstrates a solution for a specific need. ArcWeb Services offers downloadable showcase applications, but they require some level of customization.

Shuttle imaging radar (SIR) L-band radar system deployed on the Space Shuttle. Synthetic-aperture radar experiments carried aboard the NASA Space Shuttle in 1981 and 1984.

SIC codes The federal US standard for classifying establishments by their primary type of business activity. Standard Industrial Classification codes (SIC codes) are used as an identification system in business directories, publications and statistical sources. The classification system was officially replaced by NAICS in 1997, but it is still used by some organizations outside the federal government.

Side offset An adjustable value that dictates how far away from either the left or right side of a line feature an address location should be placed. A side offset prevents a point feature from being placed directly over a line feature.

Side- signal noise Signal noise is any unwanted or contaminating signal competing with the desired signal. In a SAR, two common kinds of noise are additive (receiver) noise and signal dependent noise, usually either additive or multiplicative. The relative amount of additive noise is described by the signal-to-noise ratio (SNR). Signal dependent noises, such as azimuth ambiguities or quantization noise, arise from system imperfections and are dependent on the strength of the signal itself. Speckle is sometimes considered to be a kind of signal dependent multiplicative noise in a SAR system.

Sidelap Extent of lateral overlap between images acquired on adjacent flight lines.

Sidelobes Non-zero levels in a distribution that are separated from the desired central response. Sidelobes arise naturally in antenna patterns, for example, although in general they are a nuisance, and must be suppressed as much as possible. Large side-lobes may lead to unwanted multiple images of a single feature.

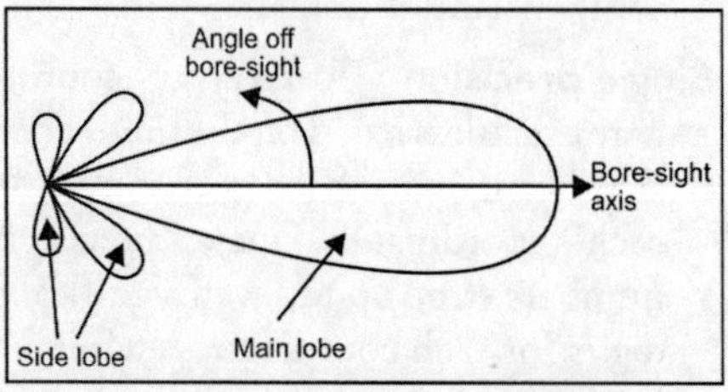

Side-looking airborne radar (SLAR) A high-resolution real aperture radar (RAR) having antennas aimed to the right or left of the flight path. Also referred to as side-looking radar. An all-weather, day/night remote sensor which is particularly effective in imaging large areas of terrain. It is an active sensor, as it generates its own energy which is transmitted and received to produce a photo-like picture of the ground. It provides high-resolution strip maps with photograph-like detail.

Side-scanning system A system that acquires images of a strip of terrain parallel with the flight or orbit path but offset to one side.

Side-shot course In Survey Analyst, used to compute a coordinate that is not a part of the main traverse course sequence.

Signal (1) Information conveyed via an electric current or electromagnetic wave. (2) Information recorded by a remote sensing system.

Signal to noise radio (S/N) (1) The ratio of the level of the signal carrying real information to that carrying spurious information as a result of defects in the system. (2) The ratio of the information content of a signal to its noninformation content (noise).

Signature The pattern of electromagnetic radiation that identifies a chemical or compound. Materials can be distinguished from one another by examining which portions of the spectrum they reflect and absorb.

Significance level In statistical testing, the probability of an incorrect rejection of the null hypotheses.

Sill A parameter of a variogram or semivariogram model that represents a value that the variogram tends toward when distances become large. Under second-order stationarity, variables become uncorrelated at large distances, so the sill of the semivariogram is equal to the variance of the random variable.

Silver halide Silver salts that are especially sensitive to visible light and convert to metallic silver when developed.

Simple edge feature In a geodatabase, a line feature that corresponds to a single network element in the logical network.

Simple feature A point, line or polygon that is not part of a geometric network and is not an annotation feature, dimension feature or custom object.

Simple junction feature In a geodatabase, a junction feature that corresponds to a single network element in the logical network.

Simple kriging A kriging method in which the weights of the values do not sum to unity. Simple kriging uses the average of the entire dataset, which is less accurate than ordinary kriging but produces a smoother result.

Simple market area An area defined by a generalized boundary drawn around the most distant set of customer points (a convex hull) that total to some value. The calculation may be unweighted (in which case every point has the same value) or weighted by a value in the underlying database, such as sales.

Simple measurement The simplest form in which measurements from COGO sources or TPS (Total Positioning System) sources can be stored.

Simple relationship A link or association between data sources that exist independently of each other.

Simple style In ArcWeb Services, a single symbol used to display the features in a layer class at all scale ranges. Users can create simple styles.

Simple temporal event In ArcGIS Tracking Analyst, an event that contains all necessary information in one temporal data message (for real-time data) or record (for fixed-time data).

Simplification A type of cartographic generalization in which the important characteristics of features are determined and unwanted detail is eliminated to retain clarity on a map whose scale has been reduced.

Simultaneous conveyance A means of defining multiple units of land in a single survey document in such a way that all their boundaries have equal legal status. A common example of simultaneous conveyance is the modern subdivision.

Single event upsets A Single Event Upset (SEU) occurs when an energetic particle travels through a transistor substrate and causes electrical signals within the transistor. This is a known phenomenon that usually occurs in near-earth orbit to spacecraft passing through the Van Allen belts, especially the northern and southern auroral zones and the south Atlantic anomaly.

Single precision A level of coordinate exactness based on the number of significant digits that can be stored for each coordinate. Single precision numbers store up to seven significant digits for each coordinate, retaining a precision of plus or minus 5 metres in an extent of 1,000,000 metres. Datasets can be stored in either single or double precision coordinates.

Single setup page One of two types of setup pages in the Survey Analyst Survey Explorer. The single setup page is used for computations that process single instrument setups.

Single threaded apartment An apartment that has only a single thread. User interface code is usually placed in an STA.

Single-coordinate precision A level of coordinate exactness based on the number of significant digits that can be stored for each coordinate. Single precision numbers store up to seven significant digits for each coordinate, retaining a precision of plus or minus 5 metres in an extent of 1,000,000 metres. Datasets can be stored in either single or double precision coordinates.

Singleton A class for which there can only be one instance in any process.

Single-user geodatabase A geodatabase that can handle a single editor and multiple readers.

Sink The location or group of locations used as the endpoint for distance analysis.

Site The physical location of an International Ground Station (IGS) or the Mission Operations Centre (MOC).

Site prospecting The process of evaluating demographic data surrounding potential locations for a business, based on a user-defined trade area or areas.

Site starter A downloadable ArcWeb Services application that users can customize and use on their site.

Sketch In ArcMap, a shape that represents a feature's geometry. Every existing feature on a map has this alternate form, a sketch, that allows visualization of that feature's composition, with all vertices and segments of the feature visible. When features are edited in ArcMap, the sketch is modified, not the original feature. A sketch must be created in order to create a feature. Only line and polygon sketches can be created, since points have neither vertices nor segments.

Sketch constraint In ArcMap editing, an angle or length limitation that can be placed on segments created using the Sketch tool.

Sketch operation In ArcMap, an editing operation that is performed on an existing sketch. Examples are Insert Vertex, Delete Vertex, Flip, Trim, Delete Sketch, Finish Sketch, and Finish Part. All of these operations are available from the Sketch shortcut menu.

Sketch tool In ArcMap, a tool that adds points, vertices or segments to create an edit sketch. Sketch points can be defined by heads-up digitizing, snapping or manually entering coordinates.

Skylab US earth-orbiting workshop that housed three crews of three astronauts in 1973 and 1974.

Skylight Component of light that is strongly scattered by the atmosphere and consists predominantly of shorter wavelengths.

Slant range In radar, an imaginary line running between the antenna and the target.

Slant-range distance Distance measured along the slant range.

Slant-range distortion Geometric distortion of a slant-range image.

Slant-range image In radar, an image in which objects are located at positions corresponding to their slant-range distances from the aircraft path. On slant-range images, the scale in the range direction is compressed in the near-range region.

SLAR Side-looking airborne radar.

Sliver polygon A small, narrow polygon feature that appears along the borders of polygons following the overlay of two or more geographic datasets. Sliver polygons may indicate topology problems with the source polygon features, or they may be a legitimate result of the overlay.

Sliver removal The act of deleting unwanted sliver polygons.

Slope The incline, or steepness, of a surface. Slope can be measured in degrees from horizontal (0–90), or per cent slope (which is the rise divided by the run, multiplied by 100). A slope of 45 degrees equals 100 per cent slope. As slope angle approaches vertical (90 degrees), the per cent slope approaches infinity. The slope of a TIN face is the steepest downhill slope of a plane defined by the face. The slope for a cell in a raster is the steepest slope of a plane defined by the cell and its eight surrounding neighbours.

Slope image A raster image that shows changes in elevation using light and shadows on terrain from a given angle and altitude of the sun.

Small circle The circle made when a flat plane intersects a sphere anywhere but through its centre. Parallels of latitude other than the equator are small circles.

Small scale Generally, a map scale that shows a relatively large area on the ground with a low level of detail.

Smart pointer A Visual C++ class implementation that encapsulates an interface pointer, providing operators and functions that can make working with the underlying type easier and less error prone.

Smart style In ArcWeb Services, a set of symbols used to display the features in a layer class at different scale ranges. Users can create smart styles.

SMIRR Shuttle Multispectral Infrared Radiometer, a non-imaging spectroradiometer carried by the NASA Space Shuttle covering ten narrow wavebands in the 0.5–2.4 m range.

SMMR Scanning Multichannel Microwave Radiometer, carried by Nimbus-7.

Smooth criterion In radar, the relationship between surface roughness, depression angle and wavelength that determines whether a surface will scatter the incident radar pulse in a smooth or intermediate fashion.

Smoothing In image processing, reducing or removing small variations in an image to reveal the global pattern or trend, either through interpolation or by passing a filter over the image.

Snap extent An ArcGIS geoprocessing option that snaps or aligns, all layers to the cell registration of a specified raster. All layers will share the lower left corner and cell size of the specified raster. Snap extent is used to resample layers to the same registration and cell size in order to perform analysis.

Snap raster An option in the Environment Setting dialog box that ensures the cell alignment of the extent will match accurately with an existing raster. This is done by snapping the lower left corner of the specified extent to the lower left corner of the nearest cell in the snap raster and snapping the upper right corner of the specified extent to the upper right corner of the nearest cell in the snap raster.

Snapping An automatic editing operation in which points or features within a specified distance (tolerance) of other points or features are moved to match or coincide exactly with each others' coordinates.

Snapping environment Settings in the ArcMap Snapping Environment window and Editing Options dialog box that define the conditions in which snapping will occur. These settings include snapping tolerance, snapping properties and snapping priority.

Snapping priority The order in which snapping will occur by layer during an ArcMap editing session, set from the Snapping Environment window.

Snapping properties In ArcMap editing, a combination of a shape to snap to and a method for determining what part of the shape will be snapped to. Snapping properties can be set to have a feature snap to a vertex, edge or endpoint of features in a specific layer. For example, a layer snapping property might allow snapping to the vertices of buildings. A more generic, sketch-specific snapping property might allow snapping to the vertices of a sketch being created.

Snapping tolerance A specified distance within which points or features within are moved to match or coincide exactly with each others' coordinates.

SnapTip In ArcMap, a user-assistance component that displays an on-screen description of the layer name or target being snapped to when the mouse pointer is paused over it. SnapTips only appear during an ArcMap edit session if they are enabled.

SNR The ratio of the information content of a signal to its noninformation content (noise).

SOAP An XML-based protocol developed by Microsoft, SAP and IBM for exchanging information between peers in a decentralized, distributed environment. SOAP allows programs on different computers to communicate independently of an operating system or platform by using the World Wide Web's Hypertext Transfer Protocol (HTTP) and XML as the basis of information exchange. SOAP is used in Web services and is now a W3C specification. SOAP was originally an acronym for Simple Object Access Protocol, but the acronym has fallen out of use.

SOC Acronym for server object container. A process in which one or more services is running, or the machine hosting this process. SOC processes are started and shut down by the SOM. The SOC processes run on the GIS server's container machines. Each container machine is capable of hosting multiple SOC processes.

Software Programs that control computer operations.

Software development kit A set of code libraries and related tools used to develop platform-specific applications.

Solution A ready-to-use application geared toward a specific audience or industry. For example, ArcWeb Services solutions can be downloaded and used with little to no customization and accessed through a Web browser.

Solver A function that performs network analysis based on a set of network data.

SOM Acronym for server object manager. A Windows service that manages a set of server objects that are distributed across one or more server object container machines, or the machine on which this service is running. When an application makes a connection to an ArcGIS Server over a LAN, it is making a connection to the SOM.

Sonar (1) Acronym for sound navigation and ranging. A system or device that measures the time lapse between emitting a sound and receiving a returned echo to determine the location, depth and shape of objects under water. Certain types of sonar consist only of a listening device that picks up sound emitted by underwater objects, such as submarines. (2) Sonar is an active form of remote sensing that employs sonic energy to image the seafloor.

Sounder An instrument that measures atmospheric profiles (e.g. temperature, pressure, moisture, etc.). Measurements can either be taken in the horizontal plane by nadir -viewing sounders or in the vertical plane by limb sounders. Limb sounders begin scanning at the limb (the horizon).

Soundex A method of phonetic spelling used for searches and address matching. Soundex uses an algorithm to represent letters and numbers with similar phonetic equivalents to facilitate searching.

Source The location or group of locations used as the starting point for distance analysis.

Source table In ArcView 3.x, one of the two tables involved in a join operation. The source table must be the inactive table; its attributes are appended to the destination (active) table.

Space coordinate system A three-dimensional, rectangular, Cartesian coordinate system that has not been adjusted for the earth's curvature. In a space coordinate system, the x- and y-axes lie in a plane tangent to the earth's surface, and the z-axis points upward.

Space segment The space-based component of the GPS system (i.e. the satellites).

Space Shuttle US manned satellite program in the 1980s, officially called the Space Transportation System (STS).

Space station A large artificial satellite used as a long-term base for manned operations in space.

Spaghetti data Vector data composed of simple lines with no topology and usually no attributes. Spaghetti lines may cross, but no intersections are created at those crossings.

Spaghetti digitizing Digitizing that does not identify intersections as it records lines.

Spatial Related to or existing within space.

Spatial adjustment An ArcMap editing function that allows transformation, rubber sheeting and edgematching of data, as well as attribute transfer.

Spatial analysis The process of examining the locations, attributes and relationships of features in spatial data through overlay and other analytical techniques in order to address a question or gain useful knowledge. Spatial analysis extracts or creates new information from spatial data.

Spatial autocorrelation A measure of the degree to which a set of spatial features and their associated data values tend to be clustered together in space (positive spatial autocorrelation) or dispersed (negative spatial autocorrelation).

Spatial bookmark In ArcMap, a shortcut created by the user that identifies a particular geographic location to be saved for later reference.

Spatial cognition The mental processes involved in gaining and using knowledge and beliefs about spatial environments. Spatial cognition includes issues of perception, memory, language, learning and problem solving, and is an object of study in humans, nonhuman animals and machines.

Spatial coverage The ground area spanned by an entire remote sensing image. Also used to describe the entire area mapped by a remote sensing project.

Spatial data (1) Any information about the location, shape of and relationships among geographic *Spatial Data Transfer Standard* – Public data formats for transferring spatial data such as DLGs, DEMs and digital orthophotos from USGS. (2) Information about the locations and shapes of geographic features and the relationships between them, usually stored as coordinates and topology.

Spatial data infrastructure A framework of technologies, policies, standards and human resources necessary to acquire, process, store, distribute and improve the use of geospatial data across multiple public and private organizations.

Spatial data transfer standard A data exchange format for transferring different databases between dissimilar computing systems, preserving meaning and minimizing the amount of external information needed to describe the data. All federal agencies are required to make their digital map data available in SDTS format upon request, and the standard is widely used in other sectors.

Spatial database A structured collection of spatial data and its related attribute data, organized for efficient storage and retrieval.

Spatial dependence A measure of the degree to which a set of spatial features and their associated data values tend to be clustered together in space (positive spatial autocorrelation) or dispersed (negative spatial autocorrelation).

Spatial domain For a spatial dataset in ArcGIS 9.1 and previous versions, the defined precision and allowable range for x- and y-coordinates and for m- and z-values, if present.

Spatial function In ArcGIS Spatial Analyst, an operation that performs spatial analysis. All Spatial Analyst tools in the Spatial Analyst toolbox and operations on the ArcGIS Spatial Analyst toolbar are classified as spatial functions. For example, distance, slope and density are examples of spatial functions.

Spatial grid A two-dimensional grid system that spans a feature class. It is used to quickly locate features in a feature class that might match the criteria of a spatial search.

Spatial index In a geodatabase, a mechanism for optimizing access to data, based on the spatial column of the business table. In most geodatabases, a system of grids is used for the spatial index. Exceptions to this are spatial indexes used in Oracle Spatial or Informix databases, which use R-tree indexes.

Spatial interpolation A process of using control points with known values to estimate values at other points.

Spatial join A type of table join operation in which fields from one layer's attribute table are appended to another layer's attribute table based on the relative locations of the features in the two layers.

Spatial modelling A methodology or set of analytical procedures used to derive information about spatial relationships between geographic phenomena.

Spatial overlay The process of superimposing layers of geographic data that cover the same area to study the relationships between them.

Spatial overlay analysis A type of analysis in which data is extracted from one layer (such as block groups) to an overlay layer (such as a trade area).

Spatial query A statement or logical expression that selects geographic features based on location or spatial relationship. For example, a spatial query might find which points are contained within a polygon or set of polygons, find features within a specified distance of a feature or find features that are adjacent to each other.

Spatial query web service A SOAP ArcWeb service that performs spatial queries and proximity searches.

Spatial reference In ArcGIS 9.2 or later, the coordinate system, tolerance and resolution used to store a spatial dataset.

Spatial reference system A reference framework consisting of a set of points, lines and/or surfaces, and a set of rules, used to define the positions of points in space in either two or three dimensions. The Cartesian coordinate system and the geographic coordinate system used on the earth's surface are common examples of coordinate systems.

Spatial resolution The area on the ground that an imaging system (such as a satellite sensor) can distinguish. The degree to which fine detail can be seen in an image. More precisely, resolution is the smallest distance between two objects that can be barely distinguished in the data. Note that if the resolution is limited by *pixel* size (which is the usual case in vegetation remote sensing), pixel size (in metres) is not the same as resolution. For example, suppose two bright sub-pixel objects on a dark background are sampled in a square grid detector array. If one object occurs in the lower left corner of a pixel, and the second object occurs in the upper right corner of a pixel to the upper right of the first pixel, these two objects are not resolved (the data would still just show a two-pixel blob against a dark background). Any further apart, however, and they would be resolved. So if the pixel size is $d \times d$ metres, the scale of the finest object that can be resolved (r) in the image is actually given by $r^2 = (2d)^2 + (2d)^2$, i.e. $r \sim 2.8d$.

Spatial statistics The field of study concerning statistical methods that use space and spatial relationships (such as distance, area, volume, length, height, orientation, centrality and/or other spatial characteristics of data) directly in their mathematical computations. Spatial statistics are used for a variety of different types of analyses, including pattern analysis, shape analysis, surface modelling and surface prediction, spatial regression, statistical comparisons of spatial datasets, statistical modelling and prediction of spatial interaction, and more. The many types of spatial statistics include descriptive, inferential, exploratory, geostatistical and econometric statistics.

Spatial weights matrix A file that quantifies spatial relationships among a set of features. Typical examples of such relationships are inverse distance, contiguity, travel time and fixed distance.

Spatial-frequency filtering The analysis of the spatial variations in DN of an image and the separation or suppression of selected frequency ranges features. This includes remotely sensed data as well as map data.

Spatialization The transformation of complex, multivariate, nonspatial data into a spatial representation located in an information space. The relative positioning of data elements within the spatial representation shows relationships between them. Spatialization is used to allow exploration of nonspatial data using spatial metaphors and spatial analysis.

Specialty In ArcGIS Network Analyst, an object used in vehicle routeing problem (VRP) analysis. A specialty is used to represent a specific capability that may be required by certain orders and supported by certain routes. Orders requiring a given specialty may only be assigned to routes that also support that specialty. For example, an order may require an electrician, so it can only be serviced by a route that also supports the electrician specialty.

Specific heat The ratio of the heat capacity of unit mass of a material to the heat capacity of unit mass of water.

Spectral hue A hue which is present in the spectral range of white light analysed by a prism or diffraction grating.

Spectral range The wavelength difference between two wavelengths in adjacent orders at the same angle of diffraction

Spectral reflectance Reflectance of electromagnetic energy at specified wavelength intervals.

Spectral resolution The range of wavelengths that an imaging system can detect.

Spectral response The relative amplitude of the response of a detector vs. the frequency of incident electromagnetic radiation.

Spectral sensitivity Response or sensitivity, of a film or detector to radiation in different spectral regions.

Spectral separation The separation of electromagnetic waves by its wavelength

Spectral signature (1) A characteristic shape to a *spectrum* which can be used as an unambiguous indicator of some chemical constituent (or vegetative component) of the area being studied. (2) The pattern of electromagnetic radiation that identifies a chemical or compound. Materials can be distinguished from one another by examining which portions of the spectrum they reflect and absorb.

Spectral vegetation index An index of relative amount and vigour of vegetation. The index is calculated from two spectral bands of AVHRR imagery.

Spectral window A band of the electromagnetic spectrum which offers maximum transmission and minimal attenuation through a particular medium.

Spectrometer (1) A photometer that measures the intensity of electromagnetic radiation as a function of its frequency. Spectrophotometers are usually used for measuring the visible portion of the spectrum. (2) An optical instrument that splits the light received from an object into its component wavelengths by means of a diffraction grating, then measuring the amplitudes of the individual wavelengths.

Spectrophotometer A photometer that measures the intensity of electromagnetic radiation as a function of its frequency. Spectrophotometers are usually used for measuring the visible portion of the spectrum.

Spectroscopy (1) The branch of science that deals with deducing as much as possible about the workings of the Universe, by studying the light emitted or otherwise modified by the objects that inhabit it. (2) The scientific study of how different chemicals and other substances absorb and reflect different parts of the electromagnetic spectrum.

Spectrum (1) A graphical presentation of the array of intensity values for data in a certain bandwidth. (2) The entire range of wavelengths (frequencies) over which electromagnetic radiation extends.

Speed of light the speed at which photons travel, approximately 3×10^8 m/s. This speed is indicated in equations by the letter c.

Spelling sensitivity In geocoding, the degree to which the spelling variation of a street name is allowed during a search for likely match candidates. The lower the value, the more likely it is that additional candidates will be retrieved and vice versa. This value is adjustable on the Address Locator Properties dialog box.

Sphere A three-dimensional shape whose centre is equidistant from every point on its surface, made by revolving a circle around its diameter.

Spherical aberration The blurring of an image that occurs when light from the margin of a lens or mirror with a spherical surface comes to a shorter focus than light from the central portion. The changing focal length is caused by deviations in the lens or mirror surface from a true sphere.

Spherical coordinate system A reference system using positions of latitude and longitude to define the locations of points on the surface of a sphere or spheroid.

Spheroid (1) A three-dimensional shape obtained by rotating an ellipse about its minor axis, resulting in an oblate spheroid or about its major axis, resulting in a prolate spheroid. (2) Mathematical figure closely approaching the geoid in form and size and used as a surface of reference for geodetic surveys. A reference spheroid or ellipsoid is a spheroid determined by revolving an ellipse about its shorter (polar) axis and used as a base for geodetic surveys of a large section of the earth (such as the Clarke spheroid of 1,866 which is used for geodetic surveys in the United States).

Spider diagram A type of market analysis that draws lines from a set of geocoded points (usually customers) to a single, central point (usually a store). Desire lines can be weighted.

Spike An anomalous data point that protrudes above or below an interpolated surface.

Spline In mathematics, a piecewise polynomial function used to approximate a smooth curve in a line or surface.

Spline interpolation An interpolation method in which cell values are estimated using a mathematical function that minimizes overall surface curvature, resulting in a smooth surface that passes exactly through the input points.

Split character A user-designated character that divides long labels into two or more lines during the labelling process. There can be more than one split character.

Split policy All attribute domains in geodatabases have a split policy associated with them. When a feature is split into two new features in ArcMap, the split policies dictate what happens to the value of the attribute with which the domain is associated. Standard split policies are duplicate, default value and geometry ratio.

SPOT Acronym for Satellite Pour l'Observation de la Terre. Earth observation satellites developed by Centre National d'Etudes Spatiales (CNES), the space agency of France. The SPOT satellites gather high-resolution imagery used in natural resource management, climatology, oceanography, environmental monitoring and the monitoring of human activities.

Spot elevation (1) An elevation measurement taken at a single location. (2) Point on a map or chart whose height above a specified datum is noted, usually by a dot or a small sawbuck and elevation value. Elevations are shown, on a selective basis, for road forks and intersections, grade crossings summit of hills, mountain.

Spot height An elevation measurement taken at a single location.

Spurious polygon A small, narrow, polygon feature that appears along the borders of polygons following the overlay of two or more geographic datasets. Sliver polygons may indicate topology problems with the source polygon features, or they may be a legitimate result of the overlay.

SQL Acronym for Structured Query Language. A syntax for retrieving and manipulating data from a relational database. SQL has become an industry standard query language in most relational database management systems.

SSL Acronym for Secure Sockets Layer. An encryption protocol for the secure transfer of private information over the Internet.

STA Acronym for single threaded apartment. An apartment that has only a single thread. User interface code is usually placed in an STA.

Stable base In cartography, any material such as a Mylar sheet or film that is more durable than paper and less likely to shrink or stretch.

Stack In computing, a data storage structure that operates on last in, first out (LIFO) protocol. As with a stack of dishes, the item placed on top of the stack last must be removed before the others may be manipulated.

Stadia Technique of distance measurement wherein the observer reads the intercept subtended on a graduated rod between two marks on the reticle of the telescope.

Staging geodatabase A normalized, fixed set of feature classes and data tables of a predefined schema from a collection of input CAD drawings.

Stand-alone application An application that runs by itself, not within an ArcGIS application.

Standard annotation Annotation that is stored in the geodatabase, consisting of geographically placed text strings that are not associated with features in the geodatabase.

Standard deviation A statistical measure of the spread of values from their mean, calculated as the square root of the sum of the squared deviations from the mean value, divided by the number of elements minus one. The standard deviation for a distribution is the square root of the variance.

Standard deviation classification A data classification method that finds the mean value, then places class breaks above and below the mean at intervals of either .25, .5, or 1 standard deviation until all the data values are contained within the classes. Values that are beyond three standard deviations from the mean are aggregated into two classes, greater than three standard deviations above the mean and less than three standard deviations below the mean.

Standard distance A measure of the compactness of a spatial distribution of features around its mean dentre. Standard distance (or standard distance deviation) is usually represented as a circle where the radius of the circle is the standard distance.

Standard generalized markup language A markup language with a predefined vocabulary and structure. SGML is used to structure information in a DTD, or Document Type Definition format, for exchanging information between different programs and machines. SGML uses a tag structure, and was standardized by ISO in 1986.

Standard industrial classification codes The federal US standard for classifying establishments by their primary type of business activity. Standard Industrial Classification codes (SIC codes) are used as an identification system in business directories, publications, and statistical sources. The classification system was officially replaced by NAICS in 1997, but it is still used by some organizations outside the federal government.

Standard line (1) A line on a sphere or spheroid that has no length compression or expansion after being projected; usually a standard parallel or central meridian. (2) Line of tangency between the projection surface and the reference globe. A standard line has no projection distortion and has the same scale as the of the reference globe.

Standard meridian A standard line that follows a meridian or line of longitude.

Standard parallel The line of latitude in a conic or cylindrical projection in normal aspect where the projection surface touches the globe. A tangent conic or cylindrical projection has one standard parallel, while a secant conic or cylindrical projection has two. At the standard parallel, the projection shows no distortion.

Standard positioning service (SPS) The normal civilian positioning accuracy obtained by using the single frequency C/A code.

Star diagram A type of diagram that consists essentially of a central point from which lines radiate outward. The central point usually represents a geographic location while the length of each line represents an attribute value or ratio. The direction of the line may represent a compass direction, a period of time or some other attribute classification. A wind rose is a common example of a star diagram.

Starter application A ready-to-use Web application that is designed for a specific purpose, such as a store locator or a map viewer. Starter applications do not require customization before use.

State (1) In programming, the current data contained by an object. (2) The process by which a Web application maintains information across a sequence of requests by the same client to the same Web application.

State plane coordinate system (1) A group of planar coordinate systems based on the division of the United States into more than 130 zones to minimize distortion caused by map projections. Each zone has its own map projection and parameters and uses either the NAD27 or NAD83 horizontal datum. The Lambert conformal conic projection is used for states that extend mostly east–west, while transverse Mercator is used for those that extend mostly north–south. The oblique Mercator projection is used for the panhandle of Alaska. (2) Coordinate systems established by the US Coast and Geodetic Survey (now the National Ocean Survey), usually one for each state, for use in defining positions of points in terms of plane rectangular (*x*,*y*) coordinates.

State tree A logical organization of all the states in the geodatabase. A state tree illustrates and maintains the logical relationship between states of a versioned geodatabase.

Stateful operation In programming, an operation that makes changes to an object or one of its associated objects, such as removing a layer from a map.

Stateless In programming, not retaining changes between calls. A stateless object or application does not store parameters or values from the last time it was called, so it is always in its original state.

Stateless operation An operation that does not make changes to an object, such as drawing a map.

Static binding A COM technique an application uses to access an object. In early binding, an object's properties and methods are discovered from an interface at compile time, instead of being checked at run time as in late binding. This difference often gives early binding performance benefits over late binding. There are two types of early binding: DispID binding and vTable binding.

Static positioning (1) Determining a position on the earth by averaging the readings taken by a stationary antenna over a period of time. (2) Location determination when the receiver's antenna is presumed to be stationary in the earth. This allows the use of various averaging techiques that improve the accuracy by factors of over 1,000.

Stationarity In geostatistics, a property of a spatial process in which all statistical properties of an attribute depend only on the relative locations of attribute values.

Stationing In the pipeline industry, another name for linear referencing. Stationing allows any point along a line feature representing a pipeline to be uniquely identified by its relative position along the line feature.

Statistic F A ratio of variances, calculated from a sample of data and used to provide information about a whole dataset. For example, statistic F may be used to provide estimates of variance or differences in a population, based on observations from two or more random samples.

Statistical surface Ordinal, interval or ratio data represented as a surface in which the height of each area is proportional to a numerical value.

Steepest path A line that follows the steepest downhill direction on a surface. Paths terminate at the surface perimeter or in surface concavities or pits.

Stephan-Boltzmann's law A law relating the surface area and temperature of an object to the power it produces via radiation: $P = 4 \pi r^2 \sigma T^4$.

Steradian The solid (conical) angle subtended at the centre of a sphere of radius r by a bounded region on the surface of the sphere having an area r squared. There are 4p steradians in a sphere.

Stereocomparator A comparator used in stereoscopically measuring images on adjacent aerial photographs.

Stereocompilation (1) A map produced with a stereoscopic plotter using aerial photographs and geodetic control data. (2) Production of a map or chart manuscript from aerial photographs and geodetic control data by means of photogrammetric instruments.

Stereogrammatic organization The presentation of features on a map in a way that implies relative importance, usually achieved with visual contrast.

Stereographic projection A tangent planar projection that views the earth's surface from a point on the globe opposite the tangent point.

Stereometer A stereoscope containing a micrometer for measuring the effects of parallax in a stereoscopic image.

Stereomodel The three-dimensional image formed where rays from points in the images of a stereoscopic pair intersect.

Stereopair Two aerial photographs of the same area taken from slightly different angles that when viewed together through a stereoscope produce a three-dimensional image.

Stereoplotter An instrument that projects a stereoscopic image from aerial photographs, converts the locations of objects and landforms on the image to x-, y- and z-coordinates and plots these coordinates as a drawing or map.

Stereoscope (1) A binocular device that produces the impression of a three-dimensional image from two overlapping images of the same area. (2) Binocular optical device for viewing overlapping images or diagrams. The left eye sees only the left image and the right eye sees only the right image.

Stereoscopic Pertaining to the use of binocular vision for observation of a pair of overlapping photographs or other perspective views, giving impression of depth.

Stereoscopic pair Two aerial photographs of the same area taken from slightly different angles that when viewed together through a stereoscope produce a three-dimensional image.

Stereoscopic vision It is the three-dimensional vision produced by the fusion of two slightly different views of a scene on each retina.

Stereoscopy The science and art that deals with the use of binocular vision or observation of a pair of overlapping photographs.

Sticky move tolerance When editing in ArcMap, a setting that defines the minimum number of pixels the pointer must move on the screen before a selected feature is moved.

Stochastic model A model that includes a random component. The random component can be a model variable, or it can be added to existing input data or model parameters.

Stop In ArcGIS Network Analyst, a network location used to determine a route in route analysis. Users can specify multiple stops, of which two must be used to represent an origin and a destination. Stops in between (known as intermediary stops) are visited en route from the first to the last stop.

Stop impedance In network analysis, the time it takes for a stop to occur, used to compute the impedance of a path or tour. For example, when a school bus drops children off or picks them up at their homes, the stop impedance might be 2 minutes at each stop.

Storage keywords A set of parameters that specify how data and indexes are stored in an ArcSDE database. Keywords are stored in a table in the ArcSDE database.

Store market analysis A type of business analysis that uses mostly data about a store or stores, rather than about customers. Examples include ring studies and analyses of equal competition areas and drive-time areas.

Store prospecting A type of business analysis that assesses the potential of a site by performing simple ring or drive-time analysis.

Storm evacuation map See: map, storm evacuation.

Straight-line allocation An ArcGIS Spatial Analyst function that identifies which cells belong to which source, based on closest proximity in a straight line.

Straight-line direction An ArcGIS Spatial Analyst function that identifies the azimuth direction from each cell to the nearest source.

Straight-line distance An ArcGIS Spatial Analyst function that calculates the distance in a straight line from every cell to the nearest source.

Stream A mode of data delivery in which objects provide data storage. Stream objects can contain any type of data in any internal structure.

Stream digitizing / Stream mode digitizing A method of digitizing in which, as the cursor is moved, points are recorded automatically at preset intervals of either distance or time.

Stream tolerance During stream mode digitizing, the minimum interval between vertices. Stream tolerance is measured in map units.

Streaming A technique for transferring data, usually over the Internet, in a real-time flow as opposed to storing it in a local file first. Streaming allows large multimedia files to be viewed before the entire file has been downloaded to a client's computer. When received by the client (local computer) the data is uncompressed and displayed using software designed to interpret and display the data rapidly.

Street network A system of interconnecting lines and points that represent a system of roads for a given area. A street network provides the foundation for network analysis; for example, finding the best route or creating service areas.

Street-based mapping A form of digital mapping that links information to geographic locations and displays address locations as point features on a map.

Stretch A display technique applied to the histogram of raster datasets, most often used to increase the visual contrast between cells.

String A set of coordinates that defines a group of linked line segments.

Structure The general pattern of z-values throughout a kriging model. The drift, or structure, forms the model's basic shape.

Structure line A line feature enforced in a TIN. There are two types of structure lines: hard and soft. Hard structure lines, also known as breaklines, represent interruptions in the slope of the surface. Soft structure lines are used to add information about the surface without implying a change in the surface behaviour across the line.

Structured query language A syntax for retrieving and manipulating data from a relational database. SQL has become an industry standard query language in most relational database management systems.

Study area The geographic area treated in an analysis.

Style An organized collection of predefined colours, symbols, properties of symbols and map elements. Styles promote standardization and consistency in mapping products.

Style manager The tool used to create new styles and edit existing ones. The Style Manager displays the contents of all the styles that are currently referenced by the map. It also contains personal and additional styles that may be used in ArcMap.

Style sheet A file or form that provides style and layout information, such as margins, fonts and alignment, for tagged content within an XML or HTML document. Style sheets are frequently used to simplify XML and HTML document design, since one style sheet may be applied to several documents. Transformational style sheets may also contain code to transform the structure of an XML document and write its content into another document.

Subfield In ArcGIS, a subset of all the fields associated with a layer or data.

Subinterval A segment of a raw wideband data interval received during a Landsat 7 contact period. Subintervals are caused by breaks in the wideband data stream due to communication dropouts and/or the inability of the spacecraft to transmit a complete observation (interval) within a single Landsat 7 contact period. The largest possible subinterval can be as long as a full imaging interval (a set of contiguous WRS scenes) transmitted during an uninterrupted contact period. The smallest possible subinterval can be as small as a set of a few contiguous ETM+ scans (a partial WRS scene). The smallest size of a subinterval is an operator modifiable parameter in LPS. If the smallest subinterval size is chosen to be as long as a full WRS scene, it will contain approximately 24 seconds worth of ETM+ data or 335 scans (without the 20 overlapping scans each .

Sublayer One of several layers that are part of a group layer in a map document.

Subscene A portion of an image that is used for detailed analysis.

Subsidence Decrease in the elevation of land surface due to tectonic, seismic, or artificial forces, without removal of surface material.

Subsumption Within a geometric network, the replacement of an orphan junction by a non-orphan junction from a user-defined feature class. The original junction is deleted from the network, and the non-orphan junction assumes the connectivity of the subsumed orphan junction.

Subtractive primary colours I n printing, the three primary colours – cyan, magenta, and yellow – that when used as filters for white light remove blue, green and red light, respectively.

Subtype In geodatabases, a subset of features in a feature class or objects in a table that share the same attributes. For example, the streets in a streets feature class could be categorized into three subtypes: local streets, collector streets and arterial streets. Creating subtypes can be more efficient than creating many feature classes or tables in a geodatabase. For example, a geodatabase with a dozen feature classes that have subtypes will perform better than a geodatabase with a hundred feature classes. Subtypes also make editing data faster and more accurate because default attribute values and domains can be set up. For example, a local street subtype could be created and defined so that whenever this type of street is added to the feature class, its speed limit attribute is automatically set to 35 miles per hour.

Suitability model A model that weights locations relative to each other based on given criteria. Suitability models might aid in finding a favourable location for a new facility, road or habitat for a species of bird.

Sunglint Bright reflectance of sunlight caused by ripples on water

Sun-synchronous orbit (1) An orbit in which a satellite is always in the same position with respect to the rotating earth at the same time of day. The satellite travels around the earth in the same direction, at an altitude of approximately 438 miles (705 kilometres). Landsat–7 is sun-synchronous, always passing overhead at approximately 10:00 am local time. (2) Earth satellite orbit in which the orbit plane is nearly polar and the altitude is such that the satellite passes over all places on earth having the same latitude twice daily at the same local sun time.

Supervised classification Digital information extraction technique in which the operator provides training-site information that the computer uses to assign pixels to categories.

Supplemental contour A contour line placed between regularly spaced contours, used when the terrain change is not large enough to be depicted with consistent contour intervals.

Surface A geographic phenomenon represented as a set of continuous data (such as elevation, geological boundaries or air pollution); a spatial distribution which associates a single value with each position in a plane, usually associated with continuous attributes.

Surface fitting Generating a statistical surface that approximates the values of a set of known x,y,z points.

Surface mode A navigation mode in ArcGlobe in which the camera target is always on the globe surface.

Surface model A representation of a geographic feature or phenomenon that can be measured continuously across some part of the earth's surface (for example, elevation). A surface model is an approximation of a surface, generalized from sample data. Surface models are stored and displayed as rasters, TINs or terrains.

Surface phenomenon Interaction between electromagnetic radiation and the surface of a material.

Surface roughness See roughness.

Surface smoothness Perpendicular or normal to the slope of the surface in 3D Analyst. As changes in slope are approached across the surface, the degree of smoothness is defined by how gradual or abrupt the normals to that surface change. A smooth surface has gradual changes in the normals to the surface; a planar surface has abrupt changes.

Surround element In digital cartography, a distinctly identifiable graphic or object in the map or page layout. For example, a map element can be a title, scale bar, legend or other map-surround element. The map area itself can be considered a map element; or an object within the map can be referred to as a map element, such as a roads layer or a school symbol.

Survey Orderly process of determining data relating to any physical or chemical characteristics of the earth. The associated data obtained in a survey. An organization engaged in making a survey.

Survey class In Survey Analyst for field measurements, a collection of survey objects of a particular type. A survey dataset contains a set of survey classes for each of the different types of measurements and computations. There is also a survey class for coordinates and a survey class for survey points.

Survey data converters In Survey Analyst for field measurements, importers that interpret operation codes and feature codes when a data collector file is imported. Users choose the converter that matches the data collector file format.

Survey dataset In Survey Analyst for field measurements, a dataset that stores survey objects.

Survey explorer In Survey Analyst for field measurements, the main interface for working with stored survey information. Data may be explored and edited directly in the Survey Explorer. Users can add lists of coordinates, measurements and computations to the Survey Explorer so they can view and analyse numerical values of measurements and coordinates created in their survey dataset.

Survey layers In Survey Analyst for field measurements, layers that are created whenever survey datasets or survey projects are added to the map. Survey layers appear in the table of contents of a map document and comprise a set of sublayers for survey points and measurements.

Survey marker An object, such as a metal disk, permanently mounted in the landscape to denote a survey station.

Survey monument An object, such as a metal disk, permanently mounted in the landscape to denote a survey station.

Survey network In Survey Analyst for field measurements, a sequence of computation dependencies – the output points of some computations are used as the input for one or more others.

Survey objects In Survey Analyst for field measurements, a collective term referring to measurements, computations, survey points and coordinates in the survey dataset.

Survey points In Survey Analyst for field measurements, named locations that are observed through various surveys. Survey points can be observed multiple times and by many surveys over time. They represent multiple coordinates, but each identifies discrete physical locations on the earth's surface.

Survey project In Survey Analyst for field measurements, a subset of the survey dataset that represents a unit of work. A survey project is used as a logical structure that owns and manages a group of measurements, points, coordinates and computations that function and belong together.

Survey station A location on the earth that has been accurately determined by geodetic survey.

Survey-aware feature classes In Survey Analyst for field measurements, the feature classes in the geodatabase that contain survey-aware features.

Survey-aware features In Survey Analyst for field measurements, features that are associated with survey data.

Surveying Measuring physical or geometric characteristics of the earth. Surveys are often classified by the type of data studied or by the instruments or methods used. Examples include geodetic, geologic, topographic, hydrographic, land, geophysical, soil, mine and engineering surveys.

SVG Acronym for scalable vector graphics. An XML-based graphics file format that describes two-dimensional vector images, including animation. SVG images scale to fit the display window without compromising quality.

SWF A vector-based Flash file format that supports text, audio, video, and end-user interaction. SWF files are often used to publish animations on the Web.

SXD Scene document. A document saved by ArcScene that has an .sxd extension.

Sybase A commercial RDBMS.

Symbol A graphic used to represent a geographic feature or class of features. Symbols can look like what they represent (trees, railroads, houses), or they can be abstract shapes (points, lines, polygons) or characters. Symbols are usually explained in a map legend.

Symbol ID code A 15-character identifier that provides the information necessary to display or transmit a tactical symbol between MIL STD–2525B compliant systems.

Symbol level drawing In ArcGIS, a setting that determines the drawing order of features based on their symbol. When symbols have more than one layer, symbol level drawing can be used to specify the order in which each layer of the symbol is drawn.

Symbol library A collection of symbols.

Symbol modifier field In MOLE, a defined area in which optional text or graphics may be entered to provide additional information about a force element or tactical graphic.

Symbolization The process of devising a set of marks of appropriate size, colour, shape and pattern, and assigning them to map features to convey their characteristics at a given map scale.

Symbology The set of conventions, rules or encoding systems that define how geographic features are represented with symbols on a map. A characteristic of a map feature may influence the size, colour and shape of the symbol used.

Synchronization In geodatabase editing, the process of applying changes made from a replica to the relative replica in a replica pair.

Synchronization version In geodatabase editing, the geodatabase version that receives changes during synchronization. The synchronization version is always a child of the replica version. With one-way and two-way replication, the synchronization version is automatically reconciled and posted with the replica version. If there are conflicts between the synchronization version and the replica version, the conflicts must be resolved before more data changes can be sent.

Synchronous (1) Occurring together, or at the same time. (2) The instantaneous alignment of two or more events in time. Events may occur at irregular intervals.

Syntax The structural rules for using statements in a command or programming language.

Synthetic aperture A synthetic aperture, or virtual antenna, consists of a long array of successive and coherent radar signals that are transmitted and received by a physically short (real) antenna as it moves along a predetermined flight or orbital path. The synthetic aperture is formed by pointing the real radar antenna of relatively small dimensions, which are restricted in size by the satellite platform, broadside to the direction of forward motion of that platform. The points at which successive pulses are transmitted can be thought of as the elements of a long synthetic array, which a signal processor will then use and process to generate a SAR image. This detailed array of radar signal data is the key to achieving high azimuth resolution. This long virtual antenna concept is the basis for synthetic aperture radar, or SAR.

Synthetic aperture time The time for a target to cross the azimuth antenna beam.

Synthetic stereo images Stereo images constructed through digital processing of a single image. Topographic data are used to calculate parallax.

Synthetic-aperture radar (SAR) A synthetic aperture radar, or SAR, is a coherent radar system that generates high-resolution remote sensing imagery. Signal processing uses magnitude and phase of the received signals over successive pulses from elements of a synthetic aperture to create an image. As the line of sight direction changes along the radar platform trajectory, a synthetic aperture is produced by signal processing that has the effect of lengthening the antenna. The achievable azimuth resolution of a SAR is approximately equal to one-half the length of the actual (real) antenna and does not depend on platform altitude (distance). High range resolution is achieved through pulse compression techniques. In order to map the ground surface the radar beam is directed to the side of the platform trajectory; with a sufficiently wide antenna beam width in the along-track direction, an identical target or area may be illuminated a number of times without a change in the antenna look angle.

A synthetic aperture radar, or SAR, is a coherent radar system that generates high resolution remote sensing imagery. Signal processing uses magnitude and phase of the received signals over successive pulses from elements of a synthetic aperture to create an image.

System Combination of components that constitute an imaging device.

System tool In ArcGIS, a geoprocessing tool. System tools are stored in system toolsets and can be copied to custom toolsets and/or toolboxes.

System toolbox In ArcGIS, a default geoprocessing toolbox that is installed with ArcGIS. System toolboxes contain system tools, organized into toolsets for ease of access.

System toolset In ArcGIS, a subset of a geoprocessing toolbox that holds system tools.

Systematic distortion Geometric irregularities on images that are caused by known and predictable characteristics.

Systematic error In Survey Analyst, one type of measurement error. Systematic error follows a mathematical or physical law, and it can be corrected to comply with a known standard.

T

Table A set of data elements arranged in rows and columns. Each row represents a single record. Each column represents a field of the record. Rows and columns intersect to form cells which contain a specific value for one field in a record.

Table of contents In ArcGIS, a tabbed list of data frames and layers (or tables) on a map that shows how the data is symbolized, the source of the data, and whether or not each layer is selectable.

Table view A representation of tabular data for viewing and editing purposes. The table view created when a table is added to ArcMap is a copy of the actual table data stored in memory.

Tabular data Descriptive information, usually alphanumeric, that is stored in rows and columns in a database and can be linked to spatial data.

Tacheometer (tachymeter) Surveying instrument designed for use in the rapid determination of distance, direction and difference of elevation from a single observation, using a short base which may be an intergraph part of the instrument.

Tactical graphic In MOLE, a type of graphic that aids in regulating the movement of force units, such as lane boundaries and obstacles.

Tactical symbol In MOLE, a type of graphic that represents a military unit (such as Company A, 1st Battalion of the 135th Infantry), equipment, or installation (such as a hospital or radar site). Force elements are also known as force units or military units.

Tag In markup languages such as XML and HTML, a set of bracketed characters that define the structural purpose of a block of content.

Tagged value A pair of keywords used to set additional properties for elements that are not part of the UML specification, such as the point, line and polygon geometry of a feature class, or the length of a field in a table.

Tangent projection A projection whose surface touches the globe's without piercing it. A tangent planar projection touches the globe at one point, while tangent conic and cylindrical projections touch the globe along a line. At the point or line of tangency, the projection is free from distortion.

Target In ArcScene and ArcGlobe, the centre point in a scene's view at which the camera is aimed. In a remote sensing investigation it is the object on the terrain of specific interest.

Target computer A computer to which an application is deployed.

Target layer In an ArcMap editing session, the layer to which edits will be applied. The target layer must be specified when creating new features and modifying existing features.

Target offset In ArcScene and ArcGlobe, the height of a target point above a surface used when calculating lines of sight and viewsheds.

Target theme In ArcView 3.x, the theme from which features are selected during a theme-on-theme selection operation. Multiple target themes may be specified for a theme-on-theme selection.

Task In ArcGIS Explorer and ArcGIS Server, a user interface combined with a particular piece of GIS functionality in a format that can be easily added to applications. Tasks make it easy for the end user of an application to perform common functions, such as querying, editing or geoprocessing. In many cases, tasks also facilitate a developer's job because they can be added to an application without requiring any code to be written.

TCP/IP Acronym for Transmission Control Protocol/Internet Protocol. The most common protocol for Internet traffic. The Transmission Control Protocol (TCP) is a communication protocol layered above the Internet Protocol (IP), which is a suite of nonproprietary communication protocols, or sets of rules, that allow computers to send and receive data over networks.

TDOP Acronym for dilution of precision. An indicator of satellite geometry for a constellation of satellites used to determine a position. Positions with a lower DOP value generally constitute better measurement results than those with higher DOP. Factors determining the total GDOP (geometric DOP) for a set of satellites include PDOP (positional DOP), HDOP (horizontal DOP), VDOP (vertical DOP), and TDOP (time DOP).

Telemeter To transmit data by radio or microwave links.

Telemetry (1) Radio signals from a spacecraft used to encode and transmit data to a ground station. (2) The science of measuring a quantity, transmitting the measured value to a distant station, and there, interpreting or recording the quantity measured.

Template In ArcMap, a kind of map document that provides a quick way to create a new map. Templates can contain data, a custom interface and a predefined layout that arranges map elements, such as north arrows, scale bars and logos, on the virtual page. Map templates have a .mxt file extension.

Temporal data Data that specifically refers to times or dates. Temporal data may refer to discrete events, such as lightning strikes; moving objects, such as trains; or repeated observations, such as counts from traffic sensors.

Temporal event In ArcGIS Tracking Analyst, a type of event used to describe observations through time of particular objects or groups of objects.

Temporal extent All the temporal information, including the earliest and last observations, of a data layer in ArcGIS Tracking Analyst.

Temporal GIS An emerging capability in GIS for integrating temporal data with location and attribute data.

Temporal object In ArcGIS Tracking Analyst, an object being observed through time.

Temporal object table The component of a complex temporal event in ArcGIS Tracking Analyst that contains feature shape and attribute information.

Temporal observation In ArcGIS Tracking Analyst, data gathered for a given object through time.

Temporal observation table The component of a complex temporal event in ArcGIS Tracking Analyst that contains feature attribute information, and possibly time and date information.

Temporal offset A function in ArcGIS Tracking Analyst that allows users to play back the temporal display of data as if it had occurred at a user-defined date and time. The temporal offset does not change the underlying data; it simply allows the data to be displayed as if it were happening at a different time.

Temporal window In ArcGIS Tracking Analyst, the time range within which data will be displayed. Temporal windows can be set at the layer level on the Symbology tab.

Temporary dataset A dataset temporarily stored on disk or in memory.

Terrain (1) An area of land having a particular characteristic, such as sandy terrain or mountainous terrain. (2) Surface of the earth.

Terrain dataset A multiresolution, TIN-based surface built from measurements stored as features in a geodatabase. Associated and supporting rules help organize the data and control how features are used to define the surface. Terrain datasets are typically derived from sources such as lidar, sonar and photogrammetric data.

Terrain dataset pyramid A data structure associated with a terrain dataset used to define a multi-resolution surface, which organizes data into different levels of detail, or pyramid levels, and serves to improve performance by enabling the terrain to access only the data required for a particular display or analysis function. Data that is over-sampled or redundant can be avoided.</SPAN>Two pyramid types can be used to build a terrain: z-tolerance and window size. Z-tolerance pyramids filter data into different pyramid levels based on their vertical significance. Window size pyramids filter the data through localized selection biased to max, min or mean height.

Terrain pyramid group In a terrain dataset, a collection of line or polygon feature classes used to represent those geographic features at different levels of detail.

Terrain pyramid level An individual level of detail in a terrain dataset pyramid. Each level has a pair of properties: resolution and scale. The resolution defines the amount of detail represented by the level, and the scale is a threshold that indicates at which display scale the level becomes active.

Terrain pyramid resolution bounds The range of pyramid levels for which polygon or polyline features will be enforced in the surface of a terrain dataset.

Terrain tiles A spatially coherent organization of terrain data facilitating efficient retrieval and editing. Tile definition is based on the average point spacing of the source data.

Tessellation The division of a two-dimensional area into polygonal tiles, or a three-dimensional area into polyhedral blocks, in such a way that no figures overlap and there are no gaps.

Text attribute table (TAT) A table containing text attributes, such as colour, font, size, location, and placement angle, for an annotation subclass in a coverage. In addition to user-defined attributes, the text attribute table contains a sequence number and text feature identifier.

Text box An entity that displays text entered by a user or derived from another source for editing purposes.

Text envelope A rectangle that bounds a text string.

Text formatting tag Tags used with text in ArcGIS that allow formatting to be modified for a portion of a text string. This allows the creation of mixed-format text where, for example, one word in a sentence is underlined. Text formatting tags adhere to XML syntax rules and can be used most places where both a text string and a text symbol can be specified. The tags are most commonly used with labels, annotation and graphic text.

Text label Text placed next to a feature on a map to describe or identify it.

Text modifier Nonspatial information about a geographic feature in a GIS, usually stored in a table and linked to the feature by a unique identifier. For example, attributes of a river might include its name, length and sediment load at a gauging station.

Text symbol A text style defined by font, size, character spacing, colour and so on, used to label maps and geographic features.

Texture Frequency of change and arrangement of tones on an image. It is digital representation of the surface of a feature.

Texture coordinate In ArcScene and ArcGlobe, a location embedded in a geometry that helps define how a texture is mapped.

Texture mapping The application of a texture to a 3D object in ArcScene and ArcGlobe.

Thematic data Thematic data layers in a data set are layers of information that deal with a particular theme. These layers are typically related information that logically goes together.

Thematic map A map designed to convey information about a single topic or theme, such as population density or geology.

Thematic mapper 'A Landsat multispectral scanner designed to acquire data to categorize the Earth's surface. Particular emphasis was placed on agricultural applications and identification of land use. The scanner continuously scans the surface of the Earth, simultaneously acquiring data in seven spectral channels. Overlaying two or more bands produces a false colour image. The ground resolution of the six visible and shortwave bands of the Thematic

Mapper is 30 meters, and the resolution of the thermal infrared band is 120 meters. Thematic mappers were flown on Landsat 4 and 5'.

Thematic mapper (TM) A cross-track scanner deployed on Landsat that records seven bands of data from the visible through the thermal IR regions.

Theme In ArcView 3.x, a set of related geographic features such as streets, parcels or rivers, along with their attributes. All features in a theme share the same coordinate system, are located within a common geographic extent and have the same attributes. Themes are similar to layers in ArcGIS 8.x and 9.0.

Theme table In ArcView 3.x, a document object linked to the set of features in a theme. It serves as an interface to the underlying database, allowing manipulation of the data.

Theme-on-theme selection In Arc View 3.x, an operation in which features in one theme are selected based on their spatial relationships with features in another theme. Theme-on-theme selection is used to find features that intersect other features, completely contain other features or are within a specified distance of other features.

Theodolite (1) A surveying instrument for measuring vertical and horizontal angles, consisting of an alidade, a telescope, and graduated circles mounted vertically and horizontally. (2) Precision surveying instrument for measuring horizontal and vertical angles.

Thermal band A general term for middle-infrared wavelengths which are transmitted through the atmospheric window at 8–13 micrometers. Occasionally also used for the windows around 3–6 micrometers.

Thermal capacity (c) See heat capacity.

Thermal conductivity (K) Measure of the rate at which heat will pass through a material, expressed in calories per centimeter per second per degree Centigrade.

Thermal crossover On a plot of radiant temperature versus time, the point at which temperature curves for two different materials intersect.

Thermal diffusivity (k) Governs the rate at which temperature changes within a substance, expressed in centimeters squared per second.

Thermal inertia (P) Measure of the response of a material to temperature changes, expressed in calories per square centimeter per square root of a second.

Thermal infrared Electromagnetic radiation with wavelengths between 3 and 25 micrometers.

Thermal IR IR region from 3 to 14 Fm that is employed in remote sensing. This spectral region spans the radiant power peak of the earth.

Thermal IR image Image acquired by a scanner that records radiation within the thermal IR band.

Thermal IR multispectral scanner (TIMS) Airborne scanner that acquires multispectral images within the 8 to 14mm band of the thermal IR region.

Thermal model Mathematical expression that relates thermal and other physical properties of a material to its temperature. Models may be used to predict temperature for given properties and conditions.

Thermography Medical applications of thermal IR images. Images of the body, called thermograms, have been used to detect tumors and monitor blood circulation.

Thiessen polygons Polygons generated from a set of sample points. Each Thiessen polygon defines an area of influence around its sample point, so that any location inside the polygon is closer to that point than any of the other sample points. Thiessen polygons are named for the American meteorologist Alfred H. Thiessen (1872–1931).

Thinning Reducing the number of points that define a line while preserving its essential shape.

THIR Temperature-Humidity Infrared Radiometer, carried by Nimbus–7.

Third normal form The third level of guidelines for designing table and data structures in a relational database. The third-normal-form guideline incorporates the guidelines of first and second normal form; in addition, it recommends removing from a table those columns that do not depend on the table's primary key. A database that follows these guidelines is said to be in third normal form.

Thread A process flow through an application. An application can have many threads.

Thread neutral apartment An apartment that has no threads permanently associated with it; threads enter and leave the apartment as required.

Three-dimensional shape A point, line or polygon that stores x-, y- and z-coordinates as part of its geometry. A point has one set of z-coordinates; lines and polygons have z-coordinates for each vertex in a shape.

Three-tier configuration A software configuration in which three software applications (commonly a client program, application server and database server) work together to accomplish a task.

Threshold ring analysis In business analysis, an operation that creates rings that contain a given population around a store or stores on a map.

Thumbnail A miniaturized version of a graphics file. A thumbnail can be used as a visual index for larger data or images.

Tic A registration or geographic control point for a coverage representing a known location on the earth's surface. Tics allow all coverage features to be recorded in a common coordinate system. Tics are used to register map sheets when they are mounted on a digitizer. They are also used to transform the coordinates of a coverage, for example, from digitizer units (inches) to the appropriate values for a particular coordinate system.

Tick marks Graphics that mark divisions of measurement on a scale bar.

Tidal datum A vertical datum in which zero height is defined by a particular tidal surface, often mean sea level. Examples of tidal surfaces include mean sea level, mean low water springs, and mean lower low water. Most traditional vertical geodetic datums are tidal datums.

Tide Periodic rise and fall of the water resulting from gravitational interactions between the sun, moon and earth. The vertical component of the particulate motion of a tidal wave. Although the accompanying horizontal movement of the water is part of the same phenomenon, it is preferable to designate this motion as tidal current.

Tie point (1) A point whose location is determined by a tie survey. (2) It is a point on the earth's surface of known or assumed position which is used to connect a survey to a particular coordinate system. Commonly used systems are the Grid Coordinate System and the Geographic Coordinate System. (3) A point on the ground which is common to two images. Several are used in the coregistration of images.

Tie survey A survey that uses a point of known location on the ground to determine the location of a second point.

Tied candidates In geocoding, two or more records that yield the same score when matching an address.

TIGER Acronym for Topologically Integrated Geographic Encoding and Referencing. The nationwide digital database developed for the 1990 census, succeeding the DIME format. TIGER files contain street address ranges, census tracts and block boundaries.

TIGER/Line files A digital database of geographic features, covering the entire United States and its territories, that provides a topological description of the geographic structure of these areas. The files are a public product created from the US Census Bureau Topologically Integrated Geographic Encoding and Referencing (TIGER) database. TIGER/Line files define the locations and spatial relationships of streets, rivers, railroads and other features to each other and to the numerous geographic entities for which the Census Bureau tabulates data from its censuses and sample surveys.

Tight coupling A high or complex degree of interconnections between the components within a program or between programs, that requires substantial overlap between methods, ontologies, class definitions and so on.

Tiling An internal subsetting of a spatial dataset (commonly raster) into a manageable rectangular set, or rows and columns of pixels, typically used to process or analyse a large raster dataset without consuming vast quantities of computer memory.

Tilt displacement A displacement in a tilted photograph that presents a slightly oblique view rather than a true vertical record.

Time mode One of two methods for displaying temporal data in ArcGIS Tracking Analyst: real-time mode or playback mode. Time modes are determined by the use of the Tracking Analyst Playback Manager.

Time window In networks, the time during which a stop can be visited. For example, on a bus route, each stop may have a time window of 15 minutes. If the bus arrives before its 15-minute time slot, it will wait until the appropriate time before proceeding. If a bus arrives after its 15-minute time slot, the stop will display a symbol to denote a time violation.

Time window violation In ArcGIS Network Analyst, a time window violation occurs when a route arrives at a network location after its associated time window has closed.

TIMS Thermal IR multispectral scanner.

TIN dataset A dataset containing a triangulated irregular network (TIN). The TIN dataset includes topological relationships between points and neighbouring triangles.

TIN layer A layer that references a set of TIN data. TIN data contains a triangulated irregular network (TIN) and includes topological relationships between points and neighbouring triangles.

TIN line type One of four types of edges that may be found in a TIN: regular lines, hard breaklines, soft breaklines and outside lines. Regular lines define the TIN's basic structure, connecting triangle nodes. Hard breaklines represent features that mark pronounced changes in slope, like roads or rivers. Soft breaklines mark milder changes in slope and sometimes artificial boundaries, such as the border of a study area. Outside lines designate parts of a TIN structure that lie beyond the TIN's zone of interpolation. Every TIN contains regular lines; other line types may or may not be present.

Tissot indicatrix A graphical representation of the spatial distortion at a particular map location. The indicatrix is the figure that results when a circle on the earth's surface is plotted to the corresponding point on a map. The shape, size and orientation of an indicatrix at any given point depend on the map projection used. In conformal (shape-preserving) projections, the indicatrix is a circle; in nonconformal projections, it is an ellipse at most locations. As a visual aid, indicatrices convey a general impression of distortion; as mathematical tools, they can be used to quantify distortion of scale and angle precisely. The indicatrix is named for Nicolas Auguste Tissot, the French mathematician who developed it.

TM Thematic mapper.

TNA Acronym for thread neutral apartment. An apartment that has no threads permanently associated with it; threads enter and leave the apartment as required.

TNTlite TNTlite is a free geospatial analysis system providing a complete fully featured Geographic Information System, RDBMS and automated image processing system with CAD, TIN, surface modelling, map layout and innovative data publishing tools. All this capability is available in a single integrated system with an identical interface, functionality and geodata structure for use on Mac OS X, Windows, Sun Solaris and popular Linux platforms. The interface, database text content, messages, map production, and all other internal aspects of TNTmips have been localized for use in numerous languages including Japanese, Chinese, Arabic, Turkish. Thai, German, Spanish, French and others. The professional version of the TNTmips is in use in over 120 nations while the free TNTlite version (restricted in project size) is used in almost every nation for educational, self learning, and small projects (e.g. archaeological sites, neighbourhood planning and precision farming). The product includes free tutorials and application booklets with step-by-step lessons.

TNTmips TNTmips is a geospatial analysis system providing a fully featured GIS, RDBMS, and automated image processing system with CAD, TIN, surface modelling, map layout and innovative data publishing tools. TNTmips has a single integrated system with an identical interface, functionality and geodata structure for use on various operating system. The interface, database text content, messages, map production and all other internal aspects of TNTmips have been localized for use in many languages, including, for example, Arabic, Thai and all romance languages. The professional version of the TNTmips is in use in over 120 nations while the free TNTlite version (restricted in project size) is used worldwide for educational, self learning and small projects (e.g. archaeological sites, neighbourhood planning, and precision farming).

Tobler's first law of geography A formulation of the concept of spatial autocorrelation by the geographer Waldo Tobler (born 1930), which states 'Everything is related to everything else, but near things are more related to each other'.

TOC In ArcGIS, a tabbed list of data frames and layers (or tables) on a map that shows how the data is symbolized, the source of the data, and whether or not each layer is selectable.

Token A line of encrypted code generated by an authenticating service that allows a client to access a Web service. In ArcWeb Services, a token must be regenerated by passing in a valid user name and password each time it expires.

Tolerance The minimum or maximum variation allowed when processing or editing a geographic feature's coordinates. For example, during editing, if a second point is placed within the snapping tolerance distance of an existing point, the second point will be snapped to the existing point.

Tone Each distinguishable shade of grey from white to black on an image.

To-node Of an arc's two endpoints, the last one digitized. From- and to-nodes give an arc left and right sides and, therefore, direction.

Tool A command that requires interaction with the GUI before an action is performed. For example, a zoom tool requires a user to use the mouse to click on or draw a box over a digital map before the tool will cause the map to be redrawn at a larger scale.

Tool Tip The description of a tool or control that is displayed on screen when the mouse is paused over it.

Toolbar A graphical user interface (GUI) with buttons that allow users to execute software commands.

Toolbox In ArcGIS, an object that contains toolsets and geoprocessing tools. It takes the form of a .tbx file on disk, or a table in a geodatabase.

Toolbox tree In ArcToolbox (part of ArcGIS 8.3 and earlier versions), a hierarchical view of toolsets and tools grouped by functionality.

Toolkit An application that software developers can use with their integrated development environment

(IDE) to make programming easier. For example, SOAP toolkits allow users to interact with SOAP Web services through various languages such as Java and .NET.

Toolset In ArcGIS, a group of geoprocessing tools that perform similar tasks.

Topo sheet A rectangular map bounded by lines of latitude and longitude, often a map sheet in either the 7.5-minute or 15-minute series published by the US Geological Survey. Quadrangles are also called topo sheets.

Topographic inversion An optical illusion that may occur on images with extensive shades. Ridges appear to be valleys and valleys appear to be ridges. The illusion is corrected by orienting the image so that the shadows trend from the top margin of the image to the bottom.

Topographic line map (TLM) Acronym for topographic line map. A map that uses line contours to show elevations and depressions of the earth's surface. Topographic line maps may be used to portray topography, elevations, infrastructure, hydrography and vegetation.
A map that uses line contours to show elevations and depressions of the earth's surface. Topographic line maps may be used to portray topography, elevations, infrastructure, hydrography and vegetation.

Topographic map A map that represents the vertical and horizontal positions of features, showing relief in some measurable form, such as contour lines, hypsometric tints and relief shading.

Topographic reversal A geomorphic phenomenon in which topographic lows coincide with structural highs and vice versa. Valleys are eroded on crests of anticlines to cause topographic lows, and synclines form ridge, or topographic highs.

Topography (1) Configuration (relief) of the land surface; the graphic delineation or portrayal of that configuration in map form, as by contour lines; in oceanography the term is applied to a surface such as the sea bottom or surface of given characteristics within the water mass. (2) The study and mapping of land surfaces, including relief (relative positions and elevations) and the position of natural and constructed features.

Topological association The spatial relationship between features that share geometry such as boundaries and vertices. When a boundary or vertex shared by two or more features is edited using the topology tools in ArcMap, the shape of each of those features is updated.

Topological feature A feature that supports network connectivity that is established and maintained based on geometric coincidence.

Topological overlay A spatial operation in which two or more maps or layers registered to a common coordinate system are superimposed, either digitally or on a transparent material, for the purpose of showing the relationships between features that occupy the same geographic space.

Topology (1) A subfield of mathematics that is applied in GIS to ensure that the spatial relationships between features are expressed explicitly.a (2) In geodatabases, the arrangement that constrains how point, line and polygon features share geometry. For example, street centrelines and census blocks share geometry, and adjacent soil polygons share geometry. Topology defines and enforces data integrity rules (for example, there should be no gaps between polygons). It supports topological relationship queries and navigation (for example, navigating feature adjacency or connectivity), supports sophisticated editing tools, and allows feature construction from unstructured geometry (for example, constructing polygons from lines).

Topology cache A temporary collection of edges and nodes used in ArcMap to query and edit the topological coincidence between features. The cache is built for the current display extent and is stored in the computer's memory.

Topology error A measured, observed, calculated or interpreted value that differs from the true value, or the value that would be obtained by a perfect observer using perfect equipment and perfect methods under perfect conditions.

Topology fix In ArcMap, a predefined method for correcting topology errors. For example, predefined topology fixes for a dangling line include snapping, trimming or extending to another line.

Topology rule An instruction to the geodatabase defining the permissible relationships of features within a given feature class or between features in two different feature classes.

Toponym A place-name.

Tour In ArcInfo Workstation, a network solver that determines the minimum-cost path to reach a series of stops, and also determines the order in which the stops are visited.

TOVS TIROS Operational Vertical Sounder.

Township (1) In the United States, a quadrangle approximately 6 miles on a side, bounded by meridians and parallels and containing 36 sections. (2) Unit of survey of the public lands of the United States, normally a quadrangle approximately 6 miles on a side with boundaries conforming to *meridians* and *parallels* within established limits, containing 36 sections. Also, in minor governmental subdivision.

TPS measurement In Survey Analyst for field measurements, an entry in an electronic or paper field book that represents observations from a theodolite. A slope distance, vertical angle, horizontal angle and a height of target define a single TPS measurement.

TPS setup In Survey Analyst for field measurements, a group of field book entries that belong together define a single set-up of the instrument. Each observation (slope distance, vertical angle, horizontal angle and a height of target) is recorded as a TPS measurement and is added to the TPS set-up. TPS is an acronym for Total Positioning System.

Tracing A function that performs network analysis on a geometric network. Specific kinds of network tracing include finding features that are connected, finding common ancestors, finding loops, tracing upstream and tracing downstream.

Track In ArcMap, ArcScene and ArcGlobe, an ordered collection of similar key frames that, when played as an animation, shows a dynamic transition between them.

Track (TRK) The direction of movement relative to a ground position.

Track identifier field In ArcGIS Tracking Analyst, a field containing a unique identifier for a given object or objects being observed. This field is used to join the components of a complex temporal event.

Tracking and data relay satellite (TDRS) Tracking and Data Relay Satellite Geostationary satellite used to communicate between ground receiving stations and satellite such as Landsat.

Tracking connection A message or Internet connection whereby real-time data can stream into ArcGIS Tracking Analyst.

Tracking data Data that specifically refers to times or dates. Temporal data may refer to discrete events, such as lightning strikes; moving objects, such as trains; or repeated observations, such as counts from traffic sensors.

Tract A small, statistical subdivision of a county that usually includes approximately 4,000 inhabitants but may include from 2,500 to 8,000 inhabitants. A census tract is designed to encompass a population with relatively uniform economic status, living conditions and some demographic characteristics. Tract boundaries normally follow physical features but may also follow administrative boundaries or other nonphysical features. A census tract is a combination of census block groups.

Trade-off As a result of changing one factor in a remote sensing system, there are compensating changes elsewhere in the system; such a compensating change is known as a trade-off.

Training area A sample of the earth's surface with known properties; the statistics of the imaged data within the area are used to determine decision boundaries in classification.

Training site Area of terrain with known properties or characteristics that is used in supervised classification.

Transaction A group of data operations that comprise a complete operational task, such as inserting a row into a table.

Transform events In linear referencing, an operation that produces a new table by copying and transforming events from one route reference to another. This allows the events to be used with a route reference having different route identifiers and/or measures.

Transformation The process of converting the coordinates of a map or an image from one system to another, typically by shifting, rotating, scaling, skewing or projecting them.

Transit Precision surveying instrument; a theodolite in which the telescope can be reversed in direction by rotation about its horizontal axis.

Transit rule A rule for adjusting the closure error in a traverse. The transit rule distributes the closure error by changing the northings and eastings of each traverse point in proportion to the northing and easting differences in each course. More specifically, a correction is computed for each northing coordinate as the difference in the course's northings divided by the sum of all the courses' northing differences. Similarly, a correction is computed for each easting coordinate using the easting coordinate differences. The corrections are applied additively to each successive coordinate pair, until the final coordinate pair is adjusted by the whole closure error amount. The transit rule assumes that course directions are measured with a higher degree of precision than the distances. Usually, observed angles are balanced for angular misclosure prior to applying a transit rule adjustment, and corrections are proportional to the x and y components of the measured line.

Translation Adding a constant value to a coordinate.

Transmissivity Property of a material that determines the amount of energy that can pass through the material.

Transmit Energy that passes through an object or material is 'transmitted'. This is in contrast to energy that may be reflected or absorbed. A window (which is not too dirty) allows light to transmit through and thus we are able to see through glass.

Transmitted energy The energy transmitted from the active sensor to the subject.

Transparency Image on a transparent photographic material, normally a positive image.

Transpiration Expulsion of water vapour and oxygen by vegetation.

Transverse aspect A map projection whose line of tangency is oriented along a meridian rather than along the equator.

Travel time In radar, the time interval between the generation of a pulse of microwave energy and its return from the terrain.

Travelling salesperson problem A Hamiltonian circuit problem in which a salesperson must find the most efficient way to visit a series of stops, then return to the starting location. In the original version of the problem, each stop may be visited only once.

Traverse (1) A predefined path or route across or over a set of geometric coordinates. (2) Sequence of lengths and directions of lines connecting a series of stations, obtained from field measurements, and used in determining positions of the stations.

Traverse course In ArcMap and Survey Analyst for field measurements, a group of observed values that define a new coordinate. A traverse course starts from a pre-existing coordinate, or a coordinate computed from the previous course.

Tree A database that stores related information in a tree-like structure, where records can be traced to parent records, which in turn can be traced to a root record.

Tree data structure A common data structure consisting of a set of nodes – basic units of data – linked hierarchically. Each node can contain one or more subordinate nodes within it, in which case it is called a parent node. The subordinate nodes are called child nodes. A node without a parent node is the root node; a node without one or more child nodes is called a leaf node. A tree data structure is used to manipulate hierarchical data and make it easily searchable.

Trend In a spatial model, nonrandom variation in the value of a variable that can be described by a mathematical function such as a polynomial.

Trend surface analysis A surface interpolation method that fits a polynomial surface by least-squares regression through the sample data points. This method results in a surface that minimizes the variance of the surface in relation to the input values. The resulting surface rarely goes through the sample data points. This is the simplest method for describing large variations, but the trend surface is susceptible to outliers in the data. Trend surface analysis is used to find general tendencies of the sample data, rather than to model a surface precisely.

Triangle Any closed, three-sided, two-dimensional polygon.

Triangulated irregular network (TIN) (1) A data model that approximates the terrain with a set of non-overlapping triangles. (2) A vector data structure that partitions geographic space into contiguous, nonoverlapping triangles. The vertices of each triangle are sample data points with x-, y-, and z-values. These sample points are connected by lines to form Delaunay triangles. TINs are used to store and display surface models.

Triangulation (1) Locating positions on the earth's surface using the principle that if the measures of one side and the two adjacent angles of a triangle are known, the other dimensions of the triangle can be determined. Surveyors begin with a known length, or baseline, and from each end use a theodolite to measure the angle to a distant point, forming a triangle. Once the lengths of the two sides and the other angle are known, a network of triangles can be extended from the first. (2) Method of extending horizontal position on the surface of the earth by measuring the angles of triangles and the included sides of selected triangles.

Trigger A piece of code in a database management system (DBMS) that executes in response to data being modified with an INSERT, UPDATE, or DELETE statement.

Trigonometric function An operator within the Raster Calculator of ArcGIS Spatial Analyst that performs various trigonometric calculations on the values in an input raster. Available trigonometric functions include Sin, Cos, Tan, Asin, Acos, and Atan.

Trilateration (1) Determining the position of a point on the earth's surface with respect to two other points by measuring the distances between all three points. (2) Method of surveying wherein the lengths of the triangle sides are measured, usually by electronic methods, and the angles are computed from the measured lengths. Compare with triangulation.

Trim guide Marks that indicate the edge of the page of a finished, printed map. Cropmarks are used as a reference for trimming excess paper after printing.

Tristimulus colour theory A theory of colour relating all hues to the combined effects of three additive primary colours corresponding to the sensitivities of the three types of cone on the retina.

True bearing A bearing measured relative to true north.

True colour A colour imaging process whereby the colour of the image is the same as the colour of the object imaged (natural colour images).

True colour image An image in which the colours represented more or less accurately correspond to the colours of the optical photons detected during the gathering of the image data. Compare *colour composite* and false colour images.

True curve A curve that is defined mathematically rather than by a series of connected vertices. A parametric curve has only two vertices, one at each end.

True north The direction from any point on the earth's surface to the geographic north pole.

True-direction projection A map projection that transforms points from a spheroid or sphere onto a tangent or secant plane. The azimuthal projection is also known as a planar or zenithal projection.

TSP A Hamiltonian circuit problem in which a salesperson must find the most efficient way to visit a series of stops, then return to the starting location. In the original version of the problem, each stop may be visited only once.

Tuple An individual row or record in a database table. Each tuple records the values for the columns defined in the table.

Turn In network analysis, a movement that explicitly models transitions between edge elements during navigation.

Turn feature class A specialized feature class that defines turn movements in an ArcGIS network dataset. Turn features explicitly model subsets of possible transitions between edge elements during navigation, and may also store the turn impedance.

Turn impedance In network analysis, the cost of making a turn at a network node. The impedance for making a left turn, for example, can be different from the impedance for making a right turn or a U-turn at the same place.

Turn table In ArcInfo Workstation and ArcView GIS 3, a table that stores information about the cost of making each turn movement in a network. A turn table identifies the edge that the turn movement comes from, the junction where the turn occurs and the edge that it turns onto.

Turn-by-turn maps A series of small maps detailing where route segments meet.

Tween In animated applications, the process of creating intermediate frames between two images so it seems like the first image merges smoothly into the second image.

Two-tier configuration A software configuration in which two software applications (commonly a client and a server) work together to accomplish a task.

Two-way replication A type of geodatabase replication that allows data changes to be sent multiple times from the parent replica to the child replica or from the child replica to the parent replica. The data on both replicas must be editable, versioned and have GlobalID columns. ArcSDE geodatabases are used to create two-way replicas.

Type inheritance In programming, a kind of inheritance in which an interface may inherit from a parent interface. A client may call the child interface as if it were the parent, as all the same members are supported.

Type library A collection of reusable classes, interfaces, enumerations and so on that can be included in programs. Type libraries usually have the extension .olb.

U

UDDI Acronym for Universal Description, Discovery and Integration. An XML-based standard for creating online directories of Web services.

UI Acronym for user interface. The portion of a computer's hardware and software that facilitates human interaction. The UI includes items that can be displayed on screen, and interacted with by using the keyboard, mouse, video, printer and data capture.

UIControl A custom button, tool, text box or combo box created with VBA.

Ultra-violet rays The portion of the electromagnetic spectrum ranging from just above the visible (about 4,000 Å) to below 400 Å, on the border of the X-ray region. In terms of wavelength, the UV spectrum ranges from 0.01Fm to 0.38 Fm.

UML Acronym for Unified Modelling Language. A modelling language that uses a series of diagrams to model the objects in a system.

UNC Acronym for Universal (or Uniform) Naming Convention. A PC format for specifying the location of resources on a local-area network (LAN). UNC follows the format \\servername\shared_resource_path

Uncertainty The degree to which the measured value of some quantity is estimated to vary from the true value. Uncertainty can arise from a variety of sources, including limitations on the precision or accuracy of a measuring instrument or system; measurement error; the integration of data that uses different scales or that describe phenomena differently; conflicting representations of the same phenomena; the variable, unquantifiable, or indefinite nature of the phenomena being measured; or the limits of human knowledge. Uncertainty is often used to describe the degree of accuracy of a measurement.

Unclosed parcel A parcel that is only partially defined or that is missing a sequence of one or more lines that would otherwise close the parcel back onto its point of beginning.

Uncommitted read The isolation level in an RDBMS specifying the minimum isolation from concurrent transactions. The transaction can read data that has been changed by concurrent transactions even before the changes are committed to the database.

Undershoot (1) A line that falls short of another line that it should intersect. (2) One type of digitizing error that results in gap between arcs.

Undevelopable surface A surface, such as the earth's, that cannot be flattened into a map without stretching, tearing, or squeezing it. To produce a flat map of the round earth its three-dimensional surface must be projected onto a developable shape such as a plane, cone or cylinder.

Undirected network flow A network state in which each edge may or may not have an associated direction of flow. In an undirected network flow, the resource that traverses a network's components can decide which direction to take, such as traffic in transportation systems.

Uniform list In Survey Analyst for field measurements, one of two types of lists in the List page. The uniform list has rows that represent survey objects of the same type.

Uninitialized flow direction A condition that occurs in a network when an edge feature is not connected through the network to sources and sinks or if the edge feature is only connected to sources and sinks through disabled features.

Union A topological overlay of two or more polygon spatial datasets that preserves the features that fall within the spatial extent of either input dataset; that is, all features from both datasets are retained and extracted into a new polygon dataset.

Unique constraint A DBMS-defined restriction specifying that each value stored in a column must be unique and that no other row can contain the same value.

Unit of measure A standard quantity used for measurements such as length, area and height.

United states geological survey A scientific agency of the US government, part of the Department of the Interior. The US Geological Survey is a fact-finding research agency that monitors, analyses, and provides scientific understanding about natural resource issues and conditions, the environment and natural hazards. The US Geological Survey is the primary civilian mapping agency in the United States. It produces digital and paper map products; aerial photography; and remotely sensed data on land cover, hydrology, geology, biology and geography.

Univariate analysis Any statistical method for evaluating a single variable, rather than the relationship between two or more variables.

Univariate distribution A function for a single variable that gives the probabilities that the variable will take a given value.

Universal description, discovery, and Integration An XML-based standard for creating online directories of Web services.

Universal kriging A kriging method often used on data with a significant spatial trend, such as a sloping surface. In universal kriging, the expected values of the sampled points are modelled as a polynomial trend. Kriging is carried out on the difference between this trend and the values of the sampled points.

Universal polar stereographic A projected coordinate system that covers all regions not included in the UTM coordinate system; that is, regions above 84 degrees north and below 80 degrees south. Its central point is either the north or south pole.

Universal soil loss equation An erosion model developed by the Agricultural Research Service of the United States Department of Agriculture that computes average annual soil loss caused by rainfall and associated overland flow. Factors used in the equation include rainfall, soil characteristics, topography, and land use and land cover. Each major factor is divided into numerous subfactors.

Universal time (UT) A timekeeping system that defines local time throughout the world by relating it to time at the prime meridian. Universal time is based on the average speed at which the earth rotates on its axis. For official purposes, universal time has been replaced by coordinated universal time; universal time is, however, still used in navigation and astronomy. Different versions of universal time correct for irregularities in the earth's rotation and orbit.

Universal transverse mercator (UTM) (1) A grid coordinate system that projects global sections onto a flat surface to measure position in specific zones. (2) A projected coordinate system that divides the world into 60 north and south zones, 6 degrees wide.

Universal transverse mercator (UTM) grid Military grid system based on the transverse Mercator projection, applied to maps of the earth's surface extending from the equator to 84 degrees north and 80 degrees south latitudes

Universal transverse mercator (UTM) Projection A widely used map projection that employs a series of identical projections around the world in the mid-latitude areas, each spanning six degrees of longitude and oriented to a meridian. This projection is characterized by its conformality; that is, it preserves angular relationships and scale and it easily allows a rectangular grid to be superimposed on it. Many worldwide topographic and planimetric maps at scales ranging between 1:24,000 and 1:250,000 use this projection.

Universe polygon In coverages, the first record in a polygon attribute table, representing the area beyond the outer boundary of the coverage.

UNIX time The number of seconds, in coordinated universal time format, since 1 January 1970 (the start of the UNIX system).

Unjoined parcel In Survey Analyst - Cadastral Editor, a parcel that has not been connected to the cadastral fabric, and that has its own local coordinate system.

Unknown points In Survey Analyst for field measurements, previously uncoordinated points.

Unmanaged raster catalog A raster catalog in which the raster datasets are not copied or altered by the geodatabase and there will only be a pointer connecting the raster catalog row to the raster datasets. Deleting a row in an unmanaged raster catalog will not delete the raster dataset from its storage location.

Unpolarized wave A polarization state in which the two perpendicular components of the electric field have equal magnitudes and a random relative phase difference.

Unprojected coordinates A measurement of a location on the earth's surface expressed in degrees of latitude and longitude.

Unsupervised classification Digital information extraction technique in which the computer assigns pixels to categories with no instructions from the operator.

Update-gram In ArcGIS, the programming object that holds updates in a delta file.

Upland Highland; ground elevation above the lowlands along rivers or between hills.

Uplink A connection through which signals transmit to a satellite.

Upper scene corners The corners associated with the trailing edge (first scan) of a scene. For descending path scenes, the upper left corner corresponds to the north-west corner of a scene and the upper right corner corresponds to the north-east corner of a scene. For ascending path scenes, the upper left corner corresponds to the south-east corner of a scene and the upper right corner corresponds to the south-west corner of a scene. These mappings hold for the band file geolocation fields and the metadata file.

UPS Acronym for Universal Polar Stereographic. A projected coordinate system that covers all regions not included in the UTM coordinate system; that is, regions above 84 degrees north and below 80 degrees south. Its central point is either the north or south pole.

Upstream In network tracing, the direction along a line or edge that opposes the direction of flow.

Urban geography The field of geography concerning the spatial and cultural patterns and processes of cities and neighbourhoods.

Urban vector map A vector-based data product in vector product format (VPF), typically at larger scales ranging from 1:2,000 to 1:25,000. UVMap data is typically collected over densely populated urban areas.

URL Acronym for uniform resource locator. A standard format for the addresses of Web sites. A URL may look like this: http://www.esri.com. The first part of the address indicates what protocol to use (such as http: or ftp:), while the second part specifies the IP address or the host name (including the domain name) where the Web site is located. An optional third part may specify the path to a specific file or resource (http://www.esri.com/products.html).

US geological survey (USGS) A bureau of the Department of the Interior. USGS was established in 1879 following several Federally sponsored independent natural resource surveys of the West and Midwest. The Department of the Interior has responsibility for most of our nationally owned public lands and natural resources. The USGS monitors resources such as energy, minerals, water, land, agriculture and irrigation. The resulting scientific information contributes to environmental-policy decision-making and public safety. For example, USGS identifies flood- and landslide-prone areas and maintains maps of the United States.

US national geodetic survey The US government agency responsible for maintaining the National Spatial Reference System (NSRS), the national coordinate system of the United States.

Usage The way in which statements in a command or programming language are actually used. In geoprocessing, usage for a tool or environment setting can be viewed at the command line.

Usage time The amount of time between when a client gets a reference to a service and when they release it.

User account administration A Web site for accessing information about an ArcWeb account. This site allows users to activate an ArcWeb Services account, view usage of ArcWeb services, manage POI and address records, and manage groups and alerts.

User interface The aspects of a computer system or program with which a software user can interact, and the commands and mechanisms used to control its operation and input data.

User name The identification used for authentication when a user logs in to a program.

User segment The component of the GPS system that includes the receivers.

USGS Acronym for United States Geological Survey. A scientific agency of the US government, part of the Department of the Interior. The US Geological Survey is a fact-finding research agency that monitors, analyses, and provides scientific understanding about natural resource issues and conditions, the environment, and natural hazards. The US Geological Survey is the primary civilian mapping agency in the United States. It produces digital and paper map products; aerial photography; and remotely sensed data on land cover, hydrology, geology, biology and geography.

USLE An erosion model developed by the Agricultural Research Service of the United States Department of Agriculture that computes average annual soil loss caused by rainfall and associated overland flow. Factors used in the equation include rainfall, soil characteristics, topography and land use and land cover. Each major factor is divided into numerous subfactors.

UT A timekeeping system that defines local time throughout the world by relating it to time at the prime meridian. Universal time is based on the average speed at which the earth rotates on its axis. For official purposes, universal time has been replaced by coordinated universal time; universal time is, however, still used in navigation and astronomy. Different versions of universal time correct for irregularities in the earth's rotation and orbit.

UTC The official timekeeping system of the world's nations since 1972. It refers local time throughout the world to time at the prime meridian, and is based on atomic clocks, but is periodically artificially adjusted so as to always remain within 0.9 seconds of universal time. The adjustment is made by the addition of leap seconds to the course of atomic time. Coordinated universal time is abbreviated 'UTC.' (The abbreviation UTC does not represent the word order of 'coordinated universal time' in either English or French. It is an extension of the 'UT*' pattern established for versions of universal time.)

Utility COM object In ArcObjects, a COM object that encapsulates a large number of fine-grained ArcObjects method calls and exposes a single coarse-grained method call. Utility COM objects are installed on a GIS server and called by server applications to minimize the round-trips between the client application and the GIS server.

Utility web service A SOAP Web service for changing the coordinate system and datum transformation of a geometry or envelope.

UTM Acronym for Universal Transverse Mercator. A projected coordinate system that divides the world into 60 north and south zones, 6 degrees wide.

UV Ultraviolet region of the electromagnetic spectrum ranging in wavelengths from 0.01 to 0.4m.

UVMap Acronym for Urban Vector Map. A vector-based data product in vector product format (VPF), typically at larger scales ranging from 1:2,000 to 1:25,000. UVMap data is typically collected over densely populated urban areas.

V

Vagueness In GIS, a state of uncertainty in data classification that exists when an attribute applies to an indeterminate quality of an object or describes an indefinite quantity. For example, the classification of an area of land as the range of golden-winged warblers (a rare species of bird) is vague for two reasons. The area populated by the birds is indefinite: it is changing constantly and can never be precisely defined. The term 'range' is also somewhat vague since the birds migrate and occupy the territory for only part of the year.

Valency In coverages, the number of arcs that begin or end at a node.

Valency table A table that lists the nodes in a data layer along with their valencies.

Valid value table (VVT) A component of the PLTS knowledge base that contains the valid attribute combinations and rendering information for the features in a feature class. VVTs are used to symbolize layers in batch mode, and provide a means to symbolize features based on more than three attributes.

Validation The process, using formal methods, of evaluating a system or software component to determine whether it functions as expected and achieves the intended results.

Validation rule A rule applied to objects in the geodatabase to ensure that their state is consistent with the system that the database is modelling. The geodatabase supports attribute, connectivity, relationship and custom validation rules.

Value A measurable quantity that may be passed to a function. Values are either assigned or determined by calculation.

Value attribute table (VAT) A table containing attributes for a grid, including user-defined attributes, the values assigned to cells in the grid, and a count of the cells with those values. Acronym for value attribute table. A table containing attributes for a grid, including user-defined attributes, the values assigned to cells in the grid and a count of the cells with those values.

Variable A symbol or placeholder that represents a changeable value or a value that has not yet been assigned.

Variable depth masking A drawing technique for hiding part of one layer using another set of features. Variable depth masking allows a layer to be drawn with gaps at specific locations without affecting other layers at these locations.

Variance A measure of the dispersion of the actual values of a variable about its mean. It is the mean of the squares of all the deviations from the mean value of a range of data.

Variance covariance matrix In surveying, the symmetric 3x3 matrix that mathematically expresses the correlation between errors in coordinates x, y, and z.

Variant A data type that can contain any kind of data.

Variogram A function of the distance and direction separating two locations that is used to quantify dependence. The variogram is defined as the variance of the difference between two variables at two locations. The variogram generally increases with distance and is described by nugget, sill and range parameters. If the data is stationary, then the variogram and the covariance are theoretically related to each other.

Variography The process of examining spatial dependence using a variogram; a set of procedures (as much art as science) for interpreting variograms.

VAS Atmospheric Sounder, carried by GEOS satellites

VB A programming language developed by Microsoft based on an object-oriented form of the BASIC language and intended for application development. Visual Basic runs on Microsoft Windows platforms.

VBA The embedded programming environment for automating, customizing and extending ESRI applications, such as ArcMap and ArcCatalog. It offers the same tools as Visual Basic in the context of an existing application. A VBA program operates on objects that represent the application and can be used to create custom symbols, workspace extensions, commands, tools, dockable windows and other objects that can be plugged in to the ArcGIS framework.

VDOP Acronym for dilution of precision. An indicator of satellite geometry for a constellation of satellites used to determine a position. Positions with a lower DOP value generally constitute better measurement results than those with higher DOP. Factors determining the total GDOP (geometric DOP) for a set of satellites include PDOP (positional DOP), HDOP (horizontal DOP), VDOP (vertical DOP), and TDOP (time DOP).

Vector A coordinate-based data model that represents geographic features as points, lines and polygons. Each point feature is represented as a single coordinate pair, while line and polygon features are represented as ordered lists of vertices. Attributes are associated with each vector feature, as opposed to a raster data model, which associates attributes with grid cells.

Vector data Vector data, when used in the context of spatial or map information, refers to a format where all map data is stored as points, lines and areas rather than as an image or continuous tone picture. These vector data have location and attribute information associated with them.

Vector data model A spatial data model that used points and their x and y coordinates to construct spatial features of points, lines and areas.

Vector data model/Vector model A representation of the world using points, lines and polygons. Vector models are useful for storing data that has discrete boundaries, such as country borders, land parcels and streets.

Vector format The expression of points, lines, and areas on a map by digitized Cartesian coordinates, directions and values.

Vector product format A vendor-neutral data format used to structure, store and access geographic data according to a defined standard. Acronym for Vector Product Format. A vendor-neutral data format used to structure, store, and access geographic data according to a defined standard.

Vectorization The conversion of raster data (an array of cell values) to vector data (a series of points, lines and polygons).

Vectorization settings GIS software settings that allow users to control the conversion of raster data to vector data. For example, vectorization settings may allow users to select which raster cells are eligible for vectorization or specify how the geometry of the output vector data will be constructed during vectorization.

Vectorization trace tool An Arc Scan tool that allows users to manually trace raster cells and generate features to be converted from raster to vector data.

Vegetation anomaly Deviation from the normal distribution or properties of vegetation. Vegetation anomalies may be caused by faults, trace elements in soil or other factors.

Vegetation index The reduction of multispectral scanning measurements to a single value for predicting and assessing vegetative characteristics such as plant leaf area, total biomass and general plant stress and vigour.

Vehicle routeing problem In ArcGIS Network Analyst, a type of network analysis for routeing a fleet of vehicles to service a set of orders with the goal of minimizing some objective (e.g. operating cost), while satisfying certain constraints. These constraints may include time windows, multiple route capacities, travel duration constraints, route zone and route seed point constraints, specialties constraints and paired order constraints.

Velocity made good (VMG) The speed you are closing in on a destination along a desired course.

Verbal scale A map scale that expresses the relationship between distance on the map and distance on the ground in words; for example, 'One inch represents 20 miles'.

Verification The process, using formal methods, of evaluating a system or software component to determine whether it satisfies the requirements imposed at the start of development.

Version In databases, an alternative state of the database that has an owner, a description, a permission (private, protected or public) and a parent version. Versions are not affected by changes occurring in other versions of the database.

Version merging The process of reconciling two versions of a dataset into a common version. If conflicting edits have been made in either of the merged versions, these conflicts are resolved, either automatically or by an interactive process.

Version reconciliation The process of updating a version of a dataset with changes made in another version. Using this technique, a version can remain up to date with changes even if it is within a long transaction lasting many months.

Vertex One of a set of ordered x,y coordinate pairs that defines the shape of a line or polygon feature.

Vertical control A basic element of a software application's GUI. Examples of controls include menus, buttons, tools, check boxes, slider bars, text input boxes and combo boxes.

Vertical control/geodetic datum A geodetic datum for any extensive measurement system of heights on, above, or below the earth's surface. Traditionally, a vertical geodetic datum defines zero height as the mean sea level at a particular location or set of locations; other heights are measured relative to a level surface passing through this point. Examples include the North American Vertical Datum of 1988; the Ordnance Datum Newlyn (used in Great Britain); and the Australian Height Datum.

Vertical coordinate system A reference system that defines the location of z-values relative to a surface. The surface may be gravity related, such as a geoid, or a more regular surface like a spheroid or sphere.

Vertical exaggeration A multiplier applied uniformly to the z-values of a three-dimensional model to enhance the natural variations of its surface. Scenes may appear too flat when the range of x- and y-values is much larger than the z-values. Setting vertical exaggeration can compensate for this apparent flattening by increasing relief.

Vertical line A line that corresponds to the direction of gravity at a point on the earth's surface; the line along which an object will fall when dropped.

Vertical photograph An aerial photograph taken with the camera lens pointed straight down.

Vertical polarization Linear polarization with the lone electric vector oriented in the vertical direction in antenna coordinates.

Vertical shift In ArcGIS, a parameter that offsets the z-origin from the surface of a vertical coordinate system. The vertical shift is similar in effect to the false easting or false northing parameters of a projected coordinate system.

Vertical transmit-horizontal receive polarization (VH) A mode of radar polarization where the microwaves of the electric field are oriented in the vertical plane for signal transmission, and where the horizontally polarized electric field of the backscattered energy is received by the radar antenna.

Vertical transmit-vertical receive polarization (VV) A mode of radar polarization where the microwaves of the electric field are oriented in the vertical plane for both signal transmission and reception by means of a radar antenna. In this case, the plane of the electric field of the microwave energy is designated by the letter V (vertical) for both transmit and receive events, i.e. VV; this transmit-receive polarity is also called like-polarized as opposed to cross-polarized (horizontal transmit - vertical receive, HV). The amount of radar backscatter received at a particular linear polarization state from a particular ground surface or object depends, in part, on the scattering mechanism and depolarization effects involved. The transmit-receive acronym is often used in conjunction with the frequency band (wavelength) designation of a particular radar system, e.g. C-VV for C-band. Several satellite SAR designs have used single-band, horizontally like-polarized systems, for example the European ERS–1 and ERS–2 (C-VV).

Vidicon An imaging device based on a sheet of transparent material whose electrical conductivity increases with the intensity of EM radiation falling on it. The variation in conductivity across the plate is measured by a sweeping electron beam and converted into a video signal. Now largely replaced by cameras employing arrays of charge-coupled devices (CCDs).

View In ArcGIS, a way to see the contents of a selected item in the Catalog tree in ArcCatalog.

Viewer In ArcGlobe and 3D Analyst, an additional window that allows a user to view 3D data in a scene from another angle. Users can have multiple viewers in a scene.

Viewing angle The acute angle between the looking axis of the sensor and the horizontal or ground elevation surface.

Viewshed The locations visible from one or more specified points or lines. Viewshed maps are useful for such applications as finding well-exposed places for communication towers, or hidden places for parking lots.

Vignetting A gradual change in overall tone of an image from the centre outwards, caused by the imaging device gathering less radiation from the periphery of its field of view than from the centre. Most usually associated with the radially increasing angle between a lens and the earth's surface, and the corresponding decrease in the light-gathering capacity of the lens.

Virtual directory A directory name, used in a URL, that corresponds to a real or actual directory on a Web server.

Virtual page The map page as seen in layout view in ArcMap.

Virtual study area The current extent of a map document in ArcGIS. The virtual study area never has a boundary.

Virtual table A logical table in a database that stores a pointer to the data rather than the data itself.

Visible band The band of the electromagnetic spectrum in which the human eye is sensitive. This band ranges from 0.4 - 0.7 micrometers.

Visible radiation The electromagnetic radiation that humans can see as colours. The visible spectrum is composed of wavelengths between 0.4 to 0.7 micrometers. Red is the longest and violet is the shortest. Landsat 7 has three visible bands in red, green, and blue.

Visible scale range A minimum and maximum value that a map scale must fall between in order for the map layers to be displayed.

VISSR Visible Infrared Spin-Scan Radiometer carried by the GOES satellites.

Visual basic A programming language developed by Microsoft based on an object-oriented form of the BASIC language and intended for application development. Visual Basic runs on Microsoft Windows platforms.

Visual basic for applications The embedded programming environment for automating, customizing, and extending ESRI applications, such as ArcMap and ArcCatalog. It offers the same tools as Visual Basic in the context of an existing application. A VBA program operates on objects that represent the application and can be used to create custom symbols, workspace extensions, commands, tools, dockable windows, and other objects that can be plugged in to the ArcGIS framework.

Visual basic virtual machine (VBVM) Acronym for Visual Basic Virtual Machine. The runtime environment used by Visual Basic code when it runs.
The runtime environment used by Visual Basic code when it runs.

Visual C++ A Microsoft implementation of the C++ programming language which is part of Microsoft's Visual Studio Development environment, used to compile and develop Windows software.

Visual centre The point on a rectangular map or image to which the eye is drawn. The visual centre lies slightly (about 5% of the total height) above the geometric centre of the page.

Visual dissonance The disturbing effect of seeing a familiar object in an unfamiliar setting or in an unexpected colour.

Visual hierarchy The presentation of features on a map in a way that implies relative importance, usually achieved with visual contrast.

Visual interpretation The interpretation of imagery by non-computerized methods.

Visualization The representation of data in a viewable medium or format. In GIS, visualization is used to organize spatial data and related information into layers that can be analysed or displayed as maps, three-dimensional scenes, summary charts, tables, time-based views and schematics.

VMap Acronym for Vector Map. A vector-based data product in vector product format (VPF) at several scales divided into groups, referred to as levels. For example, VMap Level 1 includes vector maps at a scale of 1:250,000, and VMap Level 2 includes vector maps at a scale of 1:50,000.

Volume In a TIN, the space (measured in cubic units) between a TIN surface and a plane at a specified elevation. Volume may be calculated above or below the plane.

Volume scattering In radar, interaction between electromagnetic radiation and the interior of a material.

Voronoi diagram A partition of space into areas, or cells that surround a set of geometric objects (usually points). These cells, or polygons, must satisfy the criteria for Delaunay triangles. All locations within an area are closer to the object it surrounds than to any other object in the set. Voronoi diagrams are often used to delineate areas of influence around geographic features. Voronoi diagrams are named for the Ukrainian mathematician Georgy Fedoseevich Voronoi (1868–1908).

Voxel A three-dimensional pixel used to display and rotate three-dimensional images.

VPF feature class In ArcGIS, a collection of geographic features with the same geometry type (such as point, line or polygon), the same attributes and the same spatial reference. Feature classes can be stored in geodatabases, shapefiles, coverages or other data formats. Feature classes allow homogeneous features to be grouped into a single unit for data storage purposes. For example, highways, primary roads and secondary roads can be grouped into a line feature class named 'roads'. In a geodatabase, feature classes can also store annotation and dimensions.

VPF layer A layer that references a set of VPF data. VPF data is a standard format, structure and organization for large geographic databases that are based on a georelational data model.

VRP In ArcGIS Network Analyst, a type of network analysis for routeing a fleet of vehicles to service a set of orders with the goal of minimizing some objective (e.g. operating cost), while satisfying certain constraints. These constraints may include time windows, multiple route capacities, travel duration constraints, route zone and route seed point constraints, specialties constraints and paired order constraints.

VTable binding The fastest form of early binding, during which client code that uses a COM object (such as Visual Basic) binds to a method by indexing to a virtual function table, or Vtable.

W

W3C Acronym for World Wide Web Consortium. An organization that develops standards for the World Wide Web and promotes interoperability between Web technologies, such as browsers, programming languages, and devices. Members from around the world contribute to standards for XML, SOAP, HTML, and many other Web-based protocols.

Wait time The amount of time between the time that a client requests an object from a server and the time the client receives that object.

Walk mode In ArcGlobe, a navigation mode that allows navigation on a globe close to the ground, simulating walking.

WAN Acronym for wide area network. A computer network that connects computers in a large area, such as in different cities or countries. The Internet is the most well-known example of a WAN.

War fighting element In MOLE, a real-world battle element, such as a ground force unit or a lane boundary, represented by MOLE symbols on a map.

War fighting symbol In MOLE, graphics on a map that represent battle elements such as ground troops and direction of troops. These graphics are used to plan and execute military operations in support of C4I functions. MOLE symbols fall within four main categories: force elements, tactical graphics, stacks and leaders.

Warping A procedure for adjusting the coordinates of all the data points in a dataset to allow a more accurate match between known locations and a few data points within the dataset. Rubber sheeting preserves the interconnectivity between points and objects through stretching, shrinking or reorienting their interconnecting lines.

Watch file A text file that records all dialog during an ArcInfo session. Watch files can be edited and converted to macro programs.

Waterfall model A software design methodology in which development proceeds through a top-down process of overlapping stages. First proposed in 1970, the waterfall model is a highly structured approach to a project life cycle that cascades linearly through the developmental phases of requirements analysis, design, implementation, testing, integration and maintenance.

Watershed A basin-like terrestrial region consisting of all the land that drains water into a common terminus.

Watt (W) Unit of electrical power equal to rate of work done by one ampere under a potential of one volt.

Wavelength (1) The distance between two successive crests on a wave, calculated as the velocity of the wave divided by its frequency. (2) Distance between successive wave crests or other equivalent points in a harmonic wave. Wavelengths are related to frequency: The longer the wavelength, the lower the frequency. Wavelength is indicated in equations by the Greek letter lambda (λ).

Wavelet Filter for enhancing and compressing radar images. Wavelets are operators that act both in space (azimuth and range) and in frequency (signal content). The classification and identification of texture and features in images incorporates fractal and multi-fractal models that can parameterize geophysical processes. Wavelets may be tailored to retain desired signal behaviour or to preserve smoothness. They are also used to characterize complex signals.

Wavelet compression A lossy method of data compression that uses mathematical functions and is best used in image or sound compression.

Wayfinding The mental activities engaged in by a person trying to reach a destination, usually an unfamiliar one, in real or virtual space. Wayfinding consists of acquiring information that is relevant to choosing a route or a segment of a route, and of evaluating that information in the course of travel so the route can be changed as needed. Wayfinding is the cognitive component of navigation.

Waypoint A location of interest, or a reference point on a route, stored as latitude-longitude coordinates and often captured by a GPS receiver.

Web application A software program that communicates via the World Wide Web and delivers Web-based information to the user in HTML format. Web applications are typically used to add customization and interactivity to Web pages. Web applications may also be called Web-based applications.

Web application template A file that contains a user interface as well as all the code and necessary files to use as a starting point for creating a new customized Web application. ArcGIS Server contains a number of Web application templates.

Web browser An application that allows users to access and view Web pages on their computer screens. Web browsers enable users to view HTML documents on the World Wide Web.

Web control The visual component of a Web form that executes its own action on the server. Web controls are designed specifically to work on Web forms and are similar in appearance to HTML elements.

Web coverage service (WCS) Acronym for Web Coverage Service. A standard specification for exchanging coverages over the Internet using a server. The Web Coverage Service (WCS) is the result of a collaborative effort assembled by the Open Geospatial Consortium, Inc. (OGC).
A standard specification for exchanging coverage over the Internet using a server. The Web Coverage Service (WCS) is the result of a collaborative effort assembled by the Open Geospatial Consortium, Inc. (OGC)

Web feature server specification A set of interface specifications that standardizes data manipulation and map display on the Internet. The Web Feature Server (WFS) specification is the result of a collaborative effort assembled by the Open Geospatial Consortium, Inc. (OGC).

Web form Based on ASP.NET technology, Web forms allow the creation of dynamic Web pages in a Web application. Web forms present their user interface to a client in a Web browser or other device, but generally execute their actions on the server.

Web map server specification A set of interface specifications that provides uniform access by Web clients to maps rendered by map servers on the Internet. The Web Map Server (WMS) is the result of a collaborative effort assembled by the Open Geospatial Consortium, Inc. (OGC).

Web page A page of information stored on a Web site and viewed in a Web browser. Web pages may contain text, graphics, animations, forms for data entry and links to other Web pages.

Web portal A Web resource that provides access to a broad array of related resources and services.

Web server A computer that manages Web documents, Web applications and Web services, and makes them available to the rest of the world.

Web service A software component accessible over the World Wide Web for use in other applications. Web services are built using industry standards such as XML and SOAP, and thus are not dependent on any particular operating system or programming language, allowing access to them through a wide range of applications.

Web service catalog A collection of ArcGIS Server Web services. A Web service catalog is itself a Web service with a distinct location (URL) and can be queried to obtain the list of Web services in the catalog and their URLs.

Web service description language An XML format for describing the methods, types, and connection point of a SOAP Web service.

Web site A collection of Web pages (HTML files) that are interconnected with hyperlinks and published on the World Wide Web.

Weed tolerance The minimum distance allowed between any two vertices along a line, set before digitizing. When new lines are added, vertices that fall within that distance of the last vertex are ignored. Weed tolerance applies only to vertices, not to nodes.

Weeding Reducing the number of points that define a line while preserving its essential shape.

Weight A number that indicates the importance of a variable for a particular calculation. The larger the weight assigned to the variable, the more that variable will influence the outcome of the operation.

Weight filter In geometric networks, a specification for which network features can be traced based on their weight values.

Weighted constrained adjustment In Survey Analyst for field measurements, one of two possibilities for performing a constrained adjustment. In the weighted constrained adjustment, the reference point coordinates are treated as observed measurements and their standard deviations are applied in the adjustment.

Weighted mean centre The geographic centre of a set of points as adjusted for the influence of a value associated with each point. For example, while the mean centre of a group of grocery stores would be the location obtained by averaging the stores' x,y coordinates, the weighted mean centre would be shifted closer to stores with higher sales, more square footage, or a greater quantity of some other specified attribute.

Weighted moving average The value of a point's attribute computed by averaging the values of its surrounding points, taking into account their importance or their distance from the point.

Weighted overlay A technique for combining multiple rasters by applying a common measurement scale of values to each raster, weighting each according to its importance, and adding them together to create an integrated analysis.

Weird polygon A polygon that violates topological integrity by crossing its own boundary (usually by making a small loop).

WFS A set of interface specifications that standardizes data manipulation and map display on the Internet. The Web Feature Server (WFS) specification is the result of a collaborative effort assembled by the Open Geospatial Consortium, Inc. (OGC).

WGS 1972/WGS72 Acronym for World Geodetic System 1972. A geocentric datum and coordinate system designed by the US Department of Defense, no longer in use.

WGS 1984/WGS84 Acronym for World Geodetic System 1984. The most widely used geocentric datum and geographic coordinate system today, designed by the US Department of Defense to replace WGS72. GPS measurements are based on WGS84.

WGS84 spheroid World Geodetic System reference spheroid of 1984, which is satellite determined spheroid.

WHERE clause A SQL clause that specifies a selection criterion that allows filtering of a query result set.

Whisk broom scanner A remote-sensing tool with an oscillating mirror that moves back and forth across a satellite's direction of travel, creating scan line strips that are contiguous or that overlap slightly, thereby producing an image.

White light This is the artificial construct that the human brain has developed to convey to its owner that the human eyes have detected an ensemble of photons, with a wavelength and intensity distribution approximately equal to the sun's continuum spectrum.

Wide area network A computer network that connects computers in a large area, such as in different cities or countries. The Internet is the most well-known example of a WAN.

Widget An interactive graphic component of a user interface (such as a button, scroll bar or menu bar), its controlling program, or the combination of both the component and program.

Wien's displacement law Describes the shift of the radiant power peak to shorter wavelengths as temperature increases.

Wind rose A diagram showing, for a given place and time period, how much of the time the wind blows from each direction. Wind roses have many variations, but in the typical pattern, a number of wedges (usually eight, twelve or sixteen) radiate from the centre of a circle. The width and orientation of a wedge represent the direction from which the wind blows; the length of a wedge represents the percentage of time the wind blows from that direction. More complex wind roses use colour schemes and other graphic devices to represent wind speed and related information.

Windowing The process of limiting the viewable extent of a map or data by panning and zooming.

Wireframe A three-dimensional picture of an object composed entirely of lines (wires). The lines represent the edges or surface contours, including those that would otherwise be hidden by a solid view. Wireframes make editing easier, since the screen redraws much more quickly.

Wireless application A ready-to-use Web application for a wireless client such as a handheld device, PDA, or cellular phone, designed for a specific purpose such as maps and routeing for travel assistance.

Wireless location web service A SOAP ArcWeb service that determines the location of wireless devices.

Wizard An interactive user interface that helps a software user complete a task one step at a time. Wizards are often implemented as a sequence of dialog boxes that the user can move through, filling in required details. Wizards are usually used to simplify long, difficult or complex tasks.

WMS A set of interface specifications that provides uniform access by Web clients to maps rendered by map servers on the Internet. The Web Map Server (WMS) is the result of a collaborative effort assembled by the Open Geospatial Consortium, Inc. (OGC).

Word abbreviation A shortened form of a word or phrase which represents the whole. Abbreviations are commonly a letter or group of letters taken from the complete form of the word, such as the usage of St. in place of Street.

Work order One specific task that proceeds through each stage of an organization's workflow process, including design, acceptance and construction in the field.

Workflow An organization's established processes for design, construction, and maintenance of programs, products and business objectives.

Workgroup geodatabase A Microsoft SQL Server Express database that uses ArcSDE technology to store, query and modify spatial data. Workgroup geodatabases accept up to 10 non-Web client connections and unlimited Web client connections to the SQL Server Express instance and are licensed through ArcGIS Server Workgroup edition.

Working directory A directory that indicates the appropriate location on disk to place results from analysis.

Workspace A container for geographic data. A workspace can be a folder that contains shapefiles, a geodatabase, a feature dataset, or an

ArcInfo workspace. Other multidimensional data formats such as netCDF or HDF can also be considered workspaces, and are often treated in this manner within GIS software packages.

Workstation A computer that has better graphics capabilities and more processing power than most personal computers and is able to carry out several tasks at once. Workstations usually share data and software with other computers in a network.

World file A text file containing information about where an image should be displayed in real-world coordinates. When an image has a properly configured world file, GIS software can use the information (a total of six values, including the starting coordinates, the cell size in both x- and y-dimensions, and any rotation and scaling information) to accurately overlay the image with any other data already in a projected or geographic coordinate system.

World wide web A worldwide, decentralized, public information space for sharing documents and conducting business on the Internet. Components of the World Wide Web include information in the form of HTML documents; identification tags (URLs) for the millions of computers that host this information; a set of technical specifications, called HTTP, for sending information from one computer to another; and Web browser software for accessing and displaying information.

Worldwide reference system (WRS) A global indexing scheme designed for the Landsat Program based on nominal scene centres defined by path and row coordinates.

WSDL Acronym for Web Service Description Language. An XML format for describing the methods, types, and connection point of a SOAP Web service.

W-test A type of statistical test used in surveying to detect blunders in a measurement network. The W-test is based on the assumption that the null hypothesis is rejected due to a gross error in one of the measurements and uses an alternative hypothesis to identify the erroneous measurement.

W-WAAS Wide Area Augmentation System is a FAA funded system of equipment and software that augments GPS. The WAAS provides a satellite signal for WAAS users to support enroute and precision approach aircraft navigation.

WYSIWYG - WYSIWYG In ArcGIS desktop, this term refers to the editing tools providing visual feedback of the desired end result of an editing operation while the process is still underway. For Cartographic Representations, this is the ability to see a symbolized version of the feature being operated on instead of a wireframe version of the underlying geometry of the feature.

X

x,y coordinates (x,y values) A pair of values that represents the distance from an origin (0,0) along two axes, a horizontal axis (x) and a vertical axis (y). On a map, x,y coordinates are used to represent features at the location they are found on the earth's spherical surface.

x,y event A simple coordinate pair that describes the location of a feature, such as a set of latitude and longitude degrees.

x,y,z coordinates In a planar coordinate system, three coordinates that locate a point by its distance from an origin (0,0,0) where three orthogonal axes cross. Usually, the x-coordinate is measured along the east–west axis, the y-coordinate is measured along the north–south axis, and the z-coordinate measures height or elevation.

X-axis In a planar coordinate system, the horizontal line that runs right and left (east and west of) the origin (0,0).

X-band A nominal frequency range from 12.5 to 8 GHz (2.4 to 3.75 cm wavelength) within the microwave (radar) portion of the electromagnetic spectrum. X-band is a suitable frequency for several high-resolution radar applications and has often been used for both experimental and operational airborne SAR systems, designed for military as well as civilian remote sensing applications. The corresponding wavelength for these systems is on the order of 3 cm, which has been found useful for mapping and surveillance tasks.

X-Data In AutoCAD, extra optional data attached to an AutoCAD drawing element.

XLS XML for location-based services. In ArcWeb Services, XLS is a method for encoding requests and responses and associated ADTs for the Geo-Mobility Server.

XMI Acronym for XML Metadata Interchange. A standard that specifies how to store a UML model in an XML file.

XML Acronym for Extensible Markup Language. Developed by the W3C, a standardized general purpose markup language for designing text formats that facilitates the interchange of data between computer applications. XML is a set of rules for creating standard information formats using customized tags and sharing both the format and the data across applications.

XML metadata interchange A standard that specifies how to store a UML model in an XML file.

XML recordset document In ArcGIS, an export file containing the features or records from an individual geodatabase feature class or table. Data in the file is encoded in XML and can be imported into an existing feature class or table.

XML workspace document In ArcGIS, an export file containing one or more geodatabase feature datasets, feature classes and tables. It can include schema and data or just the schema. Schema and data in the file are encoded in XML and can be imported into a geodatabase.

XPath A language used to find specific parts of an XML document and compute values from the document's content.

X-shift value applied to x coordinate readings to reduce the number of digits.

XSL Acronym for Extensible Style Language. A set of standards for defining XML document presentation and transformation. An XSL style sheet may contain information about how to display tagged content in an XML document, such as font size, background colour and text alignment. An XSL style sheet may also contain XSLT code that describes how to transform the tagged content in an XML document into an output document with another format. The W3C maintains the XSL standards.

XSLT Acronym for Extensible Style Language Transformations. A language for transforming the tagged content in an XML document into an output document with another format. An XSL style sheet contains the XSLT code that defines each transformation to be applied. Transforming a document requires the original XML document, an XSL document containing XSLT code and an XSLT processor (parser) to execute the transformations. The W3C maintains the XSLT standard.

Y

Yaw Rotation of an aircraft about its vertical axis so that the longitudinal axis deviates left or right from the flight line.

Y-axis In a planar coordinate system, the vertical line that runs above and below (north and south of) the origin (0,0). Numbers north of the origin are positive and numbers south of it are negative.

Y-shift a value applied to y coordinate readings to reduce the number of digits.

Z

Z-axis In a spherical coordinate system, the vertical line that runs parallel to the earth's rotation, passing through 90 degrees north latitude and perpendicular to the equatorial plane, where it crosses the x- and y-axes at the origin (0,0,0).

Z-axis

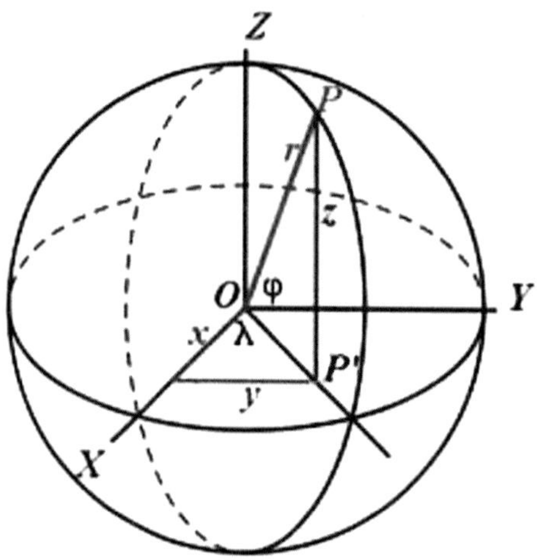

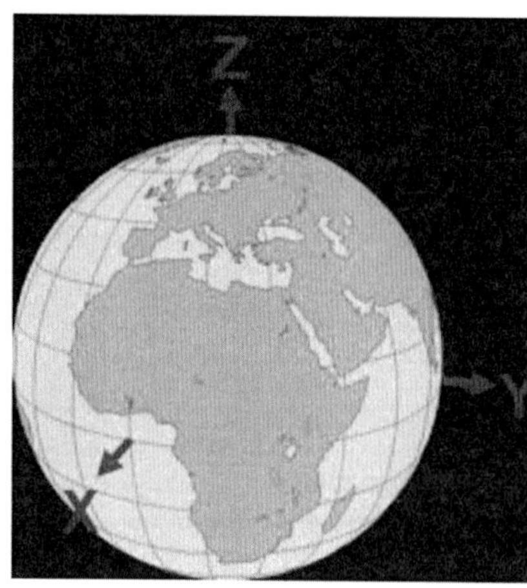

Z-coordinate The value for a given surface location that represents an attribute other than position. In an elevation or terrain model, the z-value represents elevation; in other kinds of surface models, it represents the density or quantity of a particular attribute.

Z-Coordinate

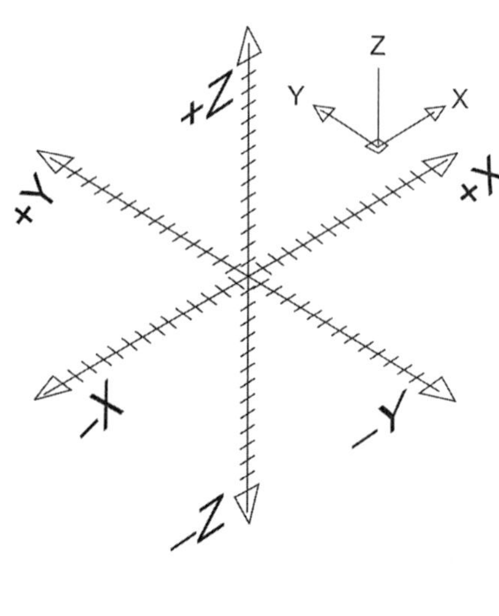

Zenith In astronomy, the point on the celestial sphere directly above an observer. Both the zenith and nadir lie on the observer's meridian (opposite the nadir); the zenith lies 180 degrees from the nadir and is observable.

Zenith

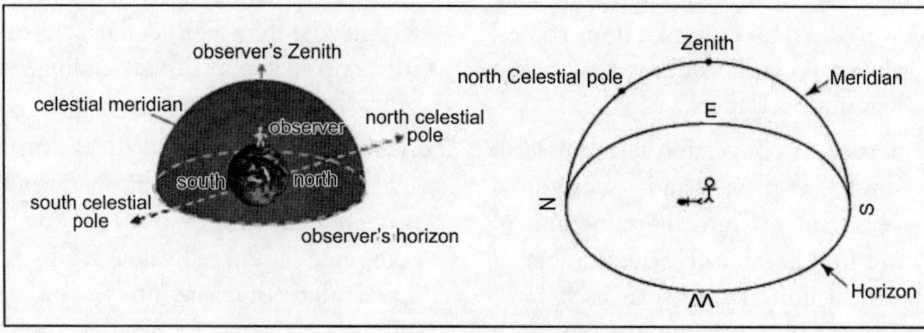

Zenith angle

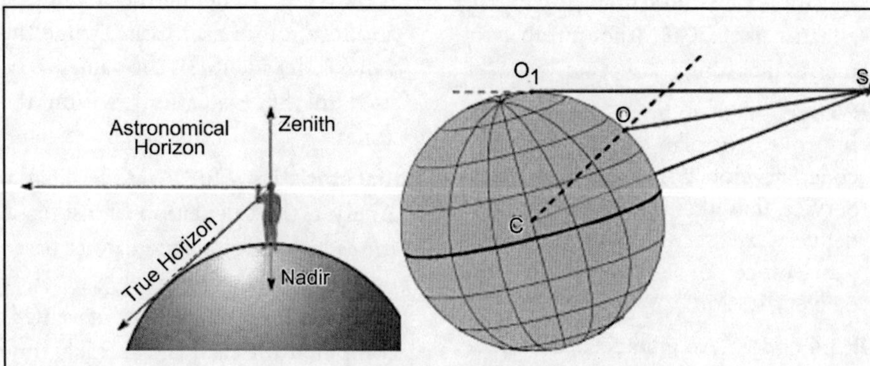

Zenith angle In Survey Analyst for field measurements, a vertical angle that is formed by the intersection of two lines in a vertical plane. Zenith angles are observed on the vertical circle of a TPS instrument.

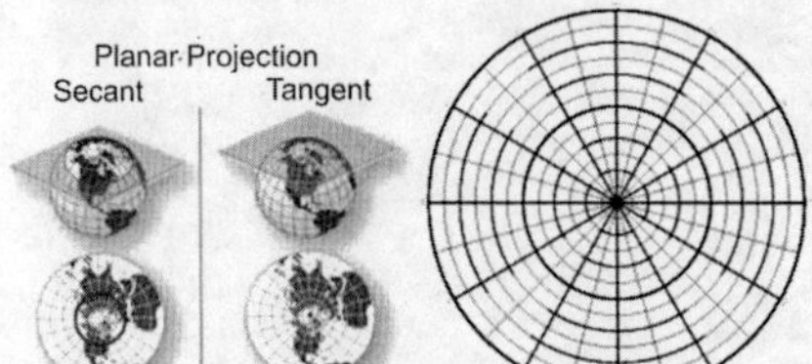

To an observer at O, the zenith angle of object S appears to be z'.
Its true zenith angle, as seen from the centre of the Earth C, is z, which is smaller.

Parallax is greatest for an observer at O_1,
where the object appears to be on the horizon.
We define the angle of parallax p by
$p = z' - z$.

Zenith telescope Instrument for observing starts near the zenith (a point on the celestial sphere directly above the observer's position).

Zenithal projection A map projection that transforms points from a spheroid or sphere onto a tangent or secant plane. The azimuthal projection is also known as a planar or zenithal projection.

Zero length line events In linear referencing, a line event whose from-measure is equal to its to-measure. A zero length line may even occur, for example, along routes, when a polygon touches a route but does not overlap it.

Z-factor A conversion factor used to adjust vertical and horizontal measurements into the same unit of measure. Specifically, the number of vertical units (z-units) in each horizontal unit. For example, if a surface's horizontal units are metres and its elevation (z) is measured in feet, the z-factor is 0.3048 (the number of metres in a foot).

ZIP code Acronym for zone improvement plan code. A five-digit code, developed by the US Postal Service, that identifies the geographic delivery area served by an individual post office or metropolitan area delivery station.

ZIP+4 code An enhanced ZIP Code that consists of the five-digit ZIP Code plus four additional digits that identify a specific geographic segment within the five-digit delivery area, such as a city block, office building, or other unit.

Zonal functions also called as zonal analysis. The creation of an output raster in which the desired function is computed on the cell values from the input value raster that intersect or fall within each zone of a specified input zone dataset. The input zone dataset is only used to define the size, shape and location of each zone, while the value raster identifies the values to be used in the evaluations within the zones.

Zonal statistics In ArcGIS Spatial Analyst, the calculation of a statistic for each zone of a zone dataset based on values from another dataset, a value raster. A single output value is computed for each cell in each zone defined by the input zone dataset.

Zenith telescope

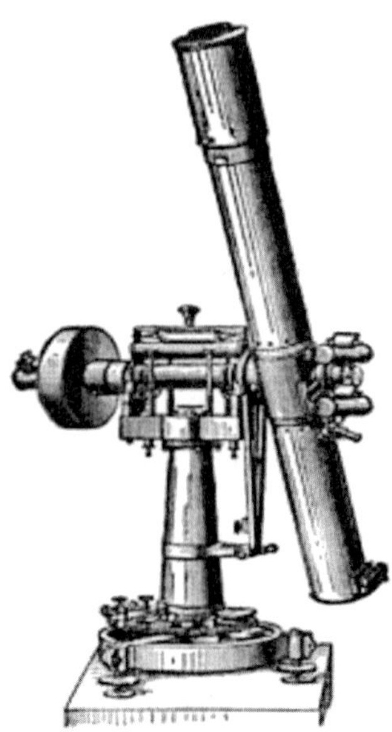

Zonal statistics

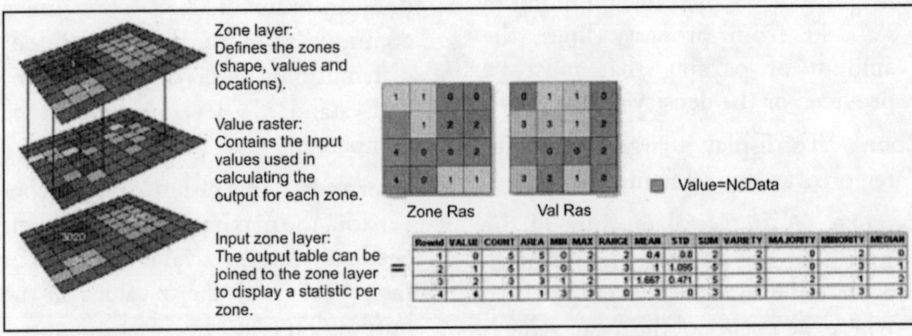

Rowid	VALUE	COUNT	AREA	MIN	MAX	RANGE	MEAN	STD	SUM	VARIETY	MAJORITY	MINORITY	MEDIAN
1	0	5	5	0	2	2	0.4	0.8	2	2	0	2	0
2	1	5	5	0	3	3	1	1.095	5	3	0	3	1
3	2	3	3	1	2	1	1.667	0.471	5	2	2	1	2
4	4	1	1	3	3	0	3	0	3	1	3	3	3

Zone

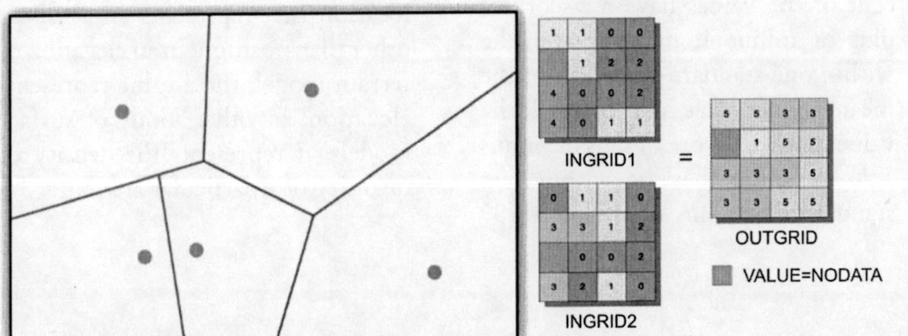

Expression: ZONALSUM (INGRID1, INGRID 2)

Zone All cells in a raster with the same value, regardless of whether or not they are contiguous.

Zone of interpolation The area in a TIN layer for which values (elevation, slope and aspect) are calculated. When a TIN layer is clipped to a smaller size to create a more focused study area, the parts that lie outside the study area remain triangulated and are represented as outside lines but they have no values. These parts are said to be outside the zone of interpolation.

Zoning The application of local government regulations that permit certain land uses within geographic areas under the government's jurisdiction. Zoning regulations typically set a broad category of land use permissible in an area, such as residential, commercial, agricultural or industrial. Zoning regulations can also set constraints on building construction within areas, which may

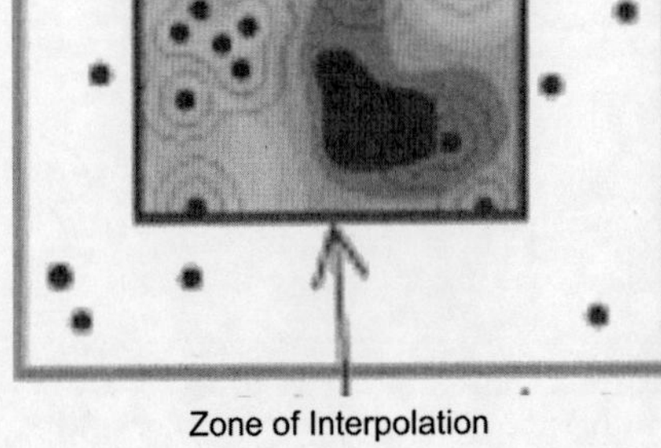

affect factors such as the maximum height of structures, minimum setbacks from property lines, the amount of parking that must be provided or the density of housing.

Zoom To display a larger or smaller region of an on-screen map or image.

Z-score A statistical measure of the spread of values from their mean, expressed in standard deviation units, where the z-score of the mean value is zero and the standard deviation is one. In a normal distribution, 68 per cent of the values have a z-score of plus or minus 1, meaning they lie within one standard deviation of the mean. Ninety-five per cent of the values have a z-score of plus or minus 1.96, meaning they lie within two standard deviations of the mean; 99 per cent of the values have a z-score of plus or minus 2.58. Z-scores are a common scale on which different distributions, with different means and standard deviations, can be compared.

Z-tolerance In raster-to-TIN conversion, the maximum allowed difference between the z-value of the input raster cell and the z-value of the output TIN at the location corresponding to the raster cell centre.

Z-value The value for a given surface location that represents an attribute other than position. In an elevation or terrain model, the z-value represents elevation; in other kinds of surface models, it represents the density or quantity of a particular attribute.